Dissipative Structures and Weak Turbulence

PERSPECTIVES IN PHYSICS
Huzihiro Araki, Albert Libchaber, and Giorgio Parisi, editors

Dissipative Structures and Weak Turbulence

Paul Manneville
*Service de Physique du Solide et
 de Résonance Magnétique
Orme des Merisiers
Gif-sur-Yvette, France*

ACADEMIC PRESS, INC.
Harcourt Brace Jovanovich, Publishers

Boston San Diego New York
London Sydney Tokyo Toronto

This book is printed on acid-free paper. ∞

Copyright © 1990 by Academic Press, Inc.
All rights reserved.
No part of this publication may be reproduced or
transmitted in any form or by any means, electronic
or mechanical, including photocopy, recording, or
any information storage and retrieval system, without
permission in writing from the publisher.

ACADEMIC PRESS, INC.
1250 Sixth Avenue, San Diego, CA 92101

United Kingdom Edition published by
ACADEMIC PRESS LIMITED
24-28 Oval Road, London NW1 7DX

Library of Congress Cataloging-in-Publication Data
Manneville, P. (Paul), (date)
    Dissipative structures and weak turbulence / Paul Manneville.
      p. cm. – (Perspectives in physics)
    Includes bibliographical references.
    ISBN 0-12-469260-5 (alk. paper)
    1. Dynamics. 2. Nonlinear theories. 3. Chaotic behavior in
systems. 4. Turbulence. I. Title. II. Series.
QA871.M33 1990
531'.11–dc20   89-48405
                                                               CIP

Printed in the United States of America
90 91 92 93   9 8 7 6 5 4 3 2 1

Table of Contents

Foreword   xv

Chapter 1. Outlook   1
1. Dissipative Structures   5
2. Transition to Temporal Chaos   10
3. Spatio-temporal Chaos and Weak Turbulence   18
4. Bibliographical Notes   23

Chapter 2. Evolution and Stability, Basic Concepts   25
1. General Framework   25
    - 1.1. Discrete versus Continuous Systems   25
    - 1.2. Autonomous versus Time-Dependent Systems   26
    - 1.3. Deterministic Evolution   27
    - 1.4. Perturbations and Stability   28
2. Global Stability   29
    - 2.1. General Viewpoint and Definitions   29
    - 2.2. Energy Method   30
    - 2.3. Different Concepts of Stability   31
3. Normal Modes, Linear and Nonlinear Dynamics   33
    - 3.1. Normal Mode Analysis   33
    - 3.2. Weakly Nonlinear Dynamics   39
4. Qualitative Dynamics   42
    - 4.1. Elements for a Phase Portrait   42
    - 4.2. Absorbing Zones and Limit Sets   45

4.3. Limit Sets and Attractors	48
4.4. Hyperbolicity, Structural Stability, and Bifurcations	49
4.5. About More General Attractors	50
5. Bibliographical Notes	53

Chapter 3. Instability Mechanisms — 55

1. Rayleigh–Bénard Convection	55
1.1. Qualitative Analysis	56
1.2. Simplified Model	57
1.3. Normal Mode Analysis	59
1.4. Linear Dynamics of Unstable Modes	62
2. Convection in Binary Mixtures	64
2.1. Stationary Mode	67
2.2. Oscillatory Mode	67
3. Thermal Convection in Nematic Liquid Crystals	70
4. Electrohydrodynamic Instabilities in Nematics	74
4.1. Carr–Helfrich Mechanism	75
4.2. Two Instability Regimes	78
5. Taylor–Couette instability	80
5.1. Centrifugal Instabilities	80
5.2. Rayleigh Instability Mechanism	82
5.3. One-Dimensional Model	83
6. Bénard–Marangoni Convection	86
6.1. Mechanism and Simplified Model	86
6.2. Marangoni versus Buoyancy-Driven Convection	90
7. Bibliographical Notes	93

Chapter 4. Thermal Convection — 95

1. Boussinesq Equations and Boundary Conditions	95
1.1. Evolution Equations	96
1.2. Boundary Conditions	99

Table of Contents   vii

2. Normal Mode Analysis   104
   2.1. Stress-Free Solution, Rayleigh (1916)   *104*
   2.2. No-Slip Solution, Pellew and Southwell (1940)   *106*
   2.3. Vicinity of the Threshold (Linear Stage)   *108*
   2.4. Approximate Determination of Threshold Curves   *109*
   2.5. Treatment of the Vertical Vorticity   *112*

3. Phenomenology of Nonlinear Convection   113
   3.1. Introduction   *113*
   3.2. Universal Secondary Modes   *115*
      3.2.1. Cross-Roll Instability   115
      3.2.2. Compression/Dilatation Instability   116
      3.2.3. Torsion Instability   117
   3.3. Specific Secondary Instabilities   *118*
      3.3.1. High Prandtl Number Instabilities   119
      3.3.2. Intermediate Prandtl Number Instabilities   121
      3.3.3. Low Prandtl Number Instabilities   122

4. Phenomenology of the Transition to Turbulence   124
   4.1. "Old" Results   *125*
   4.2. "Recent" Results, Concrete Examples   *127*
      4.2.1. Quasi-Periodicity   127
      4.2.2. Intermittency   129
      4.2.3. Sub-Harmonic Cascade   130
      4.2.4. Weakly Confined Systems   131
   4.3. Concluding Remarks   *134*

5. Bibliographical Notes   135

Chapter 5. Low-Dimensional Dynamical Systems   137

1. Dimension Reduction: A Case Study   138
   1.1. The Model and Its Normal Modes   *138*
   1.2. Elimination of Slaved Modes: Heuristic Approach   *140*
      1.2.1. Projection   140
      1.2.2. Elimination   140
      1.2.3. Formal Generalization   143

2. Center Manifold and Normal Forms — 144
2.1. Perturbative Approach to the Center Manifold — *144*
2.2. Normal Forms of Dynamical Equations — *147*
2.3. Normal Forms and Symmetries — *150*
2.4. Slightly Off the Critical Point — *151*

3. Dynamics and Bifurcations in One Dimension — 153
3.1. General Dynamics — *153*
3.2. Normal/Inverse Bifurcations — *157*
3.3. Conditional Stability, Hysteresis, and Turning Points — *159*
3.4. Transcritical Bifurcations — *160*
3.5. Imperfect Bifurcations — *162*
3.6. Mathematical Context: Unfolding of Singularities — *163*

4. Introduction to Higher Dimensional Problems — 165

5. Dynamics and Bifurcations in Two Dimensions — 169
5.1. Linear Dynamics — *169*
    5.1.1. Distinct Real Roots — 169
    5.1.2. Complex Roots — 170
    5.1.3. Double Roots — 170
5.2. Nonlinear Dynamics, an Example: The Pendulum — *172*
5.3. General Dynamics in Two Dimensions — *174*
5.4. Bifurcations in Two Dimensions — *177*
    5.4.1. Saddle-Node Bifurcation — 178
    5.4.2. Hopf Bifurcation — 179

6. Conclusion — 181

7. Bibliographical Notes — 181

Chapter 6. Beyond Periodic Behavior — 183

1. Poincaré Maps — 183
1.1. Surface of Section and First Return Map — *183*
1.2. Application to the Lorenz Model — *185*

2. Stability of a Limit Cycle — 187

Table of Contents   ix

3. Bifurcations of a Limit Cycle   191
    3.1. Normal Form in the Complex Case   *192*
    3.2. Hopf Bifurcation for Maps   *194*
    3.3. Bifurcations at Strong Resonances 1/1 and 1/2   *197*
4. Nature of Turbulence and Transition Scenarios   202
5. Sub-Harmonic Route to Turbulence   204
    5.1. The Modeling Issue   *204*
    5.2. The Sub-Harmonic Cascade   *207*
    5.3. Universality and Renormalization   *209*
    5.4. Beyond the Accumulation Point   *215*
    5.5. Concluding Remark   *217*
6. Temporal Intermittency   218
    6.1. Modeling of Intermittent Behavior   *218*
    6.2. Type I Intermittency   *219*
    6.3. Type III Intermittency   *222*
    6.4. Type II Intermittency   *226*
7. Quasi-Periodicity   228
    7.1. Introduction to the Locking Phenomenon   *228*
    7.2. The Winding Number and the Structure of Lockings   *231*
    7.3. The Breakdown of a Two-Torus   *235*
    7.4. Ruelle–Takens Scenario and n-*Periodicity*   *237*
8. Beyond "Classical" Scenarios   238
9. Bibliographical Notes   243

Chapter 7. Characterization of Temporal Chaos   247

1. Divergence of Trajectories and Lyapunov Exponents   248
    1.1. One-Dimensional Iterations   *248*
    1.2. Generalization to d-*Dimensional Maps*   *250*
    1.3. Generalization to Differential Systems   *254*
    1.4. Lyapunov Signature of Temporal Behavior   *256*

2.	**Probabilistic Approach**	257
	2.1. Entropy	*257*
	2.2. Invariant Measures	*260*
	2.3. Invariant Measures for One-Dimensional Maps	*263*
	2.4. Natural Measures	*266*
3.	**Chaos and Dimensions**	267
	3.1. Introduction	*267*
	3.2. Fractal Geometry	*267*
	3.3. Probabilistic Viewpoint	*270*
	3.4. Dimensions and Lyapunov Exponents	*273*
4.	**Experimental Approach**	275
	4.1. The Method of Time Delays	*275*
	4.2. Embedding	*276*
	4.3. Practical Problems	*277*
	4.4. Dimension Estimates	*278*
	4.5. Determination of Lyapunov Exponents	*279*
	4.6. Final Remark	*281*
5.	**Bibliographical Notes**	281

Chapter 8. Basics of Pattern Formation in Weakly Confined Systems   285

1.	**Instabilities, Confinement, and Aspect Ratios**	286
2.	**Pattern Formation**	291
	2.1. Modeling of Weakly Confined Systems	*291*
	2.2. Pattern Formation in Two Dimensions	*293*
3.	**Uniform Nonlinear Solutions**	302
	3.1. General Setting	*302*
	3.2. Steady Solutions of Two-Dimensional Models	*305*
	3.3. Rayleigh–Bénard Convection	*312*
	3.3.1. Stress-Free Solution (Malkus and Veronis, 1958)	313
	3.3.2. No-Slip Solution (Schlüter, Lortz, and Busse, 1965)	316
4.	**Modulated Structures**	318
	4.1. Systematic Expansion	*319*

Table of Contents   xi

    4.2. Roll Modulations in a Rotationally Invariant System   *326*
    4.3. Extension to Two-Dimensional Patterns   *328*
    4.4. Short-Term Stability and Early Nonlinear Selection   *330*
    4.5. Phenomenological Extension: The Example of Waves   *334*
5. Bibliographical Notes   **338**

Chapter 9. Applications of the Envelope Formalism   341

1. Phase Winding Solutions   **341**
2. Long Wavelength Instabilities   **344**
    2.1. General Formulation   *344*
    2.2. Longitudinal Perturbations and Eckhaus Instability   *345*
    2.3. Transverse Perturbations and Zigzag Instability   *346*
    2.4. Stability of Waves   *346*
3. Lateral Boundary Effects   **349**
    3.1. Rolls Parallel to a Lateral Wall   *349*
    3.2. Rolls Perpendicular to a Lateral Wall   *351*
4. Structural Defects   **352**
    4.1. Grain Boundaries   *352*
    4.2. Dislocations   *356*
5. Pattern Selection at Lowest Order   **359**
6. Bibliographical Notes   **366**

Chapter 10. Dynamics of Textures and Turbulence   369

1. Phase Diffusion in Steady Roll Patterns   **370**
    1.1. Phase Diffusion Formalism   *370*
    1.2. Phase Instabilities   *374*
    1.3. Rotationally Invariant Formulation of Phase Dynamics   *376*
2. Nonlinear Selection and Weak Turbulence   **379**
    2.1. Nonlinear Wavelength Selection Criteria   *379*

2.2. Drift Flows and the Transition to Weak Turbulence	383
3. Transition to Turbulence in Oscillating Patterns	389
4. Spatio-temporal Intermittency	392
4.1. Introduction	392
4.2. A Case Study: The Modified Swift-Hohenberg Model	394
4.3. Towards a Theory	398
5. Hydrodynamics and Turbulence	403
5.1. Introduction	403
5.2. Advection and Absolute/Convective Instabilities	405
5.3. Homogeneous, Isotropic, Developed Turbulence	408
6. Bibliographical Notes	411

Appendix 1. Macroscopic Dynamics   419

1. General Fluid Systems	419
1.1. Densities and General Balance Equations	419
1.2. Fluid Particles and the Continuity of Matter	420
1.3. Fundamental Equation of Dynamics	421
2. Thermohydrodynamics	423
2.1. Energy Continuity and First Principle	423
2.2. Entropy Source and Second Principle	425
2.3. Constitutive Equations	426
2.4. Summary of the Equations for a Simple Fluid	428
2.5. Passive Mixtures	429
2.6. Fluid Interfaces	430
3. Nematic Liquid Crystals	431
3.1. The Nematic Phase	431
3.2. Elasticity	432
3.3. Statics	433
3.4. Dynamics	435
3.5. Viscometry	437
4. Bibliographical Notes	438

Table of Contents   xiii

Appendix 2. Differential Calculus   439

1. Initial Value Problems   439
   - 1.1. Introduction   *439*
   - 1.2. Linear Maps, Matrices, and Change of Bases   *440*
   - 1.3. Eigenvalues and Invariant Subspaces   *441*
   - 1.4. Application to Linear Initial Value Problems   *443*
2. Boundary Value Problems   444
   - 2.1. Scalar Products and Adjoint Problems   *444*
   - 2.2. Adjointness for Boundary Value Problems   *447*
3. Differential Equations with Delay   450
4. Bibliographical Notes   455

Appendix 3. Numerical Simulations   457

1. Introduction   457
2. Finite Differences for Time Stepping   459
3. Treatment of Partial Differential Equations   465
   - 3.1. Finite Difference Methods   *466*
     - 3.1.1. Space Dicretization and Consistency   466
     - 3.1.2. Boundary Conditions   467
     - 3.1.3. Time Discretization and Stability   468
     - 3.1.4. Efficient Solution of Implicit Schemes   470
     - 3.1.5. Treatment of Nonlinear Terms   471
   - 3.2. Spectral Methods   *472*
4. Bibliographical Notes   474

Subject Index   477

Foreword

Important progress has been made in our understanding of the emergence and evolution of structures in macroscopic systems during the last fifteen years. Deep mathematical problems linked to the role of nonlinearities are involved, and applications have been developed ranging from the fields of natural sciences, physics, chemistry, or biology, to sociology or economics. Our main goal is to offer a text at an intermediate level, more technical than good general books for the layman that have appeared recently, but less specialized than advanced monographs or reprint collections.

We start with a long introductory chapter illustrating our purpose in some detail. The rest of the book is divided into three roughly equal parts, each with three chapters. The first part, mainly devoted to the emergence of *dissipative structures*, i.e., the instability mechanisms taking place in continuous media, will be understood as a necessary substratum to the two subsequent sections. A special emphasis is placed on *convection* which appears to be the most analyzed case study up until now.

In hydrodynamics, the word *turbulence* usually specifies a wildly irregular flow regime with enhanced mixing properties. By contrast, the dynamical regimes considered here will be termed *weakly turbulent*. Depending on geometrical constraints, two cases will be distinguished. When confinement effects are strong, the spatial structures of the instability modes are frozen and one has to deal with the dynamics of the corresponding amplitudes which play the role of generalized coordinates in some effective phase space. Part two is devoted to the study of deterministic dissipative dynamical systems, accounting for their asymptotic evolution, from steadiness to *temporal chaos*. In part three, we consider weakly confined systems where boundary effects are less stringent, so that modulation can play a primary role. The concept of *pattern* will then be the key-word and disorder will recover an irreducible *space–time* meaning. The future will tell if these two complementary viewpoints on the emergence of chaos can help us understand the nature of turbulence and improve its control.

The book itself derives directly from unpublished notes of a course on instabilities and turbulence given from 1984 to 1988 to undergraduate students preparing the "Diplome d'Etudes Approfondies de Physique des Liquides" (University of Paris VI), from lecture notes of an introductory course delivered to condensed matter physicists at a summer school in 1985 and published in *Structures et Instabilités* (Editions de Physique, Orsay-Les Ulis, 1986), and from the text of several more or less advanced seminars. The whole source material has been considerably reorganized and expanded but the style has been kept rather informal. In particular, we have always attempted to extract important ideas and to present techniques using examples worked out in detail rather than by taking a deductive abstract viewpoint. Moreover, we have tried to explain how concrete models could be derived and to show how insight into the effects of couplings could be gained by means of explicit calculations and/or numerical simulations on such simplified models.

It may be difficult to delineate the prerequisites necessary to read this book, the content of which is the result of interactions with a varied but limited panel of students. What we wanted to do was to offer some sort of a "short-circuit" between the physical and mathematical background expected from undergraduate studies and the level of current works in the field of nonlinear dynamics. With this aim, three appendices have been added. The first one presents some elements of the mechanics of continuous media, from simple fluids to liquid crystals. The second one deals with differential calculus, mostly recalling results from linear algebra. Two reasons justify the presence of the third appendix devoted to very simple numerical simulation methods applicable to ordinary or partial differential equations. First, in many respects the stability properties of numeri-

Foreword

cal schemes are comparable to those of physical systems, and second, performing numerical simulations is the best way to gain some intuition of the specificities of nonlinear systems. The literature cited is very limited since good compilations are easily accessible. With few exceptions the only articles quoted are those introducing important concepts and those illustrating our viewpoint or those from which figures have been adapted. When available, books that can serve as useful references have been mentioned. Citations are grouped at the end of each chapter with some brief comments situating their relevance.

Now, I would like to thank all those who have, at one time or another, contributed to my present understanding of instabilities, nonlinear dynamics, and turbulence. First of all, I will pay tribute to Y. Pomeau from the Ecole Normale in Paris who was at the origin of most of my personal contributions to the subject. Stimulating interactions with P. Bergé's experimental group at Saclay, M. Dubois, V. Croquette, A. Pocheau, P. Le Gal, and F. Daviaud, and the enthusiastic collaboration of J. M. Piquemal and H. Chaté are deeply acknowledged. I will also not forget E. Dubois–Violette from Orsay who initiated me into the theory of instabilities in nematics.

Let me now express my gratitude to Prof. S. Bratos from Paris VI University and to C. Godrèche from Saclay who provided me with the two courses from which the present book derives most directly. Of course, special thanks are due to Prof. A. Libchaber who was at the origin of the project and who made useful comments at several stages of the elaboration of the text, and to the Academic Press publishing board who tolerated the successive delays and helped me in the preparation of the final T$_E$X version of the manuscript.

Little would have been possible without the patience and the encouragements of my family, my wife Jacqueline and my children, Jean-Baptiste, Sébastien, Alexis, and Claire, to whom I dedicate this book.

Chapter 1
Outlook

The state of macroscopic systems can be identified by a small number of parameters (volume, number of particles,...) entering thermodynamic functions (energy, entropy,...). For example, the description of a closed system in contact with a thermal bath is contained in a free energy, $F = U - TS$ with U the internal energy, S the entropy, T the temperature, and the *equilibrium* state corresponds to a minimum of this function. This extremum property can be understood as the result of a competition between the organization induced by mechanical interactions (the energy term) and the disorder originating from the degeneracy between macroscopic states with the same energy (the entropy term). The structure of the equilibrium state is then controlled by the temperature and can be characterized by uniform macroscopic parameters, whereas the ultimate fate of fluctuations around this state amounts to an exponential relaxation well described by the linear theory of irreversible processes.

Global thermodynamic equilibrium is an exceptional situation; most often systems are out of equilibrium and evolve spontaneously to recover the lost equilibrium. In *isolated systems* displaying phase coexistence, metastability, and hysteresis, this immediately raises the problem of the nucleation and growth of the most stable phase inside the metastable phase and, at later stages, the problem of the propagation and stability of the front separating the two phases. In spite of its intrinsic interest, we will not consider such irreversible macroscopic evolution of heterogeneous systems in the following but concentrate our attention on homogeneous systems.

Out-of-equilibrium situations in isolated systems are only transient. In contrast, a system allowed to exchange matter or energy with the exterior world can be maintained permanently far from equilibrium when submitted to a gradient of intensive thermodynamic quantities. For example, a macroscopic motion is driven by a pressure gradient, a heat flux is the response to a temperature gradient, etc. In principle we need a nonlinear theory to determine the response of the system to applied stresses of arbitrary strength. In practice, however, the stresses that we are able to apply to a continuous medium are usually very weak when compared with *microscopic* interactions. We can therefore safely make an assumption of *local equilibrium* and define a *mesoscopic* scale over which this assumption holds. The corresponding size of the "infinitesimal" elements of the continuous medium must remain large when compared to the molecular dimensions but very small at a *macrosopic* scale.

Though the local equilibrium assumption is valid at this scale and molecular transport well described by a *linear response theory*, the system can be driven far from equilibrium on a *global* scale. Although solid media give many examples of strong nonlinearities (in electronics, optics, etc.), as far as physical applications are concerned, we will deal mostly with fluids. In fluid systems, the mesoscopic scale is that of the so-called *fluid particle* and macroscopic motion is governed by hydrodynamic equations as introduced briefly in Appendix 1. Most of the properties to be discussed arise from the possibility of *advection* of some physical quantity: momentum, heat, etc. In the hydrodynamic equations this is accounted for by the term $\mathbf{v} \cdot \nabla(\ )$ where $\mathbf{v}$ is the macroscopic velocity field and $\nabla(\ )$ stands for the local gradient of the quantity transported.

Extremely close to equilibrium, the quadratic convective term above can be neglected and the evolution equations are linear. As a result, the solution is unique and completely controlled by dissipative processes. It derives continuously from the equilibrium state and displays the same space-time symmetries as the driving stresses, stationary or periodic in time, uniform or periodic in space, etc. The solution is said to belong to the *thermodynamic branch*. For fluid

1. Outlook

flows, this situation corresponds to the Stokes approximation. The corresponding velocity field, called *laminar*, is entirely predictable and has the same regularity properties as the applied external forces.

Farther from equilibrium, nonlinearities are no longer negligible. In principle, one should be able to follow the solution belonging to the thermodynamic branch by extrapolation from the near-equilibrium situation. However the solution defined in that way may not be observable in practice since to be observable it must be stable, i.e., *robust to perturbations* (Chapter 2). Stability is guaranteed by an extremum principle close to thermodynamic equilibrium, but not far from it. Above a certain critical applied stress, the system can *bifurcate* toward states belonging to new branches. Along these branches some of the original symmetries of the system are broken and the typical response of an initially uniform system submitted to a time independent forcing can be time or/and space periodic. These new states have been called *dissipative structures* (Prigogine, 1955) to underline the paradoxical role of dissipation which, although naively thought to iron out fluctuations, may contribute constructively to build up organized macroscopic structures by feedback interactions involving specific *characteristic times*. The well-known example of convection will be presented in the next section and analyzed in greater detail in Chapters 3 and 4.

Usually, the state resulting from a first instability is still very regular; the system experiences a *laminar–laminar* transition. A small loss of predictability is associated with such a bifurcation since several states are equivalent under the new symmetries of the bifurcated state. An auxiliary macroscopic variable, usually a *phase variable*, is subsequently required to specify which state has been "chosen" by the system. When the forcing in increased further, the state resulting from the first or *primary* instability can itself be unstable to new *secondary*, then *tertiary*... modes, breaking new symmetries and introducing new indeterminacies. The global dynamics of the system can go from simple to very complicated and at some stage we may decide that it has become *chaotic* or *turbulent*.

According to Landau (1944), turbulence is the result of an indefinite accumulation of unstable modes, with disorder coming from the *superposition* of motions with many spatial and temporal scales and unpredictability arising from our inability to control all the phases variables associated with these modes. This picture was corrected by Ruelle and Takens (1971) who showed that the loss of long-term predictability can be the result of the nonlinear *interaction* of a finite and small number of modes. *Strange attractors* and *sensitivity to initial conditions* are the keywords of this new approach to the *laminar–turbulent* transition (Chapters 6 and 7).

The formulation in terms of *dissipative dynamical systems* at which we arrive (Chapter 5) tends to identify the *transition to turbulence* with the transition to *temporal chaos*. This indeed helps us reconcile determinism and stochasticity for systems with a small number of degrees of freedom. In contrast, turbulence is usually understood as a flow regime characterized by intense velocity fluctuations in both time and space over a large range of scales and, consequently, by highly enhanced transport properties measured by effective diffusivities much larger than their molecular counterparts. The number of excited degrees of freedom is then expected to be very large and the theory of dissipative dynamical systems no longer seems of great help. In fact, we will be concerned nearly exclusively with the processes involved in the transition to turbulence, which will be referred to as *weak turbulence*, rather than with the problem of the nature and properties of strong or *developed turbulence*.

An important question will therefore be that of the number of effective degrees of freedom, which relates to the macroscopic coherence of instability modes controlled by confinement effects. Best understood *scenarios* explored in the last decade apply to *confined systems* for which the spatial structure of the modes remains frozen (Chapters 5 to 7). When confinement effects become unable to maintain global coherence, chaos gains an irreducible *spatio-temporal* meaning which will be approached progressively in Chapters 8 to 10. Beforehand, we will examine several concrete examples of dissipative structure formation (Chapter 3) stressing on the much studied case

1. Outlook

of convection (Chapter 4) to which we now turn for a more detailed presentation.

1. Dissipative Structures

Convection is a very old notion that goes back at least to the eighteenth century when it was invoked by Lomonossov and Hadley to explain atmospheric motions. Scientific investigation begins with Bénard's work (1905) and Rayleigh's theoretical analysis (1916). This somewhat traditional topic of fluid mechanics has received considerable theoretical and experimental attention recently mainly because the primary mechanism is appealingly intuitive and because the system is sufficiently "clean" to allow a detailed comparison between theory and experiment (several other systems will be presented in Chapter 3).

The general problem is that of the stability of a fluid layer with a potentially unstable density stratification in the field of gravity: heavy fluid above light. This density stratification is obtained by heating the fluid layer from below, thermal expansion usually causing the density to decrease as the temperature is increased. To be a little more specific, we assume that in the *basic state* the fluid layer of height h is submitted to a temperature gradient $\beta = \Delta T/h$ where $\Delta T = T_b - T_t$ (> 0) is the temperature difference between the bottom (subscript "b") and top ("t") plates (Fig. 1). This regime of *pure conduction* with a fluid at rest and an unperturbed temperature field derived from the Fourier law $T_0(z) = T_b - \beta z$, belongs to the thermodynamic branch connecting the out-of-equilibrium state continuously to the state of thermal equilibrium ($\Delta T \to 0$).

The origin of the instability is easily understood by examining the evolution of a temperature fluctuation θ around the unperturbed profile $T_0(z)$, e.g., a hot droplet $\theta > 0$. Since the density ρ decreases with temperature, this droplet is lighter than the surrounding fluid at the same height and experiences a buoyancy force directed upwards; it tends to rise. But since the fluid is heated from below, the density increases with the height and the drop encounters an ever colder and

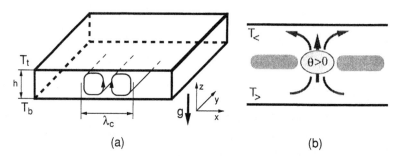

Fig. 1. Geometry and mechanism of Rayleigh-Bénard convection.

denser fluid so that it tends to rise higher; the initial fluctuation is *amplified*.

However, two dissipative processes tend to maintain the fluid in its initial rest state: friction (damping by viscosity) and heat diffusion (the warmer drop looses its heat and experiences less buoyancy). The instability develops only if the drop is accelerated sufficiently to overcome these stabilizing processes. The temperature gradient β, the natural *control parameter* of this instability, must be larger than some *critical* value β_c. This *convection threshold* can be estimated at different levels of accuracy, from the purely qualitative level in terms of *characteristic times* to a semi-quantitative level in terms of simplified models (Chapter 3) and of course to the quantitative level based on exact equations (Chapter 4).

Above the threshold, a specific structure of *organized* convection cells develops, the most unstable mode achieving some sort of optimum between the different contributions to the mechanism. A periodic modulation of the temperature and velocity fields is then expected at a well-defined wavelength called the *critical wavelength* $\lambda_c \sim 2h$ for a basically dimensional reason.

At this stage we can admit, as suggested in Fig. 1, that the dissipative structure is made up of *convection rolls* simply described by the temperature field at mid-height in the cell: $\theta(z = h/2, x, t) \sim A(t)\cos(k_c x)$, where the variable A measures the *amplitude* of the fluctuation. Therefore we have $A \equiv 0$ below the threshold and $A \neq 0$ above (*order parameter* in the sense of the Landau theory

1. Outlook

of phase transitions). For $\beta \sim \beta_c$ and A small enough, the evolution of the fluctuation should be governed by a linear differential equation $dA/dt = sA$, where the growth rate s is negative below threshold (damping) and positive above (amplification). It is then reasonable to assume $s = r/\tau_0$, where $r = (\beta - \beta_c)/\beta_c$ is the relative distance from the threshold and τ_0 the natural time scale for the process.

However, beyond the threshold the fluctuation cannot grow indefinitely since the dissipation increases while the driving force decreases. To account for this self-limitation we assume that the "bare" amplification rate s is replaced by an "effective" rate $s_\text{eff} = (r - gA^2)/\tau_0$ independent of the sign of A (with $g > 0$ so that self-limitation really occurs). This yields a *nonlinear* evolution equation for the amplitude:

$$\tau_0 \frac{dA}{dt} = rA - gA^3 \qquad (1)$$

which correctly describes the *bifurcation* from the conduction regime toward the convection regime. The conduction regime is associated with the trivial solution $A_{f,t} = 0$ of the *fixed point equation* $dA/dt \equiv 0$. This solution exists for all r but it is unstable for $r > 0$. On the other hand, the convection regime is represented by the two equivalent nontrivial solutions $A_{f,nt}^{(\pm)} = \pm\sqrt{r/g}$, which only exist above the threshold ($r > 0$).

Notice however that self-limitation that actually holds for convection may not take place in other cases. For example, if g is negative higher order terms must be included to ensure saturation (see Chapter 5 for a detailed analysis) but even wilder transitions can happen when the perturbative approach does not converge, possibly indicating a direct transition to strong turbulence.

Since the instability discussed up to now is the first to develop, it is called the *primary instability*. However, the convection cells that emerge can themselves be considered as a new basic state, the stability of which has to be studied. The characteristics of *secondary instability* modes that can appear are usually more difficult to understand since the new starting point is much more complicated than the initial basic state. In convection, their nature depends on the

Fig. 2. In a parallelepipedic box with $\Gamma_x = 2.0$ and $\Gamma_y = 1.2$, convection can set in as two short rolls parallel to the small side of the box. Thermal gradients in the fluid are visualized by differential interferometry (after Bergé and Dubois, 1981).

physical properties of the fluid and especially on the relative importance of the two stabilizing processes, the viscous dissipation and the thermal diffusion. When the temperature field controls the nonlinear dynamics, secondary instabilities involve processes confined in thermal boundary layers close to the horizontal plates. In the opposite case, the role of velocity field becomes essential and secondary instabilities involve delicate feedback loops of truly hydrodynamic origin coupling the convection rolls to a large scale horizontal flow component. Nontrivial time dependence can introduce itself at this stage or only after a *tertiary instability* but, in all cases, turbulence sets in when β is further increased (a more detailed review will be presented in Chapter 4).

This description may seem to confirm the Ruelle-Takens picture alluded to earlier since a finite and small number of steps are required before the occurrence of a turbulent regime. However, the experiments supporting this view were performed in wide containers implying the presence of many rolls, and the observed convection patterns were most often strongly disordered. Moreover, a slow residual time dependence due to defect motion in the pattern could interfere with the actual process of transition to turbulence. Experimental work performed in the last decade has been marked by the recognition of the role of *confinement effects*, which help us control this structural disorder and decide whether it is parasitic or intrinsic.

In fact, the number of effectively interacting modes is related

1. Outlook 9

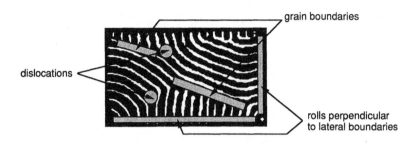

Fig. 3. Pattern observed in convection at large aspect ratios; thick lines mark the place of sinking fluid; the *texture* is made of a large *grain* with a slowly varying roll orientation bounded by smaller grains and lateral walls; notice the presence of line and point defects called *grain boundaries* and *dislocations* (after Pocheau and Croquette, 1984).

not to the number of independent physical processes (always very limited) but rather to geometric characteristics of the experiment. From the physical dimensions of the container, *aspect ratios* can be defined that give a dimensionless measure of the lateral extension ℓ in terms of the spatial scale intrinsic to the instability mechanism λ_c: $\Gamma = \ell/\lambda_c$. When Γ is small, the spatial structure of the modes is frozen by the lateral confinement effects, e.g., a pair of counter-rotating cells as in Fig. 2. Moreover, their instability thresholds are well apart from each other so that a small number of modes can easily be excited while others are kept stable. In this case, usually called *confined*, the problem is then to determine the *effective dynamics* of the unstable modes, which we will examine at the qualitative level in the next section and in greater detail in Chapter 5. The corresponding routes to *temporal chaos*, the so-called *scenarios* will be analyzed in Chapter 6.

In the opposite case, $\Gamma \gg 1$, called *weakly confined* or *extended*, the convection field is ordered at a local scale only. The overall structure is better understood in terms of *patterns* or *textures* as illustrated in Fig. 3. Their description raises specific problems associated with the large *position degeneracy* of the convection cells and the occurrence of *structural defects*. This will be introduced in Section 3 as a premise of the developments in Chapters 8 to 10.

2. Transition to Temporal Chaos

In Section 1, we introduced the amplitude A as a *generalized coordinate* serving to characterize the state of the layer as a whole. It can be understood as the *degree of freedom* associated with the main convection instability mode. In addition, we derived the nonlinear equation governing the time dependence of A by a heuristic argument yielding a cubic self-interaction term. However, several instability modes corresponding to different spatial structures are likely to interact, each introducing its own amplitude. Equation (1), which governs a single amplitude, can be generalized to a system of equations involving a set of degrees of freedom sufficiently large to account for the global dynamics of the system beyond the first instability threshold.

As already noted, in confined geometry, modes are isolated, and by playing with the control parameters we can excite some of them while keeping the others unexcited. The procedure that eliminates irrelevant modes and leads to the reduced system is best illustrated using an example involving only two modes, one is taken close to its linear instability threshold and governed by

$$\frac{dA_1}{dt} = s_1 A_1 + f_1(A_1, A_2)$$

with $s_1 \sim 0$, positive or negative, the other is taken strongly stable, i.e.,

$$\frac{dA_2}{dt} = s_2 A_2 + f_2(A_1, A_2)$$

with s_2 negative and "large". Owing to the existence of this wide gap in the spectrum ($|s_2| \gg |s_1|$) the slow variable A_1 is easily seen to drive the evolution of the fast variable A_2. Indeed, for fluctuations at the rate s_1, $|dA_2/dt| \sim |s_1| A_2$ is negligible when compared with $|s_2| A_2$ in the equation for A_2, which then follows the variations of f_2 *adiabatically*. The stable mode is said to be *slaved* to the unstable mode and we simply have $|s_2| A_2 = f_2(A_1, A_2)$, or, solving for A_2,

$$A_2 = g_2(A_1),$$

1. Outlook

which defines a curve in the two-dimensional phase space spanned by A_1 and A_2. Further inserting this expression into the equation for A_1, we obtain

$$\frac{dA_1}{dt} = s_1 A_1 + f_1(A_1, g_2(A_1)) = s_1 A_1 + g_1(A_1),$$

which governs the effective dynamics of the remaining degree of freedom. This can be generalized to systems having m nearly neutral modes (*central modes*) provided that a sufficiently large gap exists between them and the remaining stable modes, which yields an effective dynamics for a m-dimensional reduced system governing modes constrained to 'live' on some *center manifold*, see Chapter 5.

When the system resulting from this reduction procedure is one-dimensional, e.g., Equation (1), there can only be a relaxation toward *fixed points*. A general proof of this property stems from the fact that for a single variable, $dA/dt = F(A)$ can always be written in the form $dA/dt = -\partial G/\partial A$, where $G = -\int F(A)\,dA$ is the potential from which the evolution equation derives. The dynamics then merely amounts to a search for local minima of G, but this variational property does not hold in general for systems in higher dimensional spaces so that there is room for nontrivial time-dependent asymptotic regimes. Skipping the two-dimensional case that typically yields at most periodic behaviors associated with stable closed trajectories or *limit cycles* in phase space (Chapter 5), we directly turn to a three-dimensional example generating complex time-behavior: the celebrated Lorenz model (1963) defined as

$$\frac{dA_1}{dt} = \sigma(A_2 - A_1),$$
$$\frac{dA_2}{dt} = rA_1 - A_2 - A_1 A_3,$$
$$\frac{dA_3}{dt} = -bA_3 + A_1 A_2.$$

Though deterministic, this system possesses trajectories that are unpredictable in the long term for certain values of its parameters (Fig. 4).

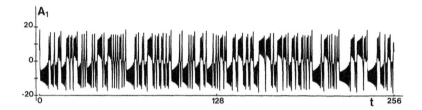

Fig. 4. Typical long time series of variable A_1 for the Lorenz model for $\sigma = 10$, $b = 8/3$, and $r = 28$.

The next issue is therefore the understanding of the transition from regular to complex temporal behavior that takes place when the control parameters are varied. The study of the different *scenarios* of transition to turbulence is precisely the subject of *bifurcation theory*, in the largest possible sense (Chapter 6). At this stage, the introduction of the so-called *Poincaré section technique* is essential (see Fig. 5). Indeed, to study bifurcations beyond periodic behavior we must detect perturbations hidden by the periodic driving by performing a *stroboscopic* analysis of the system. In phase space, this amounts to taking successive sections of trajectories with some surface of section and then defining a *first return map*. The initial time-continuous dynamical system is therefore converted into an iterated map called a *discrete-time dynamical system* which serves to analyze subsequent instabilities. For example, the birth of a doubly periodic regime with two incommensurate periods gives rise to a *2-torus* easily viewed as the direct product of a limit cycle losing stability by a circle drawn on the Poincaré surface.

According to Landau (1944), the transition to turbulence is the result of the accumulation of oscillatory instabilities with unrelated frequencies. The corresponding attractor is then an n-torus and the system is turbulent at the limit $n \to \infty$. The loss of predictability comes from the fact that phases introduced at each bifurcation remain unknown functions of initial conditions. However, if this picture were to hold, motions in phase space would remain correlated in the long term since phase differences would be fixed once for all, in contrast with the strong "mixing" property attributed to turbu-

1. Outlook

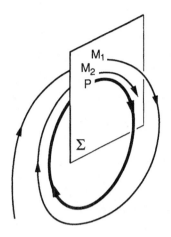

Fig. 5. Poincaré section of a limit cycle by a transverse surface Σ is carried out and the dynamics is reduced from continuous to discrete time by considering the associated first return map: M_1, $M_2, \ldots \Rightarrow M_{n+1} = \Phi(M_n)$.

lence. The scenario proposed by Landau to explain the "nature of turbulence" by a superposition of independent modes is in a sense "too linear". In fact, modes interact in a nonlinear way and, as a nontrivial consequence, the Ruelle–Takens picture emerges: the resulting dynamics becomes uncorrelated in the long term owing to the *instability of trajectories* belonging to *strange attractors*.

As implied by the discussion above, the transition to chaos can be studied using first return maps and, in practice, much has been learned from the consideration of model iterations. As an example of nontrivial properties of simple nonlinear maps, Fig. 6 displays the bifurcation diagram of the so-called *logistic map*

$$X_{n+1} = 4rX_n(1 - X_n).$$

In this figure, the *attractor*, i.e., the set of points asymptotically visited by the trajectories, is plotted as a function of the control parameter r in the interval $[1/2, 1]$. A whole cascade of *sub-harmonic bifurcations* with increasing periods 2^m takes place with bifurcation thresholds converging in a geometric progression toward an accumulation point r_∞ beyond which chaos sets in. This first example of *scenario* displays universal features (Feigenbaum, 1978) of great interest for the interpretation of the transition to temporal chaos, which will make up part of the subjects treated in Chapter 6.

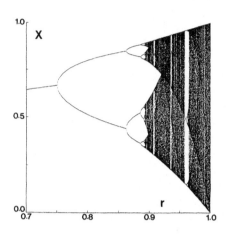

Fig. 6. Bifurcation diagram of the period-doubling scenario: the single fixed point bifurcates at $r = 3/4$ toward a period-2 cycle stable up to $r = 0.862\ldots$ where a period-4 cycle appears, etc.; the cascade accumulates at $r_\infty = 0.892\ldots$; beyond r_∞ chaos can be present (densely covered X-intervals at given r); notice the windows of periodic behavior.

Beyond the threshold for aperiodic motion, trajectories are unpredictable in the long term. This unpredictability, which manifests itself in the *decay of correlations*, comes from an *instability of trajectories in phase space*. Roughly speaking, two trajectories that are issued from neighboring points in phase space do not stay close to each other but fly apart so that, after a while, knowing one instantaneous position does not help to predict the other. This instability can be viewed as the result of some permanent *stretching* in phase space along one or more directions. However, for dissipative systems, after transients have decayed, trajectories belong to attractors that are bounded in phase space and have vanishing volume owing to phase space contraction. To make this compatible with the divergence of trajectories (stretching), some sort of *folding* or a similar process must occur at given places in phase space. As a result, chaotic attractors usually display a *fractal* structure, locally the product of a continuous manifold by a Cantor set as illustrated in Fig. 7 on the Hénon model (1976):

$$X_{n+1} = Y_n,$$
$$Y_{n+1} = 1 - aY_n^2 + bX_n$$

which extends the one-dimensional logistic map to two dimensions consistently.

1. Outlook

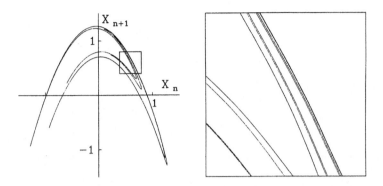

Fig. 7. The Hénon attractor is displayed for the "classical" values $a = 1.4$ and $b = 0.3$ (left) and enlargement of region $[0.4, 0.8] \times [0.4, 0.8]$ showing the locally self-similar transverse fractal structure (right).

The divergence of trajectories related to the behavior "along" the continuous component of the attractor can be best understood by considering a reduced map simply forgetting the "transverse" Cantorian component. As an illustration, let us consider the simplest possible expanding map given by

$$X_{n+1} = 2X_n \bmod(1)$$

(*diadic map*) and visualize the divergence of trajectories by plotting $Z_{n+1} = Z_n + \exp(2\pi i X_n)$ in the complex plane. Figure 8a displays two trajectories with initial conditions that differ by 2×10^{-7}. At the resolution of the picture they can be distinguished easily after about 15 iterations, and after 18 iterations they live their own life, which gives an idea of the way experimental measurements can be affected by the instability of trajectories. In Fig. 8b, the same representation is used for a single trajectory followed during a long time interval. Both *short term determinism* and *long term unpredictability* are easily identified in this figure since pieces of trajectory form defined patterns associated with $X_n \sim 0$ on a local scale, whereas on a global scale we remain unable to predict in which region of the plane the trajectory will move. For comparison, Fig. 8c displays the *random walk* corresponding to a series $\{Z_n\}$ defined as previously but with $\{X_n\}$ given by a random number generator.

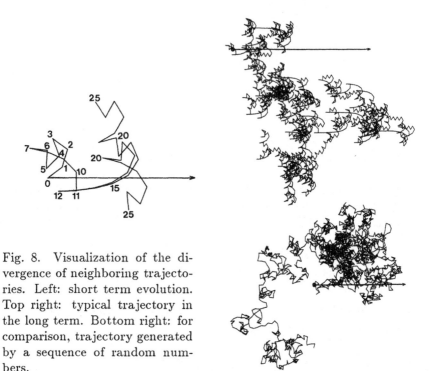

Fig. 8. Visualization of the divergence of neighboring trajectories. Left: short term evolution. Top right: typical trajectory in the long term. Bottom right: for comparison, trajectory generated by a sequence of random numbers.

A quantitative characterization of the chaotic behavior can be obtained from the measurement of the average rate of divergence of trajectories. For an iteration of a single variable $X_{n+1} = f(X_n)$ this rate, called the *Lyapunov exponent* of the map, is given by

$$\gamma = \lim_{n\to\infty} \frac{1}{n} \sum_{i=0}^{n-1} \log(|f'(X_i)|),$$

where f' denotes the derivative of f with respect to X. For higher dimensional systems, a whole Lyapunov spectrum can be determined and the existence of at least one positive Lyapunov exponent testifies to the presence of *deterministic chaos*.

Hints on the transverse *fractal* structure of an attractor, e.g., in Fig. 7, can be obtained form the consideration of the classical triadic *Cantor set* obtained by cutting out the central part]1/3, 2/3[

1. Outlook

of a reference interval $\mathsf{U} = [0,1] = \mathsf{U}_0$, then repeating the process on each of the two sub-intervals $\mathsf{U}_1^1 = [0, 1/3]$ and $\mathsf{U}_1^2 = [2/3, 1]$, and continuing it iteratively on the 2^n segments U_n^j kept at level n. Though made of isolated points, the Cantor set seems to fill space more than an ordinary set of points owing to its internal *self-similarity* (parts are similar to the whole after a blow-up by a factor of 3^n). This property can be measured in terms of a *fractal dimension* defined as

$$d_f = \lim_{\epsilon \to 0} \left[\log \left(\mathcal{N}(\epsilon) \right) / \log(\epsilon^{-1}) \right],$$

where $\mathcal{N}(\epsilon)$ is the number of balls of radius ϵ covering the set. For the triadic Cantor set, this definition yields $d_f = \log(2)/\log(3) = 0.631\ldots$, well between 0 and 1, the dimensions (in the usual sense) of a finite set of isolated points and of a continuous segment, respectively. Chapter 7 will be devoted to the study of various means of characterizing these different facets of strange attractors by extending this simple introduction of Lyapunov exponents and fractal dimensions.

At this point, we should note that the success of the dynamical systems approach in explaining temporal chaos is essentially conceptual and that difficulties can appear in the interpretation of experimental data. In fact, the actual structure of the dynamical system governing effective modes is not known *a priori*, so that the best understood situations correspond to cases where everything can be made local in phase space, which is most favorable to theoretical analysis. Chaos then grows in a well controlled and rather continuous way and, as a consequence, the universal features of "classical" scenarios can be predicted reliably (at least in limited regions of parameter space). When the phase-space structure is involved in a more "global" manner more complicated situations occur with attractor coexistence, *crises*, and *intermittency*. This can be grasped theoretically only at the price of stringent modeling assumptions, the amount of universality left then decreasing rapidly.

3. Spatio-temporal Chaos and Weak Turbulence

Chaos has a strictly temporal meaning as long as the spatial structure remains frozen. This leads us to a new limitation of the dynamical systems approach: What happens when the degrees of freedom have an undoubtedly spatial significance and how does the local/global dilemma translate from phase space to physical space? We will begin this program in Chapter 8 by presenting examples of pattern evolution from numerical simulations to illustrate the different stages of nonlinear selection. In the subsequent chapters we will be interested first in setting a proper framework able to deal with the occurrence of slow spatio-temporal disorder (Chapter 9). Then we will examine specific aspects of the transition to turbulence at the large aspect ratio limit and some possible implications for more developed turbulence (Chapter 10).

When the system is extended, the best starting point to understand its nonlinear behavior is the limit of the laterally unbounded layer. Normal modes are then continuously indexed by the wavelength and, as already noticed, they are quasi-degenerate so that many of them can become unstable even slightly above the threshold. At the linear level, normal modes are independent. The heuristic introduction of nonlinearities in Section 1 involved a single mode (Fig. 9a). When many modes with comparable wavelength can be present with nearly equal amplitude we have to face the problem of their mutual interferences inducing large scale modulations (Fig. 9b). For patterns made of rolls, as in convection, symmetry considerations suggest that these modulations could result from the combination of two basic ingredients: wavelength dilatation/compression and torsion (Fig. 9c, d). In practice, things are more complicated owing to the feedback effects of large scale secondary flows induced by the modulations themselves and to the existence of localized perturbations associated with structural defects. In this introductory section we will forget these sources of difficulty and suppose most of the time that modulations are present only in the direction of the wavevector of the structure (no torsion).

The basic idea is to extend the notion of amplitude $A(t)$ to

1. Outlook

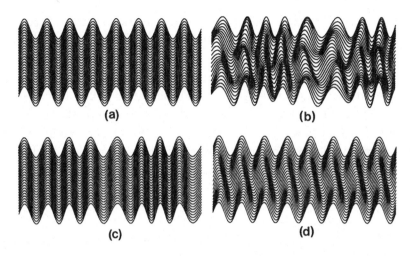

Fig. 9. a) reference structure made of a single Fourier mode; b) modulated structure due to interferences within a wave-packet of modes belonging to linearly unstable band; c) dilatation/compression of the wavelength; d) torsion of the rolls.

account also for spatial dependence $A(x,t)$. At the limit of the laterally infinite system we can assume a "condensation" of the system into a reference state of perfect rolls aligned along the y-axis: $\theta \sim A \cos[k_c(x - x_0)]$, where A measures the "intensity" of convection. However, comparing with the expression introduced in Section 2, we see that the term x_0 has been added serving to locate the rolls, the positions of which are not fixed in the laboratory frame by the instability mechanism. This is made necessary by the translation invariance property of the problem at the limit of infinite size. At this stage, it turns out to be more convenient to rewrite the basic *ansatz* in terms of a complex amplitude: $\theta \sim \frac{1}{2}(A \exp(i\,k_c x) + \text{c.c.})$, therefore assuming $A = |A| \exp(i\phi)$ with $\phi = -k_c x_0$. The evolution equation for the (now) complex amplitude A then reads

$$\tau_0 \frac{dA}{dt} = rA - g|A|^2 A \qquad (1')$$

instead of (1). The evolution equation for the intensity $|A|$ and the phase ϕ are easily obtained from (1'). We get

$$\tau_0 d|A|/dt = r|A| - g|A|^3 \qquad \text{and} \qquad \tau_0 |A|\, d\phi/dt = 0\,.$$

The second equation yields $\phi = \phi_0$, constant, whereas the first equation is strictly equivalent to (1) and gives $|A| \to \sqrt{r/g}$ when $t \to \infty$ for r positive.

Modulations to the reference state are best described by a spatially varying amplitude $A \to A(x,t)$, hereafter called the *envelope*. At least heuristically, it is reasonable to admit that the coherence implied in the instability mechanism will be at the origin of a diffusive relaxation of inhomogeneities. This assumption leads us to complete equation (1') by a term $\xi_0^2 \partial_{x^2} A$, which yields

$$\tau_0 \partial_t A = rA + \xi_0^2 \partial_{x^2} A - g|A|^2 A. \tag{2}$$

As will be shown in Chapter 9, this equation is expected to hold for any well-behaved stationary roll instability close to threshold (apart from possible intricacies alluded to previously).

Some immediate properties can be derived directly from (2) by turning it into a parameterless form. The relevant space and time modulation scales are then seen to diverge as $\xi_x = \xi_0 r^{-1/2}$ and $\tau = \tau_0 r^{-1}$ close to the threshold ($r \to 0$). In this context, "extended geometry" obviously means lateral extension such that $l \gg \xi_x$ and "slow dynamics" means evolution on time scales much longer than τ. Further direct consequences of (2) and its extension to two-dimensional modulations (i.e., for $A(x,y,t)$) will be studied in Chapter 9: the stability of roll patterns, the dynamics of defects at lowest order or the gross features of pattern selection, which will enable us to understand most of the features displayed in Fig. 3.

By analyzing the evolution of the modulus and phase of the envelope we can notice great disparity of behaviors since uniform intensity modulations relax at the finite rate τ whereas uniform phase changes are obviously neutral. At finite but long wavelength, the spontaneous relaxation of intensity modulations is therefore expected on time scales of the order of τ but the phase relaxation should be much slower. Accordingly, there should exist a nonlinear regime sufficiently far from threshold where intensity modulations are slaved to the phase fluctuations and can be eliminated adiabatically. This elimination is not a simple matter (see the first part of

1. Outlook

Chapter 10) but it results in an effective equation for the phase that, in the simplest case, reads

$$\partial_t \phi = D_{\text{eff}}\, \partial_{x^2} \phi$$

where D_{eff} is an effective phase diffusivity.

The calculation of D_{eff} is the first step of the analysis of the nonlinear phase dynamics. A change of sign of the effective diffusivities as a function of the control parameter and the wavevector of the underlying pattern is the symptom of a *phase instability*. In fact, more general modulations have to be considered to account for global properties of patterns, especially when large scale flows are involved, as in convection when dominant physical processes relate to the velocity field rather than to the temperature field. The behavior of the phase is then found more subtle than merely diffusive, and generalized phase equations are required.

This approach in terms of *phase dynamics* is extremely important since it stresses the fact that phase variables are slow variables associated with continuous invariance properties of the pattern (translation, rotation, ...). The associated transition to weak turbulence has gained an irreducible space-time dimension for which the term *phase turbulence* has been coined (Kuramoto and Tsuzuki, 1976). The extension to different physical situations, such as oscillatory instabilities, especially those yielding propagating waves, have received considerable attention recently.

The underlying assumption to the theoretical approach sketched above is that some sort of perturbation expansion remains possible, that is to say that the primary bifurcation is continuous (or weakly discontinuous). Then the envelope formalism, the phase expansion, etc., yield a description of global disorder with reference to some remaining local order and leads to a reasonable understanding of a class of phenomena at the origin of turbulence in large aspect ratio systems. When this approach is valid the problems left should be only technical, not conceptual, even when the level of turbulence increases as the control parameters are varied.

However, this simple picture no longer holds first when the structure, though stable to infinitesimal unlocalized perturbations, is unstable to localized finite amplitude disturbances. We then have to face more serious problems since the perturbative approach may fail owing to the absence of convergence when nothing stable exists "close to" the basic state or, even worse, when the basic state is linearly stable for all values of the control parameter so that the starting point for the expansion is completely lacking (e.g., Poiseuille flow in a circular pipe). The transition to turbulence may then be abrupt locally in space, which can be considered as the spatial generalization of the strictly temporal discontinuous case alluded to above. Unfortunately, nontrivial finite amplitude nonlinear solutions to the unperturbed problem that could serve as a starting point for a theory are not known in general. A somewhat simple possibility remains, which corresponds to a transition to turbulence *via spatio-temporal intermittency* (second main part of Chapter 10). This regime is characterized by the coexistence of patches of turbulence immersed in the rest of structure still in the laminar state; the transition scenario then amounts to a progressive increase of the turbulent fraction by a process akin to a contact contamination (Pomeau, 1986). The account of such a type of transition most naturally involves tools borrowed from statistical physics and no longer from dynamical systems theory, even extended to include the envelope formalism. However in that case we can still say that turbulence is "weak" since, though possibly strong locally, it affects only a part of the system that can be very small, close to the spatio-temporal intermittency threshold.

Up until now, the discussion has been implicitly restricted to the case of closed flows, i.e., flows that take place in laterally bounded systems. However turbulence is more often understood as a dynamical regime taking place in open flows, i.e., flow through pipes, jets, wakes, boundary layer flows, shear layers, etc., all situations where a global transport downstream takes place and where the advection term $\mathbf{v}\cdot\nabla()$ of the hydrodynamic equations is expected to contribute in a specific way. The abstract approach to the transition to turbulence is much less developed in this case owing to an addi-

1. Outlook

tional difficulty that comes from the presence of this global transport. As discussed in the third part of Chapter 10, this leads to a basic distinction between so-called *absolute* and *convective* instabilities, according to whether the unstable mode is strong enough to develop everywhere in spite of the downstream flow or just grows while being carried downstream, leaving a quiescent medium behind it.

Beyond the understanding of the transition to turbulence, including this source of complication, the problem remains of the nature of *developed turbulence* characterized by an energy cascade from large scale coherent structures where energy is injected to small-scale intermittent structures where it is dissipated (Kolmogorov, 1941). Though a global picture is still lacking, we may hope to gain a deeper insight into this problem by combining recent advances in the theory of space-time chaos with our understanding of dissipative dynamical systems, the first being perhaps more adapted to small scales and the second more relevant for the description of the large scales.

4. Bibliographical Notes

General aspects of the generation of non-equilibrium structures are discussed for example in:

[1] G. Nicolis and I. Prigogine, *Self-organization in nonequilibrium systems, from dissipative structures to order through fluctuations* (Wiley, New York, 1977).

[2] H. Haken, *Synergetics*, 3rd Edition (Springer, New York, 1983).

A good introduction to temporal chaos is given in:

[3] P. Bergé, Y. Pomeau, and Ch. Vidal, *Order within chaos* (Wiley, New York, 1987).

At a more technical level, one finds:

[4] H. G. Schuster, *Deterministic chaos* (Physik-Verlag, Weinheim, 1984).

Those interested in original works can consult one of the several collections of reprints that have appeared, notably:

[5] P. Citanović, ed., *Universality in chaos* (Adam Hilger, Bristol, 1984).

[6] Hao Bai-lin, ed., *Chaos* (World Scientific, Singapore, 1984).

Indispensable elements of fluid dynamics with insight in instabilities and turbulence can be found in:

[7] D. J. Tritton, *Physical fluid dynamics*, 2nd Edition (Clarendon Press, Oxford, 1988).

and many interesting illustrations in:

[8] M. Van Dyke, *An album of fluid motion* (The Parabolic Press, Stanford, 1982).

More on instabilities leading to chaos can be found in:

[9] H. L. Swinney and J. P. Gollub, eds., *Hydrodynamic instabilities and the transition to turbulence*, Topics in Applied Physics Vol. 45 (Springer, Berlin, 1981).

whereas a classical starting textbook on developed turbulence is:

[10] H. Tennekes and J. L. Lumley, *A first course in turbulence* (MIT-Press, Cambridge, 1972)

We are not aware of monographs giving a sufficiently broad coverage of spatio-temporal chaos at the level of this introductory chapter. Reference to more specialized items will be given later.

Chapter 2

Evolution and Stability, Basic Concepts

1. General Framework

In this chapter, we discuss the theoretical description of the dynamics of a given system at a formal level. We simply assume that the physical system can be reduced to a well-defined mathematical model in terms of a set of state variables governed by evolution equations depending on parameters. The state variables or *degrees of freedom*, further denoted as $V \equiv \{V\} \equiv \{V_n; n = 1, 2, \ldots\}$, are supposed to form a complete set of generalized coordinates in some relevant *phase space* and the external working conditions applied to the system are specified by an additional set of *control parameters*, $r \equiv \{r\} \equiv \{r_p; p = 1, 2, \ldots\}$. The evolution equations can be written symbolically as

$$\frac{dV}{dt} = F_r(V;t). \tag{1}$$

1.1. Discrete versus Continuous Systems

Several cases can be distinguished according to the nature of the system's description. Firstly, *discrete systems* are described by a finite set of scalar functions of time, as in classical mechanics where a *degree of freedom* traditionally refers to a pair of variables {generalized coordinate, conjugate momentum} rather than to a single variable as will be meant in the following. Other examples can be found, e.g., in electronics with discrete components where variables are electric charges, intensities, and potentials at various points of a given circuit. The evolution of such systems is governed by sets of *ordinary differential equations*, which require the specification of *initial conditions*. To these *time-continuous* systems we can oppose *discrete-time* systems whose variables are defined at a sequence of times ($t_n = n\Delta t; n = 1, \ldots$). The evolution is then described by *iterations*. In population dynamics, for example, the most relevant variables may be the number of members of given species at a given period of the year. Though differential equations can be seen as the limit of difference equations and therefore time-continuous systems as limiting cases of discrete-time systems, it should be kept in mind that, as suggested by the above example, useful reductions to discrete time systems involve some kind of internal clock that legitimates a *stroboscopic* analysis of the system.

Secondly, as can be easily understood from the local-equilibrium assumption, the description of *continuous media* with local properties involves fields governed by *partial differential equations*. Degrees of freedom are no longer scalars but functions of space: the values of fields at points in some domain of physical space, e.g., the three components of the velocity in a fluid. In addition to initial conditions, we have to specify *boundary conditions*. The phase space is then functional and therefore infinite-dimensional. This can be visualized by replacing differential operators by finite differences and increasing progressively the resolution (for practical illustrations of such numerical methods see Appendix 3). More complicated situations can involve time delays (for ordinary differential equations with delay, see Appendix 2, Section 3) and/or nonlocal properties leading to integro-differential equations, all cases that involve an infinite number of degrees of freedom but can be tackled by finite dimensional approximations, at least at a preliminary level.

1.2. Autonomous versus Time-Dependent Systems

The system is said to be *autonomous* when time t does not appear explicitly on the r.h.s of (1). A nonautonomous system can however be made autonomous at the price of a change of independent

2. Evolution and Stability, Basic Concepts

variable increasing its dimension. Indeed, considering $W = (V, t)$ as coordinates in an enlarged phase space and denoting the new independent variable as u, we can write $dV/dt = F_r(V;t)$ as $dV/du = F_r(V;t)$ with the additional equation $dt/du = 1$. The resulting set, $dW/du = \widetilde{F}_r(W)$, is then autonomous. The equation for t is trivially integrated to give $t = u + t_0$, where t_0 controls the phase of the system with respect to the forcing. In practice, only the case of periodic forcing can be treated in this way.

Another case of interest is when the explicit time dependence in equation (1) arises from the presence of "external noise":

$$\frac{dV}{dt} = F_r(V) + \zeta(t), \qquad (1')$$

where $\zeta(t)$ is a small random term with well defined statistical properties. Equation (1') is called a *Langevin equation*. In the presence of noise, only statistical properties can be predicted, the instantaneous values of observables losing their interest. One has then to turn from stochastic evolution equations to functional equations of the Fokker–Planck type that govern the probability densities of the variables under study.

1.3. Deterministic Evolution

Though periodic forcing and external noise are of interest, in the following and except when explicitly stated, we assume that the system is deterministic and autonomous:

$$\frac{dV}{dt} = F_r(V). \qquad (2)$$

The very notion of determinism is intimately associated with the property of existence and uniqueness of the *trajectory* that passes through a given point in phase space at a given time. For ordinary differential equations, this property holds if the *vector field* F_r is differentiable with continuous first derivative (class $\mathcal{C}^1$). However, existence and uniqueness is usually only *local* in time since the possibility of a singularity at some finite time is not discarded (for example, consider $dA/dt = A^2$ on R, which yields $1/A - 1/A_0 = -t$ with $A = A_0$ at $t = 0$; the solution then diverges at $t_* = 1/A_0 > 0$ when A_0 is positive). However, when the phase space is *compact*, i.e., closed and bounded, say a circle or a n-dimensional torus T^n, the possibility of a divergence is excluded so that the existence-and-uniqueness property becomes *global* in time. An *evolution operator* can be defined, mapping the phase space onto itself: $A(t) = U_t(A_0)$, so that:

$$A(t_1 + t_2) = U_{t_1+t_2}(A_0) = U_{t_2}(A(t_1))$$
$$= U_{t_2}(U_{t_1}(A_0)) = [U_{t_2} \circ U_{t_1}](A_0).$$

The distinction between the *transient* and *permanent regime* is essential: in particular, the "steady-state regime" must be understood as the regime that is reached asymptotically, long after transients have decayed (long time limit). Permanent regimes can be simple, i.e., time-independent, time-periodic, or more complicated, i.e., quasi-periodic or even chaotic. The fundamental theoretical problem is then to "predict" the characteristics of the permanent regime reached by a given system as a function of available initial conditions and control parameters.

1.4. Perturbations and Stability

Let $V_0(t)$ correspond to some permanent regime, hereafter called the *basic state*. Usually, it can be thought of as deriving continuously from thermodynamic equilibrium along the *thermodynamic branch*, therefore complying with the space-time symmetries of the external constraints (boundary conditions and applied forces, spatially uniform or not, time independent or not). Here, we consider time-independent external constraints yielding an autonomous problem. The basic state then fulfills

$$\frac{dV_0}{dt} = F_r(V_0), \tag{3}$$

which defines V_0 as an implicit function of r.

2. Evolution and Stability, Basic Concepts

A permanent regime is observable if it is *stable*, i.e., robust to perturbations. Let V be another solution and define the perturbation V' as
$$V' = V - V_0\,.$$
The evolution of V' is governed by

$$\frac{dV'}{dt} = F_r(V_0 + V') = \widetilde{F}_{r,V_0}(V')\,, \qquad (4)$$

which depends on V_0 implicitly as recalled by the second subscript to $\widetilde{F}$. Now, $V' = 0$ is a solution of (4), since V_0 is a solution of (3). losely speaking, the basic solution V_0 is *stable* if perturbation V' remains under control.

2. Global Stability

2.1. General Viewpoint and Definitions

The stability of thermodynamic equilibrium is analyzed in terms of the extrema of a *potential* (the free energy for a purely thermal system that can exchange energy with the exterior world). In the same vein, sufficiently close to equilibrium, fluxes remain proportional to thermodynamic forces and the linear response regime can be discussed from the minimization of a quantity called the *entropy production*. Farther from equilibrium, the basic idea is then to try to analyze the stability of the system in terms of the extrema of a function which generalizes the entropy production. The aim is thus the determination of such a function, called a *Lyapunov function*, to get stability criteria not restricted, if possible, to an infinitesimal neighborhood of the basic state.

Let $V_0(t)$ be a reference trajectory initiated at $V_0(0)$ and $V(t)$ be another trajectory starting at $V(0)$, both governed by (2), and let $|\cdots|$ denote a distance between two points in phase space; then:

- the basic state is said to be *uniformly stable* if the distance between the basic solution and the perturbed solution can be kept arbitrarily small for all times, or more formally, if, ϵ being given, δ can be found such that

$$|V(0) - V_0(0)| < \delta \quad \Rightarrow \quad |V(t) - V_0(t)| < \epsilon \quad \text{for} \quad t > 0;$$

- it is *asymptotically stable* if, in addition, the distance $|V(t) - V_0(t)|$ tends to zero when time increases.

Let $G(V)$ be a function of V, continuous with continuous first order partial derivatives. Assume further that it is *positive definite*, i.e., such that

$$G(V = 0) = 0 \quad \text{and} \quad G(V \neq 0) > 0.$$

If, as a function of t through V, G is decreasing ($dG/dt \leq 0$) it is called a *Lyapunov function* for the system.

If such a Lyapunov function exists, then, considering level curves of G, one can show that the basic state $V = 0$ is stable. Moreover, if the time derivative dG/dt is negative definite, the basic state is asymptotically stable (Lyapunov's theorems, see Bibliographic Notes for references).

2.2. Energy Method

The problem is therefore to find recipes for constructing a useful G. One such recipe is given by the *energy method*. Consider a particle submitted to a viscous friction governed by $m\,d\mathbf{v}/dt = -\eta\,\mathbf{v}$. The kinetic energy $K = \frac{1}{2}m\mathbf{v}^2$ is positive definite and

$$\frac{dK}{dt} = m\mathbf{v}\frac{d\mathbf{v}}{dt} = \mathbf{v}(-\eta\mathbf{v}) = -\eta\mathbf{v}^2$$

is negative definite so that solution $\mathbf{v} = 0$ is stable as expected. We are then looking for a generalization of this quantity for systems such as (2) and we can try the "kinetic energy" $K = \frac{1}{2}\sum_n V_n^2$ and study $dK/dt = \sum_n V_n\,dV_n/dt = \sum_n V_n F_n$ (here the subscript n has been restored to distinguish between the different degrees of freedom; in the continuous case the sum is replaced by an integral).

2. Evolution and Stability, Basic Concepts

2.3. Different Concepts of Stability

Methods based on the search for a Lyapunov function lead to the definition of different levels of stability. We say that we have:

- *asymptotic stability* if $dK/dt < 0$ holds in the vicinity of the basic state;
- *global monotonic stability* if the "energy" contained in arbitrary perturbations can only decrease, e.g., in the immediate vicinity of thermodynamic equilibrium;
- *asymptotic stability in the mean* if we have global stability but, before decreasing, the energy contained in some perturbations can grow at the beginning (a less stringent condition);
- *conditional stability* if we have asymptotic stability only for perturbations with an energy smaller than some upper bound (still less stringent);
- *linear instability* if we have instability even against infinitesimal perturbations.

The stability of a given basic state usually depends on control parameters, which leads to the determination of *stability criteria* resulting in a *stability diagram* of the type displayed in Fig. 1. The situation can be summarized as follows:

- Global methods (especially the energy method) yield *sufficient* conditions of *stability*: $r < r_g \Rightarrow$ stability. Bounds obtained in that way are usually very conservative. Below r_g, a threshold for "monotonous" stability r_m can sometimes be determined that marks the difference between regions I and II.
- Linear stability analysis yields a *sufficient* condition of *instability*: indeed $r > r_c$ implies instability. It is a local method and the energy of relevant perturbations goes to zero.
- The global condition of stability and the local condition of instability are both not necessary; in case of *conditional stability*, the basic flow can remain stable for $r > r_g$ or become unstable for $r < r_c$ if some control on the amplitude of the perturbations can be kept (region III and IV).

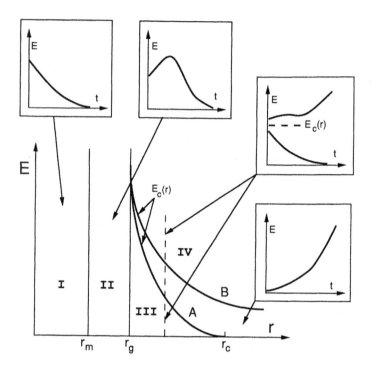

Fig. 1. Stability of time-asymptotic regimes of a given system as a function the "energy" of the perturbations and the control parameters.

The conditional stability threshold can often be obtained by a perturbation expansion around the linear instability threshold ("weakly nonlinear theory", extremity of curve A close to r_c). The situation is more delicate when the basic flow is linearly stable at all values of r since then the starting point of the perturbation expansion is missing (curve B, e.g., Poiseuille flow in a pipe with a circular section). In this latter case, though stable against all (unlocalized) infinitesimal perturbations, the basic state can be unstable to finite amplitude localized perturbations and a direct transition to turbulence can be observed (see Chapter 10).

Different kinds of perturbations can be distinguished according to whether they can or cannot be controlled to some extent by improving the experimental setup or/and reducing the external

2. Evolution and Stability, Basic Concepts

noise. Specific perturbations with well-defined space-time behavior and controlled amplitude can even be introduced to excite internal instability modes. In any case, infinitesimal perturbations are unavoidable so that *linear instability* actually means *unconditional instability*; if the basic solution is linearly unstable, it cannot be observed.

3. Normal Modes, Linear and Nonlinear Dynamics

Global methods do not tell much about the physical origin of instabilities mainly because they assume little about the perturbations responsible for them. To get more information, we must restrict the class of disturbances considered to those evolving from infinitesimal fluctuations which, being governed by linear equations, are easier to handle. We now turn to this *local* approach to stability analysis.

3.1. Normal Mode Analysis

Evolution equations for infinitesimal perturbations can be obtained in a systematic way by inserting $V = V_0 + \epsilon V'$ in the primitive equations $dV/dt = F_r(V)$ and expanding them in powers of ϵ:

$$\epsilon \frac{dV'}{dt} = \epsilon L_r V' + \epsilon^2 N_2(V', V') + \ldots ;$$

here L_r is a linear operator, the formal derivative of F_r with respect to the variable V around V_0: $L_r = \delta F_r/\delta V|_{V_0}$; N_2 gathers terms quadratic in V', etc. (terms of order ϵ^0 vanish identically). Neglecting all equations beyond first order, we get the linearized problem

$$\frac{dV'}{dt} = L_r V'. \tag{5}$$

Stability analysis is basically an initial value problem. As can be shown easily (see Appendix 2), the general solution of (5) is formally given by $V'(t) = U_t V'(0)$ where the linear evolution operator U_t is given by

$$U_t = \exp(L_r t) = \sum_k \frac{t^k}{k!} L_r^k.$$

In principle, the general solution of the linearized problem can be obtained as a *superposition* of infinitesimal perturbations isolated by *projection* onto the eigenvectors of L_r. In practice, this may not be simple because L_r cannot always be written in strictly diagonal form, which can happen when some eigenvalues are degenerate. Generalized eigenvectors have then to be introduced leading to Jordan normal forms and secular terms (see Chapter 5, Section 5.1 and Appendix 2). For simplicity, we consider here only the case of a nondegenerate spectrum. Operator L_r can then always be put in strictly diagonal form and, in the one-dimensional eigen-subspace associated with the root s_n of the characteristic equation $\det(L_r - sI) = 0$, we simply have $\exp(L_r t) = \exp(s_n t)$.

The evolution of an arbitrary *superposition* $V' = \sum_n A_n X_n$, where the coefficient A_n is called the *amplitude* of the mode X_n, can be discussed in terms of the algebraic properties of the linearized problem. When the problem is self-adjoint, eigenvalues are all real; when it is not, eigenvalues may be complex, but since the primitive problem is real, they always form complex conjugate pairs. Let us make this less restrictive assumption and write $s_n = \sigma_n + i\omega_n$. The time dependence of the amplitude A_n is then given by

$$\exp(s_n t) = \big(\cos(\omega_n t) + i\sin(\omega_n t)\big) \exp(\sigma_n t).$$

The real part σ_n of s_n is called the *growth rate*:

- $\sigma_n < 0$: the perturbation decays, the mode X_n is *stable*, (more strictly speaking, the basic state is stable against mode X_n);
- $\sigma_n > 0$: the perturbation grows, the mode is *unstable*;
- $\sigma_n = 0$: the perturbation neither grows nor decays, the mode is *marginal* or *neutral*.

Note that in case of nondiagonalizable degeneracy, secular terms are of the form $(\alpha_0 + \alpha_1 t + \alpha_2 t^2 \ldots) \exp(s_n t)$ so that the marginal stabil-

2. Evolution and Stability, Basic Concepts

ity condition stated above does not exclude a less-than-exponential growth of perturbations at the linear stage.

A linear instability takes place when at least one mode is unstable. Eigenmodes can be ordered by decreasing values of the growth rate: $\sigma_1 > \sigma_2 > \ldots$; the most unstable or least stable mode, here X_1, is often called the "most dangerous" mode.

The imaginary part ω_n of s_n controls the rest of time dependence:

- $\omega_n \neq 0$: the mode is *oscillating*;
- $\omega_n = 0$: the mode is *stationary*.

The simplest case corresponds to a fully discrete system, e.g., an electronic circuit with discrete components. Operator L_r is then merely a constant matrix and we are left with a purely algebraic problem. Eigenmodes are linear combinations of basic variables and the spectrum of L_r is discrete. When the system is a continuous medium, modes X_n are still functions of space. They are solutions of a *homogeneous boundary value problem*.

Consider for example a fluid governed by Navier–Stokes equations; nonlinearities are contained in the advection term $\mathbf{v}\cdot\nabla\mathbf{v}$, which for $\mathbf{v} = \mathbf{v}_0 + \mathbf{v}'$ gives at lowest order $\mathbf{v}'\cdot\nabla\mathbf{v}_0 + \mathbf{v}_0\cdot\nabla\mathbf{v}'$. The linearized problem is thus a set of partial differential equations whose coefficients depend in general on space through $\mathbf{v}_0$ and $\nabla\mathbf{v}_0$. In addition, since $\mathbf{v}_0$ fulfills all boundary conditions at order ϵ^0, boundary conditions on $\mathbf{v}'$ are homogeneous ($\mathbf{v}'|_b \equiv 0$).

When the system is unbounded in some directions the spectrum displays a continuous part. In fact, a whole series of continuous branches depending on separation parameters are obtained. For example, perturbations in an unbounded horizontal fluid layer between two horizontal plates with uniform boundary conditions (viz. the ideal Rayleigh–Bénard experiment) are best analyzed using Fourier modes of the form

$$X_n(x, y, z) = \exp(i\mathbf{k}_h \cdot \mathbf{x}_h)\widehat{X}_n(z),$$

where the two-dimensional horizontal wavevector $\mathbf{k}_h = (k_x, k_y)$ is the separation parameter.

Partial derivatives ∂_x and ∂_y are then converted into ik_x and ik_y, respectively, so that the partial differential problem in x, y, and z is reduced to an ordinary differential problem in z. Denoting the derivative with respect to z as ∂ (without subscript) we have $L_r = L_r(\partial_x, \partial_y, \partial_z) = L_r(ik_x, ik_y, \partial)$ so that the eigenvalue problem reads $sX = L_r(ik_x, ik_y, \partial)X$. Eigenvalues are then labeled by the wavevector $\mathbf{k}_\mathrm{h}$ in addition to the discrete index n associated with the confinement direction z:

$$s_n = s_n(r, \mathbf{k}_\mathrm{h}) = \sigma_n(r, \mathbf{k}_\mathrm{h}) + i\omega_n(r, \mathbf{k}_\mathrm{h}).$$

If the system is quasi-one-dimensional, e.g., a narrow channel or a pipe, there is only one wavevector component but two discrete indices. In the following, we no longer specify which situation we refer to and simply denote the separation parameter as k, further dropping the subscript h.

Usually, experiments are performed by increasing r without controlling k. At a given r, for each branch we can determine

$$\tilde{\sigma}_n(r) = \sup_k \sigma_n(r, k)$$

and find out which $\tilde{\sigma}_n$ goes though zero first, i.e., which mode (k, n) becomes unstable first (Fig. 2a).

In fact, it is customary to present the results in a different way by drawing lines of constant σ in the plane (r, k), and focusing on regions corresponding to unstable or marginal modes ($\sigma \geq 0$, see Fig. 2b). The *marginal stability condition* of the most unstable mode is then given by

$$\sigma_1(r, k) = 0,$$

and the *threshold curve* is obtained by solving this relation for r:

$$r = r_1(k).$$

This curve typically displays a minimum at $k = k_\mathrm{c}$ for $r = r_\mathrm{c} = r_1(k_\mathrm{c})$, which defines the *instability threshold*. The instability is

2. Evolution and Stability, Basic Concepts

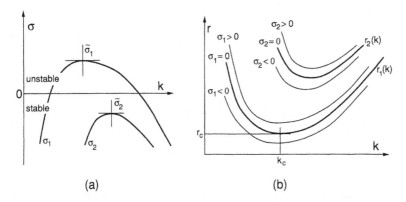

Fig. 2. a) Eigenmodes can be ordered according to the value of the real part of their growth rate σ_n as a function of the wavevector k at given control parameter r. b) Solving conditions $\sigma_n(r, k) = \sigma$ for r yields a different presentation of the linear stability problem; the critical wavevector k_c and the threshold r_c are given by the minimum of the marginal stability curve $\sigma = 0$ corresponding to the most unstable mode.

called *cellular* when the minimum occurs at $k_c \neq 0$ and *homogeneous* when $k_c = 0$. Instabilities in continuous media can therefore be classified according to the value of ω_c and k_c: stationary ($\omega_c = 0$) or oscillatory ($\omega_c \neq 0$), homogeneous ($k_c = 0$) or cellular ($k_c \neq 0$).

Notice that the critical mode can be highly degenerate owing to symmetry. Considering more specifically the case of a layer with $k \to \mathbf{k}_h = (k_x, k_y)$, we easily understand that if the instability mechanism is isotropic in the horizontal plane, continuous rotational invariance implies that only the modulus k_c of $\mathbf{k}_h$ is fixed by the marginal stability condition, not the orientation (this is the case of Rayleigh–Bénard convection in isotropic liquids which is cellular, $k_c \neq 0$, and stationary, $\omega_c = 0$). In contrast, anisotropy in the horizontal plane implies discrete symmetry properties, e.g., reflection with respect to some given vertical plane, so that a finite set of critical wavevectors is to be found, typically a single pair $\mathbf{k}_c = \pm k_c \hat{\mathbf{x}}$ where $\hat{\mathbf{x}}$ is the unit vector in the horizontal direction singled out by the instability mechanism (case of convection in nematic liquid crystals, see Chapter 3).

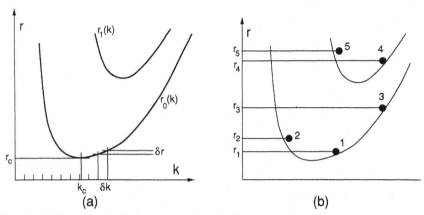

Fig. 3. a) In extended geometry, $\ell \gg 2\pi/k_c$, normal modes form a quasi-continuum. b) In confined geometry, $\ell \sim 2\pi/k_c$, normal modes are isolated.

However, systems are always laterally bounded and we have to find the solution of a fully three-dimensional homogeneous boundary value problem. This results in an infinite series of discrete eigenvalues and associated eigenmodes with well-defined spatial structures. Two distinct situations arise according to the value of the *aspect ratio*, either large or small. Assuming for simplicity that the typical length-scale of the instability $\lambda_c = 2\pi/k_c$ is finite, i.e., $k_c \neq 0$, we define the aspect ratio as $\Gamma = \ell/\lambda_c$ where ℓ is a typical measure of the lateral size. Then, when lateral boundaries are set at a large distance, $\Gamma \gg 1$, the actual spectrum is discrete but remains quasi-continuous. For periodic boundary conditions, possible wavevectors are of the form $k_n = 2\pi n/\ell$ so that $k_{n+1} - k_n = 2\pi/\ell \ll k_c$. Neighboring modes have comparable structure and differ only by the precise value of their wavevector. Moreover, as illustrated in Fig. 3a, they are quasi-degenerate and many of them can be destabilized even close to the threshold. The opposite limit, $\Gamma \sim 1$, is illustrated in Fig. 3b. A discrete eigenproblem has then to be solved yielding a fully discrete spectrum. The very concept of wavevector loses its meaning and critical modes are better indexed by integers. We have no right to draw the marginal stability curve, except perhaps to remember the physical origin of the normal modes.

3.2. Weakly Nonlinear Dynamics

At the linear stage, the evolution of perturbations amounts to an indefinite exponential growth or decay. The linear theory is likely to fail early for growing modes and we must worry about the evolution of perturbations that contain unstable components. Since much work has already been done to understand the eigenvalue problem, we must try not to lose the information gained but to take advantage of it for representing the general nonlinear solution using the amplitudes introduced earlier (the most natural coordinates of the system). Here we consider the case of a fully discrete spectrum, postponing the necessary adaptations to the case of a quasi-continuous spectrum to the final chapters.

Considering the evolution of a disturbance corresponding to a pure mode X_n with initial amplitude A_0, we can write

$$V'(t) = A_0 \exp(s_n t) X_n = A_n(t) X_n,$$

that is to say $A_n(t)$ is a solution of

$$\frac{dA_n}{dt} = s_n A_n \quad \text{with} \quad A_n(0) = A_0. \tag{6}$$

Such an equation can be obtained directly by projecting (5) onto the basis of eigenvectors, which implies using the algebraic properties of the linearized problem (scalar product). At this step, amplitudes of modes belonging to different eigenspaces are still uncoupled. The required extension of (6) including nonlinear interactions is derived by projection of the complete problem (4).

The linear problem is not self-adjoint in general. By definition, the adjoint problem is obtained from

$$\langle \widetilde{V} | L_r V \rangle = \langle L_r^\dagger \widetilde{V} | V \rangle = \overline{\langle V | L_r^\dagger \widetilde{V} \rangle}$$

where $\langle \cdot | \cdot \rangle$ denotes the scalar product (see Appendix 2, Section 2.2). Eigenvectors of $L_r^\dagger$ are obtained by solving

$$L_r^\dagger \widetilde{V} = s \widetilde{V}.$$

This results in a bi-orthogonal series of eigenfunctions $\{\widetilde{X}_n, X_n\}$ such that
$$(\bar{s}_m - s_n)\langle \widetilde{X}_m | X_n \rangle = 0,$$
that is to say $\widetilde{X}_m$ and X_n are orthogonal if s_m and s_n are not conjugate complex.

Substituting $V' = \sum_n A_n X_n$ in system (4), performing the scalar product by $\widetilde{X}_n$ to isolate the equation governing the amplitude A_n of mode X_n, we get
$$\frac{dA_n}{dt} = s_n A_n + \langle \widetilde{X}_n | N(\Sigma_m A_m X_m) \rangle,$$
where N gathers all nonlinear contributions (i.e., $\sum_p N_p$ where N_p is formally of order p). In order to include also the nondiagonalizable case, we should use generalized eigenvectors and rewrite this equation with s_n replaced by the restriction L_n of L_r to the generalized eigenspace corresponding to that eigenvalue.

In hydrodynamics, nonlinearities are given by a bilinear form so that we expect
$$\frac{dA_n}{dt} = L_n A_n + \sum_{m,p} g_{n,mp} A_m A_p,$$
where coefficients $g_{n,mp}$ denote the scalar products involving $\widetilde{X}_n$ and X_m, X_p and their spatial derivatives. More generally, this projection procedure yields a series
$$\frac{dA_n}{dt} = L_n A_n + \sum_{mp} g_{n,mp} A_m A_p + \sum_{mpq} g_{n,mpq} A_m A_p A_q + \ldots,$$
where the structure of the coupling terms stems from spatial resonance properties between modes X_n.

Let us illustrate this on a schematic example
$$\partial_t v + v\, \partial_x v = Lv \qquad (7)$$
with $LX_n = s_n X_n$ for periodic perturbations with period ℓ: $X_n = \sin(2\pi n x/\ell)$ in the x-direction. The set $\{X_n\}$ can be taken as a basis

2. Evolution and Stability, Basic Concepts

in the functional state space. The general solution is then expanded on this basis: $v = A_1 X_1 + A_2 X_2 + \ldots$, which defines the set of amplitudes $\{A_n\}$. Inserting this expansion into the evolution equation and Fourier analyzing the nonlinear term (the projection step) we obtain a set of ordinary differential equations in the announced form. If, for simplicity, we truncate the expansion and keep only two modes, the fundamental mode $n = 1$ and its first harmonic $n = 2$, we get:

$$\frac{dA_1}{dt} = s_1 A_1 + \frac{k}{2} A_1 A_2 ,$$
$$\frac{dA_2}{dt} = s_2 A_2 - \frac{k}{2} A_1^2 . \tag{8}$$

If, in addition, we assume that the fundamental mode is close to the threshold $2\pi/\ell \sim k_c$, i.e., $s_1 \sim 0$, and the first harmonic is stable and far from critical, then we can proceed to the adiabatic elimination of A_2 to get an effective equation for A_1 alone, as will be discussed in depth in Chapter 5.

This possibility of distinguishing between stable and unstable modes, with the assocoated *adiabatic elimination of slaved modes* leading to a small set of *central modes* in effective interaction, is precisely what makes this approach so important; otherwise the progress from an initial formulation in terms of partial differential equations (primitive problem) to an infinite set of nonlinear ordinary differential equations could seem minor. Unfortunately, the plain projection procedure sketched above can be performed in a straightforward way only when the analytical expression of eigenmodes can be handled easily (e.g., Rayleigh–Bénard convection with stress-free boundary conditions, see Chapter 4). More frequently, it has to be combined with some sort of perturbation expansion, which obscures the picture a little, especially in the case of a quasi-continuous spectrum requiring special techniques.

4. Qualitative Dynamics

4.1. Elements for a Phase Portrait

At this stage, it is interesting to introduce some vocabulary from the theory of *dissipative dynamical systems* which deals with *qualitative* properties of trajectories describing the evolution of a given system in its phase space. This will be done mainly in a pictorial way using system (8) which, by a proper rescaling of its variables, will be rewritten as

$$\frac{dA_1}{dt} = rA_1 + A_1 A_2,$$
$$\frac{dA_2}{dt} = -A_2 - A_1^2, \tag{9}$$

where the control parameter r ($= s_1/|s_2|$) is the only remaining variable coefficient.

The *vector field* corresponding to System (9) is sketched in Fig. 4. It is parallel to the A_2 axis when dA_1/dt vanishes, that is to say $rA_1 + A_1 A_2 = A_1(r + A_2) = 0$, i.e., $A_1 = 0$ and $A_2 = -r$. In the same way, it is parallel to the A_1 axis when $A_2 = -A_1^2$. At the intersections of these lines, we have both $dA_1/dt = 0$ and $dA_2/dt = 0$. The trajectories initiated at such points reduce trivially to the points themselves since the system does not evolve. These conditions define the *fixed points*, also called the *singular, critical* or *equilibrium* points, of the vector field. In contrast, trajectories initiated at any other point, called *regular*, do not reduce trivially to a point.

Let us consider again the fixed points. For $r > 0$, we have three of them, the first one remains permanently at the origin $A_1 = A_2 = 0$; the two others are nontrivial, with $A_2 = -r$ and $A_1 = \pm\sqrt{r}$. Let (A_1', A_2') be an infinitesimal perturbation around one of these fixed points. They belong to the *tangent space* at the corresponding point and evolution equations linearized around that point define the *tangent operator* (note that a tangent space and a tangent operator can be defined at any point and not only at fixed points).

2. Evolution and Stability, Basic Concepts

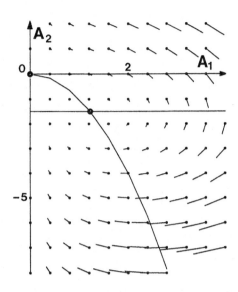

Fig. 4. Vector field defined by system (9) for $A_1 \geq 0$ and $r = 1.5$; complete by symmetry for $A_1 < 0$.

At the origin, the tangent system reads

$$\frac{dA'_1}{dt} = rA'_1, \qquad \frac{dA'_2}{dt} = -A'_2,$$

i.e., the operator is diagonal with eigenvalues r and -1. The corresponding eigenvectors are directed along the A_1 axis (unstable since we assume $r > 0$) and the A_2 axis (stable), respectively. The tangent space at the origin is split into a stable subspace and an unstable subspace, both one-dimensional. Such a point is called a *saddle*.

The case of the other fixed points is less trivial. For example, at the point $(\sqrt{r}, -r)$, inserting $A_1 = \sqrt{r} + A'_1$ and $A_2 = -r + A'_2$ into (9), we obtain

$$\frac{dA'_1}{dt} = \sqrt{r}A'_2, \qquad \frac{dA'_2}{dt} = -A'_2 - 2A'_1\sqrt{r},$$

so that the tangent operator is no longer diagonal. The eigenvalues are the roots of the characteristic equation $s(s+1) + 2r = 0$, i.e., $s_+ = -2r + \mathcal{O}(r^2)$ and $s_- = -1 + 2r + \mathcal{O}(r^2)$. The eigenvectors associated with s_+ and s_- are $(1, -2\sqrt{r})$ and $(-\sqrt{r}, 1)$, respectively. This fixed point is stable in all directions; it is called a *sink*. The

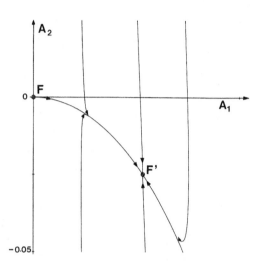

Fig. 5. Phase portrait of system (9) in the neighborhood of its fixed points for $r = 0.025$; stable and unstable manifolds are the special trajectories arriving at or emerging from the fixed points and tangent to the eigenvectors of the linear stability problem at these points.

opposite case of a completely unstable fixed point would be called a *source*.

Now, suppose that we take an initial condition with $A_1 = 0$; then we have $A_1(t) \equiv 0$ and $A_2(t) = A_2(0)\exp(-t)$, which tends to zero when $t \to +\infty$. This special trajectory is an example of *stable manifold* of a fixed point. More mathematically, the stable manifold is formed by the set of trajectories that are tangent to the stable linear subspace when $t \to +\infty$. In the same way, the *unstable manifold*, which can be naively understood as the extrapolation of a trajectory starting infinitely close to the origin along the unstable direction, is better defined as the set of trajectories that are tangent to the unstable subspace when $t \to -\infty$. In Fig. 5, the unstable manifold of the origin is seen to reach the nontrivial fixed point along its least stable direction.

The stable manifold of the nontrivial fixed point is two-dimensional. However, we can define two submanifolds tangent to each stable eigenvector. The submanifold tangent to the least stable direction has already been obtained as the unstable manifold of the origin. The other can be obtained most easily as an unstable manifold of the time-reversed system.

Stable manifolds of different fixed points cannot intersect owing

2. Evolution and Stability, Basic Concepts

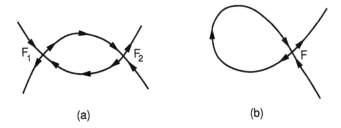

Fig. 6. Example of heteroclinic (a) and homoclinic (b) trajectories.

to the property of uniqueness of trajectories passing through a given point in phase space since the crossing point would have two possible futures. The same holds for unstable manifolds of different fixed points (use the time-reversed system). In contrast, stable and unstable manifolds of different fixed points can intersect each other along special sets of trajectories called *heteroclinic trajectories*. Such trajectories come asymptotically from one fixed point along its unstable manifold and reach asymptotically the other along its stable manifold. In the same way, nothing forbids the existence of a *homoclinic trajectory* that starts from a fixed point along its unstable manifold and comes back to the same fixed point along its stable manifold (see Fig. 6). The positions of the stable and unstable manifolds thus put severe constraints on the global aspect of the *phase portrait* of the system.

4.2. Absorbing Zones and Limit Sets

Let us now turn to the characterization of the asymptotic regime that develops when t tends to infinity. For ideal mechanical systems, Liouville's Theorem ensures the conservation of volumes in phase space under the evolution. The number of accessible states remains constant and the possibility of a chaotic behavior emerges from the interplay between this conservation and the instability of certain configurations, as realized first by Poincaré at the turn of the century. In contrast, dissipation is expected to reduce the number of accessible asymptotic states, e.g., only the rest state for a real free pendulum with viscous friction. In phase space, the contraction of volumes is measured by the divergence of the vector field. For system (9), we have

$$\frac{\partial}{\partial A_1}(rA_1 + A_1 A_2) + \frac{\partial}{\partial A_2}(-A_2 - A_1^2) = r - 1 + A_2,$$

from which we see that contraction is not uniform but effective only for $A_2 < 1 - r$. However, it turns out that the expanding region plays a role only during the transient since, whatever the initial condition, the trajectory is rapidly brought into the contracting region. To set this fact on firmer ground and prove the confinement of trajectories in phase space, we can adapt the energy method presented in Section 2, introducing first $K = \frac{1}{2}(A_1^2 + A_2^2)$ and computing

$$\frac{dK}{dt} = \frac{d}{dt}\left(\frac{1}{2}\left(A_1^2 + A_2^2\right)\right) = rA_1^2 - A_2^2. \tag{10}$$

The absence of cubic terms in (10) stems from the structure of the nonlinear term in (7), $v\,\partial_x v$. This important property is characteristic of systems of "hydrodynamic type" whose nonlinearities derive from the advection term $\mathbf{v}\cdot\nabla\mathbf{v}$ which conserves the energy (in the ordinary sense, i.e., $\frac{1}{2}\mathbf{v}^2$, here simply $\frac{1}{2}v^2$).

Let us consider first the case $r < 0$. The quadratic form on the r.h.s. being negative definite, the energy decays in all circumstances so that all trajectories approach the origin which, as we already know, is a stable fixed point. This decrease of the energy can be interpreted as the shrinking with time of a disc of radius $A = \sqrt{A_1^2 + A_2^2}$ centered at the origin. Such a domain of phase space is called an *absorbing zone*. More generally, an absorbing zone is a region at the boundary of which the vector field always points inwards.

When $r > 0$, no firm conclusion can be drawn from dK/dt which is no longer definite negative. However, it is easily seen that adequately centered, sufficiently large discs remain absorbing. Indeed, performing a translation of the origin, $A_i = \tilde{A}_i + A'_i$ ($i = 1, 2$),

2. Evolution and Stability, Basic Concepts

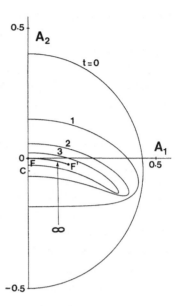

Fig. 7. Shrinking in time of a circular domain centered at C for model (9) with $r = 0.025$; complete by symmetry for $A_1 < 0$.

which yields

$$\frac{dA'_1}{dt} = \left(r\tilde{A}_1 + \tilde{A}_1\tilde{A}_2\right) + \left(r + \tilde{A}_2\right)A'_1 + \tilde{A}_1 A'_2 + A'_1 A'_2,$$

$$\frac{dA'_2}{dt} = -\tilde{A}_2 - \tilde{A}_1^2 - A'_2 - 2\tilde{A}_1 A'_1 - {A'_1}^2,$$

and computing the time derivative of $K' = \frac{1}{2}({A'_1}^2 + {A'_2}^2)$, we obtain

$$\frac{dK'}{dt} = \left(r\tilde{A}_1 + \tilde{A}_1\tilde{A}_2\right)A'_1 - \left(\tilde{A}_2 + \tilde{A}_1^2\right)A'_2$$
$$+ \left(r + \tilde{A}_2\right){A'_1}^2 - \tilde{A}_1 A'_1 A'_2 - {A'_2}^2.$$

Taking $\tilde{A}_i$ ($i = 1, 2$) so that the quadratic form on the r.h.s. is negative definite will prove the result since dK'/dt is dominated by its quadratic term at sufficiently large distances. This condition results in the inequality $\tilde{A}_2 < -r - \tilde{A}_1^2/4$, and assuming $\tilde{A}_1 = 0$ and $(r + \tilde{A}_2) = -Q < 0$ we obtain

$$\frac{dK'}{dt} = (Q+r)A'_2 - Q{A'_1}^2 - {A'_2}^2.$$

As an example, Fig. 7 displays the time evolution of a disc centered at $(0, -2r)$ showing that trajectories cannot escape to infinity.

4.3. Limit Sets and Attractors

From the picture just given, we see that when time proceeds the system always enters the region where contraction is effective. The surface of any domain taken in the absorbing zone therefore shrinks to zero. In fact, a domain containing the three fixed points for $r > 0$ is seen to collapse on a curve segment going from one nontrivial fixed point to the other and passing through the origin as time goes on. Such a set is called an *attracting part* for the dynamical system. It contains some of its *limit sets*. A limit set relative to some initial condition is the set of accumulation points of the corresponding trajectory. For model (9), when r is positive, we have obviously three limit sets: the origin and the two nontrivial fixed points. All trajectories reach one or the other nontrivial fixed point except those starting with $A_1 = 0$ and reaching the origin. The latter is an example of *exceptional limit set* since the corresponding set of initial conditions has a vanishing measure (a line embedded in a two-dimensional phase space) and therefore cannot be picked "at random" with nonzero probability. On the other hand, the *inset* of the nontrivial fixed points (the set of initial conditions that lead to it) has a finite surface; accordingly, randomly chosen initial conditions will lead to one or the other with nonzero probability. We can even see that all initial conditions taken in a sufficiently small (topologically) *open* neighborhood of each of the fixed points converge to them. Each of them is an *attractor* of the flow. Notice that an attractor must be *indecomposable*: it cannot be split into several independent components corresponding to distinct asymptotic behaviors; here the two nontrivial fixed points belong to the same attracting set but form two distinct attractors.

The *basin of attraction* of a given attractor is the set of all points in phase space that tend asymptotically to it as time increases. From Fig. 7 we see that the basin of attraction of each nontrivial fixed point is an open half-plane, either $A_1 > 0$ or $A_1 < 0$. Their common boundary is the stable manifold of the origin, an exceptional limit set.

That the inset of an attractor must contain an open neighborhood actually means that no trajectory emerges from the attractor,

2. Evolution and Stability, Basic Concepts

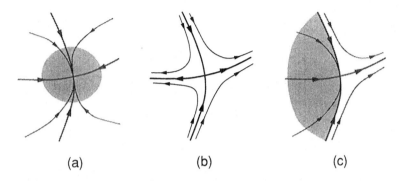

Fig. 8. a) An attractor is a limit set with an open inset. b) An exceptional limit set can be reached only for initial conditions taken right on its stable manifold (the inset has zero measure). c) A vague attractor is an unstable limit set whose inset has a nonvanishing measure.

i.e., it has no *outset*. This condition, which is another manner to express the stability of the set, is best appreciated when considering an intermediate case where the limit set has an inset with finite measure but has a nonvoid outset, i.e., is unstable. Indeed, taking for example $dA_1/dt = A_1^2$, $dA_2/dt = -A_2$, we see that the fixed point at the origin is a limit set for all initial conditions with $A_1 \leq 0$ but that trajectories starting with $A_1 > 0$ all go to infinity. The outset of the origin is the whole positive half-A_1-axis. The inset is not topologically an open neighborhood of the fixed point since the latter lies right on the boundary. The origin cannot be an attractor but, to acknowledge the fact that it still attracts a whole set of trajectories, it is called a *vague attractor*. From its very definition, an attractor is stable against small perturbations, whereas a vague attractor is obviously not since it is particularly sensitive to perturbations; in the example given, any infinitesimal fluctuation with $\delta A_1 > 0$ diverges immediately.

4.4. Hyperbolicity, Structural Stability, and Bifurcations

When control parameters are varied, the phase portrait may change qualitatively. As shown in Fig. 9, the shape of trajectories of system (9) close to the origin changes from parabolic ($r < 0$) to hyperbolic ($r > 0$). When $r = 0$ the origin is marginally stable. This

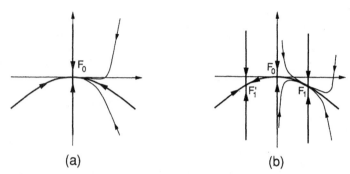

Fig. 9. Qualitative aspect of the phase portrait of system (9) for $r < 0$ (a) and $r > 0$ (b).

condition defines a *bifurcation point* in parameter space. At such a point, the dynamics are *not robust* to perturbations brought to the system since even a small change in its analytical expression changes the *qualitative* aspect of the trajectories. The system is said to be *structurally unstable*. In contrast, at finite r, small perturbations to the system only lead to a slight *quantitative* change of the position of the fixed points but does not change their number, nor the relative position of their stable and unstable manifolds. The general aspect of trajectories in their vicinity remains the same, while at large distances, the trajectories are not sensitive to the precise value of r anyway.

A fixed point that is marginally stable is said to be *nonhyperbolic*. The term *hyperbolic* comes from the aspect of trajectories close to a saddle point (the origin in Fig. 9b) but also includes the case of the sink (the origin in Fig. 9a) or that of a source and applies also to more general limit sets. A structurally stable system has only hyperbolic limit sets, and bifurcations take place when some limit sets become nonhyperbolic.

4.5. About More General Attractors

In the example presented above, attractors were fixed points and the asymptotic time dependence was trivial. Even in two dimensions more complicated attractors can exist. An example is given by the

2. Evolution and Stability, Basic Concepts

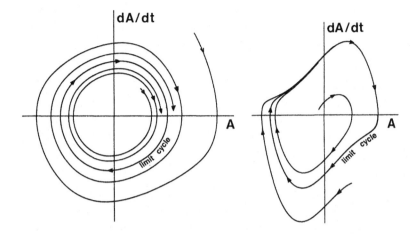

Fig. 10. Example of limit cycles and transient trajectories tending to them for the van der Pol oscillator. Left: nearly harmonic oscillations for $\epsilon = 0.05$. Right: relaxation oscillations for $\epsilon = 0.5$. For $\epsilon < 0$ trajectories spiral toward the origin.

van der Pol oscillator defined by

$$\frac{d^2 A}{dt^2} - (2\epsilon - A^2)\frac{dA}{dt} + A = 0\,.$$

This is nothing but a harmonic oscillator ($d^2A/dt^2 + A = 0$) with a damping term (coefficient of dA/dt) which is a function of A itself. Forgetting for a while the nonlinear term, we see that for ϵ negative, the damping is 'normal'. In a phase space parametrized by A and dA/dt, this results in trajectories spiraling toward the origin. When ϵ is positive, close to the origin the effective damping is "anomalous", which makes the origin unstable against a pair of complex modes with eigenvalues $s \simeq \epsilon \pm i$. On the other hand, far from the origin, say for $A^2 \gg \epsilon$, trajectories should spiral inwards; the problem of what happens between is solved experimentally in Fig. 10: trajectories converge toward closed loops called *limit cycles*, either nearly circular for ϵ small (Fig. 10a) or more "rectangular" for ϵ large. Time series of the variable A would show damped, nearly harmonic, or strongly anharmonic oscillations.

Beyond this example of periodic attractors represented by limit cycles in phase space, other less trivial possibilities exist, which require some additional definitions. First the transient part of a trajectory is expected to wander in phase space, whereas some sort of recurrence is expected when the asymptotic regime is reached. Time independence associated with a fixed point is of course the simplest possibility, and periodicity is already slightly less trivial. In view of more complicated cases, we shall say that a point is *nonwandering* if trajectories starting in its neighborhood (i.e., in an open set surrounding it) visit this neighborhood again and again, indefinitely. In contrast, a *wandering point* is a point whose neighborhood is visited at most a finite number of times, i.e., only during the transient. Finally, we will say that, in the general case, an attractor is a nonwandering attracting set (i.e., with an open inset) which contains a dense orbit (i.e., an orbit that visits the neighborhood of all its points) to fulfill the indecomposability requirement.

The existence and properties of periodic and more complicated attractors will be the subject of forthcoming chapters. Let us just say that stability properties of a time-dependent limit set have to be appreciated from the evolution of infinitesimal perturbations after removal of the trivial time dependence forced by the limit set itself. For example, the stability of a limit cycle such as those appearing in Fig. 10 can be determined after averaging over one period. When the limit set is apparently more complicated than strictly periodic, a first piece of information is obtained from the Fourier spectrum of time series that reveals more than a single family of spectral lines with one fundamental and its harmonics. The divergence away from the considered trajectory has then to be appreciated in the long term, that is to say after averaging over a time-window whose width tends to infinity. The instability of trajectories belonging to stable limit sets will be understood as the very "nature" of *temporal chaos* in so far as long term divergence of orbits in phase space will prevent detailed predictability in spite of determinism, as will be further analyzed in Chapter 7.

2. Evolution and Stability, Basic Concepts

5. Bibliographical Notes

An introductory presentation of stability problems can be found in:

[1] C. Normand, Y. Pomeau, and M.G. Velarde, *Convective instability: a physicist approach*, Rev. Mod. Phys. **49**, 581 (1977).

Lyapunov's approach is discussed in:

[2] J. La Salle and S. Lefschetz, *Stability by Liapunov's direct method*, (Academic Press, New York, 1961).

For the entropy production theorem, see e.g.,

[3] I. Prigogine, *Introduction to thermodynamics of irreversible processes* (Wiley, New York, 1967).

A presentation of the different stability criteria related to the energy method applied to the hydrodynamic flows is developed in:

[4] D.D. Joseph, *Stability of fluid motion*, Springer Tracts in Natural Philosophy, Vol. 27 (Springer, Berlin, 1976).

For a pictorial introduction to phase space dynamics, see:

[5] R.H. Abraham and D.C. Shaw, *Dynamics: the geometry of behavior; Part 1: periodic behavior; Part 2: chaotic behavior* (Aerial Press, Santa Cruz, 1983).

Finally, note that a rather self-contained analysis of problems involving noise, Langevin equations, Fokker–Planck equations evoked at the end of the introduction can be found in:

[6] C.W. Gardiner, *Handbook of stochastic methods for physics, chemistry, and natural sciences* (Springer, Berlin, 1983).

Chapter 3

Instability Mechanisms

Discrete systems can be treated directly within the framework of dynamical systems theory but continuous systems first need a reduction process that relies on the cooperation of fluctuations on a macroscopic scale. In convection, the origin of this cooperation has been briefly introduced in Chapter 1. However, the processes involved are often much less intuitive and some physical insight into instability mechanisms is more easily obtained in terms of simplified models. In this chapter, starting with the simplest case corresponding to stationary plain convection in Section 1, we give several examples of how such simplified models can be derived, complicating the picture progressively to account for new phenomena, e.g., double diffusion, Section 2; anisotropy, Section 3; centrifugal forces, Section 5; or new circumstances such as a time periodic excitation, Section 4, or instability competition, Section 6.

1. Rayleigh–Bénard Convection

Here we develop first the qualitative analysis in terms of *characteristic times*, then we make it more precise by building a simplified model. The normal mode analysis of this model is shown to give insight into the different processes involved in the instability mechanism and the role of relevant dimensionless parameters. The full quantitative analysis yielding the accurate prediction of the threshold is postponed until Chapter 4.

1.1. Qualitative Analysis

As was already apparent from the introductory discussion in Chapter 1, convection is the response of a system submitted to an unstable stratification of the density. The basic destabilizing force is the differential buoyancy experienced by a fluid particle subjected to a temperature fluctuation. Let us estimate the typical acceleration due to this differential buoyancy; α being the thermal expansion coefficient (i.e., $-\rho^{-1}\,\partial\rho/\partial T$), g the gravity acceleration, ρ some average density, the order of magnitude of density fluctuations is $\rho\alpha\,\Delta T$, where ΔT is the temperature difference between the top (cold) and the bottom (hot) plates. The potential buoyancy force *per* unit volume is then $\rho\alpha g\,\Delta T$, which allows the definition of the characteristic time $\tau_{\rm B}$ through

$$\rho\alpha g\Delta T = \text{force} = \rho \times \text{acceleration} = \rho\,\frac{h}{\tau_{\rm B}^2}$$

(apart from numerical factors, $\tau_{\rm B}$ is the time required for a hot bubble to rise, or a cold bubble to sink, over the distance h).

Damping is expected from the irreversible trend to uniformity: relaxation of velocity gradients *via* viscous friction and relaxation of temperature gradients *via* heat diffusion. Both processes are governed by a law of the form

$$\partial_t(\cdots) \propto \nabla^2(\cdots),$$

where the proportionality coefficient is the diffusivity of the quantity considered, dimensionally $[l]^2[t]^{-1}$ viz. cm^2/s. Here we have two diffusivities: the kinematic viscosity $\nu = \eta/\rho$, where η is the dynamical viscosity, and the heat diffusivity $\kappa = \chi/C$, where χ is the heat conductivity and C the heat capacity *per* unit volume.

Characteristic times τ_v and τ_θ are then defined from

$$\nu = \frac{h^2}{\tau_v} \qquad \kappa = \frac{h^2}{\tau_\theta}.$$

The Rayleigh number

$$R = \frac{\alpha g\,\Delta T\,h^3}{\kappa\nu}$$

3. Instability Mechanisms

can then be understood as the ratio $\tau_v \tau_\theta / \tau_B^2$, where τ_B contains explicitly the control parameter ΔT.

When $R \ll 1$, i.e., $\tau_B \gg \tau_v \tau_\theta$, the buoyancy force is insufficient to make the hot (cold) bubble rise (sink) sufficiently quickly. Damping processes, especially thermal diffusion, iron out the fluctuation so that the layer remains at rest. On the contrary, when $R \gg 1$, i.e., $\tau_B \ll \tau_v \tau_\theta$, the buoyancy is expected to be strong enough to overturn the layer. The convection threshold should then correspond to some "intermediate" value of R that remains to be estimated.

1.2. Simplified Model

The previous analysis suggests that a simplified approach in terms of a temperature fluctuation θ coupled to the vertical motion v_z should be sufficient to account for the main features of the instability mechanism. Moreover, since only the differential buoyancy between fluid particles at the same altitude is involved, a model including only the horizontal dependence should be sufficient. Such a model is termed *one-dimensional*. The price to be paid for this simplification will appear at the end of the argument.

In the basic state, the fluid submitted to the temperature difference $\Delta T = T_b - T_t$ remains at rest. Assuming a thermal diffusivity independent of the temperature, we get a purely conducting (Fourier) temperature profile with a constant gradient $\beta = \Delta T / h$. With our definition, $T_0(z) = T_b - \beta z$ with β positive when the fluid is heated from below.

Let us introduce infinitesimal fluctuations. For the temperature field we assume $T = T_0(z) + \theta$, and since we start from the rest state ($\mathbf{v}_0 \equiv 0$), the total velocity field is merely that corresponding to the fluctuation.

We consider first the equation for the vertical velocity that involves the buoyancy force. The temperature distribution generates a density distribution

$$\rho_0(z) = \rho(T_0(z)) = \rho_b(1 + \alpha \beta z),$$

where ρ_b is the reference density at the bottom, $z = 0$. The differential buoyancy then reads

$$(-g)(\rho - \rho_0) = g(\rho(T_0) - \rho(T_0 + \theta)) = \rho_b \alpha g \theta.$$

(In the first term on the l.h.s. we have $(-g)$ since the gravity is directed downwards.)

If we neglect the vertical dependence of the fluctuations, we get the simplified motion equation

$$\partial_t v_z = \nu \partial_{x^2} v_z + \alpha g \theta. \qquad (1a)$$

The first term on the r.h.s. accounts for viscous diffusion along the x-direction only (one-dimensional model). The second term is the acceleration due to the buoyancy force and it is easily checked that $\partial_t v_z$ as the same sign as θ, i.e., upwards acceleration for a positive temperature fluctuation. The absence of the pressure term is related to the absence of a horizontal velocity component since in the incompressibility condition $\partial_x v_x + \partial_z v_z = 0$ we consider the first term irrelevant at this stage.

Some care is needed to treat the heat equation since the thermal evolution of a given fluid particle is given by

$$C \frac{dT}{dt} = \chi \partial_{x^2} T,$$

but the total derivative with respect to time appears on the left hand side since the fluid particle must be followed during its motion. The advection term $\mathbf{v} \cdot \nabla T$ then contributes nontrivially at first order in perturbation since, when applied to $T = T_0(z) + \theta$ $\mathbf{v} \cdot \nabla$, reduced to $v_z \partial_z$ in the one-dimensional model, yields $-\beta v_z$. Finally, we get

$$C(\partial_t \theta - \beta v_z) = \chi \partial_{x^2} \theta$$

or, with $\kappa = \chi/C$ as defined earlier,

$$\partial_t \theta = \kappa \partial_{x^2} \theta + \beta v_z. \qquad (1b)$$

Rayleigh's mechanism therefore involves a linear coupling between the equation for v_z and the equation for θ through the buoyancy term

3. Instability Mechanisms

in Equation (1a) and the term βv_z in Equation (1b). At thermal equilibrium, when $\beta = 0$, fluctuations are left uncoupled.

1.3. Normal Mode Analysis

Linear stability analysis is, in fact, an *initial value problem* for infinitesimal perturbations. Since the resulting mathematical problem is linear, a *superposition principle* holds. The perturbations can be expanded on a basis adapted to the geometry of the problem. Here, since we assume an unbounded horizontal layer, we must choose *Fourier modes* $\{v_z, \theta\} = \{V, \Theta\} \cos(kx)$. Coefficients V and Θ remain a function of time so that system (1) now reads

$$\begin{aligned} \frac{dV}{dt} &= -\nu k^2 V + \alpha g \Theta \,, \\ \frac{d\Theta}{dt} &= -\kappa k^2 \Theta + \beta V \,. \end{aligned} \quad (2)$$

This linear differential equation is easily solved by assuming an exponential time dependence involving a yet unknown growth rate s: $\{V, \Theta\} \to \{V, \Theta\} \exp(st)$, where $\{V, \Theta\}$ are now pure numbers. This converts the time differential problem into an algebraic 2×2 homogeneous linear system

$$\begin{aligned} (s + \nu k^2)V - \alpha g \Theta &= 0 \,, \\ -\beta V + (s + \kappa k^2)\Theta &= 0 \,. \end{aligned} \quad (3)$$

A perturbation will exist if this system has a nontrivial solution, which can be the case only if its determinant vanishes:

$$\begin{aligned} 0 &= (s + \nu k^2)(s + \kappa k^2) - \alpha g \beta \\ &= s^2 + (\nu k^2 + \kappa k^2)s + \kappa \nu k^4 - \alpha g \beta \,. \end{aligned} \quad (4)$$

This *compatibility condition* relates the growth rate s to the wavevector k of the perturbation and contains the control parameter β.

Equation (4) is quadratic in s. It can have two roots, $s_\pm$, with

$$\begin{aligned} s_+ + s_- &= \Sigma = -(\nu k^2 + \kappa k^2) \,, \\ s_+ s_- &= \Pi = \nu \kappa k^4 - \alpha g \beta \,. \end{aligned}$$

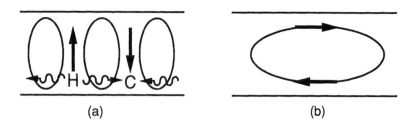

Fig. 1. a) Thermal diffusion and viscous damping associated with the vertical shear are the dominant dissipative processes for wavelengths shorter than $2h$. b) Viscous damping associated with the horizontal shear implied by the continuity condition is the basic stabilizing mechanism at wavelengths longer than $2h$.

s_+ and s_- can be real or complex according to the sign of the discriminant

$$\Delta = (\nu k^2 + \kappa k^2)^2 + 4\left(\alpha g \beta - (\nu k^2)(\kappa k^2)\right)$$
$$= (\nu k^2 - \kappa k^2)^2 + 4\alpha g \beta.$$

Δ can be negative only when β is negative, i.e., when the layer is heated from above. When β is negative and sufficiently large, the two roots are conjugate complex but their real part $\Sigma/2$ remains negative so that the corresponding perturbations are damped; there cannot be unstable oscillatory modes.

When β is positive, Δ remains positive and the roots are real but their Σ stays negative. In order to detect the instability, we need only to discuss the sign of the product Π. When it is positive both roots are negative and we have two stable modes. This occurs when β is small, especially when $\beta \to 0$, i.e., close to thermal equilibrium. When Π is negative, root s_- remains negative but s_+ becomes positive; one of the modes grows exponentially and the layer is unstable. This happens when β is large so that the release of gravitational energy by buoyancy can overcome dissipation losses.

Marginal stability corresponds to $s_+ = 0$, that is to say $\Pi = 0$:

$$\alpha g \beta_m = \nu \kappa k^4. \tag{5}$$

3. Instability Mechanisms

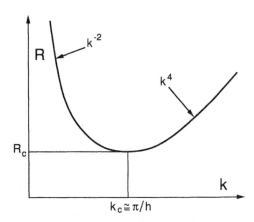

Fig. 2. Sketch of the marginal stability curve as derived from the qualitative analysis.

This relation suggests that β_m increases as k^4 for large k. A detailed calculation shows that this is indeed the case, which could be expected from the fact that for a short spatial period, the induced vertical shear $(\partial_x v_z)$ and horizontal temperature gradient $(\partial_x \theta)$ are large and the associated dissipation is dominant (see Fig. 1a).

However, the decrease at small k predicted by (5) is incorrect. This discrepancy is easy to understand since damping by the horizontal shear $(\partial_z v_x)$ has been neglected while the presence of the horizontal velocity field is made necessary by the closing of flow lines not included in the one-dimensional model. When k goes to zero this neglected contribution is expected to be dominant as suggested in Fig. 1b.

To get an idea of the behavior of β_m in the small k limit we balance the buoyancy release on the l.h.s. of (5) by a single effective damping involving the horizontal shear instead of the two dissipative processes introduced initially. The continuity equation $\partial_x v_x + \partial_z v_z = 0$ gives us an order of magnitude of v_x in terms of v_z, viz. $kv_x \sim v_z/h$. The gradient involved in the dissipative process is $\partial_z v_x \sim v_x/h$, which has to be expressed in terms of v_z, the variable actually involved in the coupling. We get $(v_x/h) \sim (v_z/kh^2)$, which suggests a $1/k$-enhancement of the gradient and a $1/k^2$ contribution to the dissipation on the r.h.s. of (5) leading to the marginal stability curve sketched in Fig. 2.

Unstable modes with wavevectors that are neither too large nor too small realize a kind of optimum minimizing β_m. In terms of wavelengths, the optimal diameter of the individual cells is expected to be of the order of the height h of the layer, which makes $\lambda_c \sim 2h$, and thus $k_c = 2\pi/\lambda_c \sim \pi/h$. The instability threshold β_c is then given by the corresponding value of the marginal gradient:

$$\alpha g \beta_c = \kappa \nu \left(\frac{\pi}{h}\right)^4,$$

or, in terms of the Rayleigh number defined by $R = \alpha g \beta h^4 / \kappa \nu$,

$$R_c \sim \pi^4.$$

Let us remark that this estimate is basically dimensional and, therefore, correct to within some numerical factor. The exact calculation performed in Chapter 4 will show that this factor is of the order of 10 to 20 according to the precise kind of boundary conditions considered. Though not very good, this estimate is however relatively satisfactory when taking into account the crudeness of the model. The argument, which correctly predicts the stationary nature of the instability, can be carried further to give insight into the nature of unstable modes.

1.4. Linear Dynamics of Unstable Modes

In order to extract more from model (1), let us first write it in dimensionless form; the most natural length unit is the height h of the layer. For the time unit, we may choose either the thermal diffusion time over the thickness h, $\tau_\theta = h^2/\kappa$, or the viscous diffusion time, $\tau_v = h^2/\nu$. In the following, we take τ_θ so that the velocity unit is $h/\tau_\theta = \kappa/h$. Finally, from the expression of the Rayleigh number, it can be checked that $\kappa\nu/\alpha g h^3$ has the dimension of a temperature. This will be our temperature unit, more suitable than ΔT that is the natural control parameter in the problem. Inserting $x = \hat{x}h$, $t = \hat{t}h^2/\kappa$, $v_z = \hat{v}_z \kappa/h$, and $\theta = \hat{\theta}\kappa\nu/\alpha g h^3$ into system (1), dropping the carets, and after simplification, we get

$$\partial_t v_z = P\left(\partial_{x^2} v_z + \theta\right), \qquad (6a)$$

$$\partial_t \theta = R v_z + \partial_{x^2}\theta, \qquad (6b)$$

3. Instability Mechanisms

where x, t, v_z, and θ are now dimensionless quantities.

The parameter P, which appears in the first equation, is the Prandtl number defined as

$$P = \nu/\kappa.$$

It is characteristic of the given fluid and can be understood as the ratio of the two disspative times: $P = \tau_\theta/\tau_v$. It does not appear in the calculation of the threshold but controls the physical nature of the unstable modes.

System (3) for normal modes now reads

$$sV = P(-k^2 V + \Theta),$$
$$s\Theta = RV - k^2\Theta.$$

Here the wavevector k is dimensionless and directly assumed to be of the order of π from the start. Condition (4) can then be written as

$$s^2 + \pi^2(P+1)s + P(\pi^4 - R) = 0. \tag{7}$$

Recalling that we are interested only in the real roots close to the threshold, we find it preferable to search for their approximate values from order of magnitude comparisons. Since the growth rate s_+ of the near marginal mode is close to zero, it can be obtained by neglecting s^2, which yields

$$s_+ \sim \frac{P}{P+1} \frac{R - \pi^4}{k^2}. \tag{7'}$$

The other root s_- is strongly negative and since the last term in (7) is nearly vanishing, it can be neglected so that s_- can be obtained from the comparison between the two first terms. This gives

$$s_- \sim -\pi^2(P+1).$$

When P is much greater than unity, the mode corresponding to the second root, $s_- \sim -\pi^2 P$, is essentially viscous. This is shown by restoring the physical dimension, that is to say $s_- \to \tau_\theta^{-1} s_- \sim \nu(\pi/h)^2$. In contrast, the unstable mode is mostly thermal since $P/(P+1) \to 1$ when $P \to \infty$ so that $s_+ \sim \tau_\theta^{-1}$ times a purely numerical factor that changes its sign when the threshold is crossed.

Let us go further with the large P limit. Returning to System (6) we see that when P tends to infinity the l.h.s. of (6a) becomes negligible, so that the velocity field is simply given by

$$\partial_{x^2} v_z + \theta = 0\,;$$

i.e., it follows adiabatically the (slow) variations of the temperature field; thus, $\pi^2 v_z \sim \theta$ which can be inserted in (6b) to get an effective equation for θ:

$$\partial_t \theta = \left(\frac{R}{\pi^2} - \pi^2\right)\theta\,.$$

Expression (7′) is then obviously recovered at the limit $P \to \infty$.

When P is much smaller than unity, the situation is reversed and the temperature field can be shown to follow the velocity fluctuations. The fact that the two cases do not appear symmetrical is merely due to the choice of the time scale. Qualitative arguments are better developed in terms of the slowest variables and τ_θ is more adapted to the large P limit while τ_v is more suitable when $P \to 0$. Though the instability threshold is independent of P, the physical nature of the unstable modes plays an important role at the nonlinear stage above threshold.

2. Convection in Binary Mixtures

Density variations can have other sources than thermal expansion that leads to the standard Rayleigh–Bénard instability. In a mixture, concentration fluctuations can induce differential buoyancy forces able to drive convection by a comparable mechanism. Here we shall restrict ourselves to the case of a layer of binary mixture characterized by a single concentration N and submitted to a concentration gradient in addition to the temperature gradient, the so-called

3. Instability Mechanisms

thermo-haline problem. New effects can be expected from the richness brought by a second variable and its associated control parameter. In the present case, they will emerge from the difference between the molecular and thermal diffusivity, the latter being several orders of magnitude larger than the former (because the diffusion of a given chemical species implies a microscopic displacement of corresponding molecules, one by one, which makes a lower bound to diffusion, whereas thermal diffusion governing energy fluctuations is faster since it involves all kinds of molecules and all possible energetic processes such as vibrations).

The experimental situation is the same as for plain Rayleigh–Bénard convection except that in addition to being isothermal, the plates are also permeable. The fluid layer is therefore in contact with baths at temperatures T_b and T_t, which are also solute reservoirs at concentrations N_b and N_t. The rest state is thus characterized by linear temperature and concentration distributions

$$T(z) = T_b - \beta z \quad \text{and} \quad N(z) = N_b - \beta' z,$$

where $\beta = \Delta T/h$ and $\beta' = \Delta N/h$ are the applied gradients.

The coupling with the vertical velocity is again due to differential buoyancy but it can now have two origins, thermal and chemical. Let us write the equation of state as

$$\rho = \rho_b(1 - \alpha\theta + \alpha' n),$$

where θ and n are the fluctuations of T and N around the rest state distribution. The thermal expansion now denoted as α is positive as usual but α' can be positive or negative according to the overall composition of the mixture (solute and solvent).

Within the framework of a one-dimensional model, the new force equation reads

$$\partial_t v_z = \nu \partial_{x^2} v_z + g(\alpha\theta - \alpha' n).$$

The thermal diffusion equation is unchanged:

$$\partial_t \theta - \beta v_z = \kappa \partial_{x^2} \theta;$$

and the molecular diffusion equation has an analogous structure:

$$\partial_t n - \beta' v_z = \kappa' \partial_{x^2} n,$$

where the second term on the l.h.s. accounts for the convective transport of the solute due to the concentration gradient. κ' is the molecular diffusion coefficient.

The normal mode analysis is straightforward; we assume now that $\{v_z, \theta, n\} = \{V, \Theta, N\} \cos(kx) \exp(st)$, which leads to the 3×3 homogeneous system

$$\begin{aligned}
(s + \nu k^2)V & \quad -\alpha g \Theta & \quad +\alpha' g N & = 0, \\
-\beta V & \quad +(s + \kappa k^2)\Theta & & = 0, \\
-\beta' V & & \quad +(s + \kappa' k^2)N & = 0,
\end{aligned}$$

and to the compatibility condition

$$\begin{aligned}
0 = &(s + \nu k^2)(s + \kappa k^2)(s + \kappa' k^2) \\
&+ (s + \kappa k^2)\alpha' g \beta' - (s + \kappa' k^2)\alpha g \beta,
\end{aligned}$$

which can be expanded as

$$s^3 - a_2 s^2 + a_1 s - a_0 = 0 \qquad (8)$$

with

$$\begin{aligned}
a_2 &= -(\nu k^2 + \kappa k^2 + \kappa' k^2), & (9a) \\
a_1 &= (\kappa k^2)(\kappa' k^2) + (\kappa' k^2)(\nu k^2) + (\nu k^2)(\kappa k^2) \\
&\quad + \alpha' g \beta' - \alpha g \beta, & (9b) \\
a_0 &= -(\nu k^2)(\kappa k^2)(\kappa' k^2) - \kappa k^2 \alpha' g \beta' + \kappa' k^2 \alpha g \beta. & (9c)
\end{aligned}$$

This cubic equation can have either three real roots or one real root and a pair of complex conjugate roots so that oscillatory behavior is possible. Let us denote the three roots as s_1, s_2, and s_3; then

$$a_2 = s_1 + s_2 + s_3, \qquad a_1 = s_1 s_2 + s_2 s_3 + s_3 s_1, \qquad a_0 = s_1 s_2 s_3. \qquad (10)$$

3. Instability Mechanisms

2.1. Stationary Mode

The threshold of a stationary instability is obviously given by $a_0 = 0$. Dividing (9c) by $(\nu k^2)(\kappa k^2)(\kappa' k^2)$, we obtain

$$1 = \frac{\alpha g \beta}{\kappa \nu k^4} - \frac{\alpha' g \beta'}{\kappa' \nu k^4}.$$

Defining the two dimensionless Rayleigh numbers

$$R = \frac{\alpha g \beta h^4}{\kappa \nu}, \qquad R' = \frac{\alpha' g \beta' h^4}{\kappa' \nu}$$

and assuming as before by that $k_c \sim \pi/h$ we get

$$R - R' = \pi^4. \tag{11}$$

The instability that takes place when the l.h.s. is larger than the critical value, here roughly estimated as π^4, can be understood as the result of a simple superposition of two parallel mechanisms, each involving a particular kind of fluctuation. When boundary conditions on the concentration are similar to those for the temperature, e.g., $n = \theta = 0$ at $z = 0, h$, it can be shown easily by algebraic manipulations of the primitive equations that the threshold of the stationary instability is exactly given by condition (11) except for the replacement of π^4 by the exact value.

2.2. Oscillatory Mode

Oscillatory convection at threshold, implies, for example, s_1 real and $s_2 = i\omega$, $s_3 = -i\omega$. Upon insertion into (10), we get:

$$a_2 = s_1, \qquad a_1 = \omega^2 \geq 0, \qquad a_0 = s_1 \omega^2. \tag{12}$$

The threshold condition is then obtained by eliminating s_1 and ω between these three equations, which gives the relation

$$a_0 = a_2 a_1. \tag{13}$$

The explicit form of condition (13) is complicated:

$$-(\nu k^2)(\kappa k^2)(\kappa' k^2) - (\kappa k^2)\alpha' g\beta' + (\kappa' k^2)\alpha g\beta$$
$$= -(\nu k^2 + \kappa k^2 + \kappa' k^2)\Big((\kappa k^2)(\kappa' k^2) + (\kappa' k^2)(\nu k^2)$$
$$+ (\nu k^2)(\kappa k^2) + \alpha' g\beta' - \alpha g\beta\Big).$$

Using the previous definitions of R, R', introducing the Schmidt number $S = \nu/\kappa'$ in addition to the Prandtl number $P = \nu/\kappa$, and assuming again $k \sim \pi/h$, we obtain

$$-\pi^4 - R' + R = -\left(1 + \frac{1}{P} + \frac{1}{S}\right)\left(\pi^4(1+P+S) + PR' - SR\right)$$

or, after simplification

$$S\left(1 + \frac{1}{P}\right)R - P\left(1 + \frac{1}{S}\right)R' = (P+S)\left(1 + \frac{1}{P}\right)\left(1 + \frac{1}{S}\right)\pi^4. \tag{14}$$

As illustrated in Fig. 3, the graphs of conditions (11) and (14), which also read

$$R = R' + \pi^4,$$
$$R = \frac{P^2}{S^2}\frac{1+S}{1+P}R' + \frac{(1+S)(P+S)}{S^2}\pi^4,$$

are straight lines that intersect at a point whose coordinates are

$$R' = \pi^4 \frac{1+P}{S-P}, \quad R = \pi^4 \frac{1+S}{S-P}.$$

Since molecular diffusion of the solute is expected to be slower than thermal diffusion, i.e., $\kappa \gg \kappa'$ or $\nu/\kappa = P \ll \nu/\kappa' = S$, the slope of the second line is smaller than unity and the intersection takes place for $R' > 0$. The fluid layer is stable for values of (R, R') below the lines. The instability that develops first, stationary or oscillatory, depends on which line is crossed first when coming from the stable region. We remark that the condition $a_1 \geq 0$ (from the second relation in (12)), which can be written as

$$SR - PR' \leq \pi^4(1+P+S), \tag{15}$$

3. Instability Mechanisms

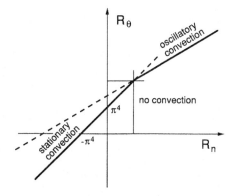

Fig. 3. Different types of instability expected for thermo-haline convection as a function of the thermal and solutal Rayleigh numbers R and R'.

is automatically fulfilled in the region of the plane where the oscillatory instability is expected.

It should be emphasized that by simultaneously controlling two parameters, namely R and R', we can manage to have the system unstable against the two different instability modes. This is an example of what is called a "codimension-2" point: the degeneracy is here between the stationary mode and an oscillatory mode, the frequency of which goes to zero since this particular point lies on the line where inequality (15) becomes an equality, i.e., $\omega = 0$ (note however that the assumption $k_c = \pi/h$ for both modes can be invalidated by a more accurate calculation).

Up until now, cross-diffusion effects have been neglected. However, in a mixture, a temperature gradient induces a flux of the solute (Soret effect) and a concentration gradient induces a heat flux (Dufour effect). In the present situation, fluxes induced by these cross-effects remain negligible when compared with fluxes induced by the gradients imposed from the outside. This is no longer the case when the plates are impervious so that the concentration gradient has to adjust to the temperature gradient. The Soret effect has then to be taken into account in the determination of the basic state. A nontrivial concentration gradient builds up at order zero which modifies the linear stability analysis. In any case, the presence of the concentration variable introduces important modifications in the mechanism of "plain" Rayleigh–Bénard convection due to the

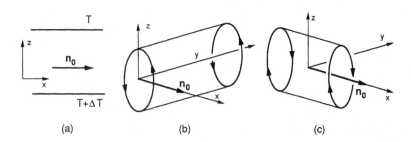

Fig. 4. a) Convection experiment in planar configuration, the unperturbed director field $\mathbf{n}_0$ is parallel to the plates; the anisotropy of nematics is expected to have some effect when the axis of the convection roll is perpendicular to $\mathbf{n}_0$ (b), not when it is parallel to $\mathbf{n}_0$ (c).

long-lived character of concentration fluctuations (for complements, see Bibliographic Notes).

3. Thermal Convection in Nematic Liquid Crystals

Nematic liquid crystals are ordered fluids usually made of elongated molecules. As such, they display anisotropic properties both at and out of equilibrium. In this section we simply need to know: (i) that a supplementary orientational field called the *director* (a unit vector) is involved in their continuous description; (ii) that the anisotropy implies a focalization of fluxes; (iii) that the orientation can be controlled by anchoring at boundaries; and (iv) that dynamic equilibrium requires a balance of bulk torques applied to the director, mainly the viscous torque induced by shearing and the elastic torque tending to restore the director orientation. Complements can be found in Appendix 1 that reviews most of what is needed in practical applications.

Let us consider the convection experiment described in Fig. 4a. The nematic layer is confined between horizontal plates which are treated to favor a *planar* orientation along the x axis, i.e., $\mathbf{n}_0 = \hat{\mathbf{x}}$. An immediate consequence is expected: the orientation of convec-

3. Instability Mechanisms

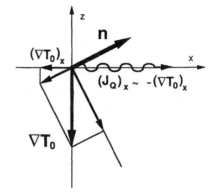

Fig. 5. Owing to the anisotropy of the thermal conductivity, a horizontal heat flux is produced as soon as the applied temperature gradient is not strictly perpendicular to the director.

tion rolls is no longer degenerate in the (x,y) plane but must be related to the unperturbed director orientation due to coupling between orientation and flow. Indeed when the roll axis is parallel to the director, the shear due to convective motions is applied perpendicularly to the principal axis of the molecules so that, by symmetry, no heat focalization effects are expected (Fig. 4b). On the contrary, when rolls are perpendicular to the director, the coupling of the orientation and the velocity field should be at a maximum (Fig. 4c). This is what we now have to examine.

Within the framework of a one-dimensional model where perturbations depend only on x and t, we assume that orientation fluctuations remain confined to the (x,z) plane:

$$\mathbf{n} = \mathbf{n}_0 + n_z \widehat{\mathbf{z}}.$$

The most important contribution to the destabilizing part of the Rayleigh mechanism is due to heat focalization (Fig.5). Let T be the total temperature field $T_0(z) + \theta$. The general expression of the heat flux is given by

$$\mathbf{J}_Q = -\chi_\| \nabla T - \chi_a (\mathbf{n} \cdot \nabla T)\mathbf{n}, \tag{16}$$

where $\chi_\|$ is the conductivity along $\mathbf{n}$ and $\chi_a = \chi_\| - \chi_\perp$ is the anisotropic part.

The modified heat equation then reads

$$\partial_t \theta - \beta v_z = \kappa_\| \partial_{x^2} \theta - \kappa_a \beta \partial_x n_z. \tag{17a}$$

The l.h.s. is obviously left unmodified when compared with the case of isotropic liquids ($-\beta v_z$ accounts for the advection of the fluid particle at lowest order). A new term appears on the r.h.s. which comes from the lowest order contribution of the second term in the heat flux, i.e., the focalization effect that vanishes when the orientation is uniform ($\partial_x \to 0$) and, of course, when the anisotropic part of the conductivity vanishes. The parameters $\kappa_\|$ and κ_a are the diffusivities obtained from the conductivities after dividing by the heat capacity.

The vertical component of the force equation is only slightly modified:

$$\rho \, \partial_t v_z = \eta_2 \, \partial_{x^2} v_z + \rho g \alpha \theta + \alpha_2 \, \partial_{xt} n_z; \tag{17b}$$

η_2 and α_2 are viscosity coefficients corresponding to a shear perpendicular to the unperturbed orientation. The first one accounts for the standard response of the velocity field to the applied shear, the second is related to the motion induced by a rotating director (hence the ∂_t term).

The newest contribution arises from the torque balance equation and describes the dynamical equilibrium of the director. It contains the feedback action of the flow on the orientation *via* a viscous torque:

$$\mathbf{\Gamma}_{\text{total}} = 0 = \mathbf{\Gamma}_{\text{visc.}} + \mathbf{\Gamma}_{\text{elast.}} \tag{18}$$

(note that supplementary restoring or destabilizing torques can be applied by submitting conveniently oriented external electric or magnetic fields). Here only the y component of the torque is nontrivial and we get:

$$\Gamma_{\text{total},y} = 0 = (-K_3 \partial_{x^2} n_z) + (\alpha_2 \partial_x v_z + \gamma_1 \partial_t n_z). \tag{17c}$$

The first term is the elastic restoring torque involving the Frank constant relative to a bend deformation. The viscous part is made of

3. Instability Mechanisms

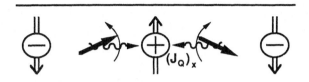

Fig. 6. Schematic illustration of the modified Rayleigh–Bénard mechanism. The supplemetary horizontal heat flux induced by the disorientation reinforces the temperature fluctuation (heat focalization).

two contributions: a source term involving the vertical shear which is expected to make the molecules rotate and a damping term describing the viscous friction of the director. Forgetting the source term, it is easily seen that what remains is a diffusion equation for the director, the orientational diffusivity being given by K_3/γ_1. K_3 and γ_1 are both positive, whereas α_2 is expected to be negative by examination of the effect of the shear.

The outcome of this simplified analysis is a system of three equations for three unknowns θ, v_z, and n_z or rather the curvature $\psi = \partial_x n_z$. A summary of the modified mechanism is given in Fig. 6.

Analyzing system (17) using normal modes of the form $\{v_z, \theta, \psi\} = \{V, \Theta, \Psi\}\cos(kx)\exp(st)$ and assuming that the instability is stationary ($s = 0$), we obtain the threshold condition from

$$\begin{aligned} -\kappa_\| k^2 \Theta &\quad +\beta V &\quad -\kappa_a \beta \Psi &= 0, \\ \rho \alpha g \Theta &\quad -\eta_2 k^2 V &\quad &= 0, \\ &\quad -\alpha_2 k^2 V &\quad +K_3 k^2 \Psi &= 0, \end{aligned}$$

which leads to

$$(K_3 k^2)(\kappa_\| \eta_2 k^4 - \alpha \rho g \beta) + \alpha_2 k^2 \kappa_a \alpha \rho g \beta = 0$$

or

$$\frac{\alpha \rho g \beta}{\kappa_\| \eta_2 k^4}\left(1 - \frac{\alpha_2 \kappa_a}{K_3}\right) = 1.$$

To conclude the argument, it remains to add $k \sim \pi/h$ as usual. Introducing the Rayleigh number

$$R = \frac{\rho \alpha g \beta h^4}{\kappa_\| \eta_2}$$

we get

$$R_c = \frac{\pi^4}{1+A},$$

where $A = -\alpha_2 \kappa_a / K_3$ contains the correction brought by the heat focalization (through κ_a). A is expected to be positive since α_2 is negative and κ_a is positive in general ($K_\| > K_\perp$). The elastic strength of the director field being comparatively small, typical values of A can be very large, say a few hundreds. Recalling that π^4 is the estimate obtained in the same conditions for an isotropic liquid with similar properties, we see that the heat focalization process can lower the instability threshold by a large factor.

Focalization effects can be at the origin of exotic situations, e.g., instability by heating from above when the unperturbed director is perpendicular to the plates (*homeotropic configuration*). Most of them can be analyzed by similar simple arguments but one-dimensional models may sometimes be insufficient.

4. Electrohydrodynamic Instabilities in Nematics

A dielectric fluid inserted between the two plates of a capacitor submitted to a potential difference can become unstable against a convection like-instability mode. Indeed, under usual conditions an unstable stratification of the charge density builds up resulting from charge injection processes at the electrodes, and a mechanism comparable to that of plain convection can operate leading to a convective state.

Here we consider rather what happens when charge injection can be neglected and the fluid is a nematic. The instability mechanism is then analogous to the modified thermal mechanism examined

3. Instability Mechanisms

above. The new variable introduced is the electric charge density ρ_e and the control parameter is the potential difference applied to the capacitor. Owing to its interest for display systems, this instability (Carr, Helfrich, 1969) has been much studied at the beginning of the '70s. It is presented here mainly because most experiments involve a time-periodic excitation, which leads us to analyze a nonautonomous system for the first time.

4.1. The Carr–Helfrich Mechanism

The geometry of the experiment is described in Fig. 7. The plates are usually covered with thin oxide layers orienting the molecules in planar configuration and preventing charge injection. In the unperturbed state, the nematic is at rest ($\mathbf{v} \equiv 0$) and uniformly aligned along the x direction ($\mathbf{n}_0 \equiv \hat{\mathbf{x}}$). Moreover, the overall charge density vanishes and the electric field is aligned along the z axis ($\mathbf{E}_0 = E_0 \hat{\mathbf{z}}$ with $E_0 = U_0/h$, where U_0 is the applied potential and h is the height of the cell). We consider here a one-dimensional linear model coupling the charge density fluctuations ρ_e to the vertical velocity v_z and the orientation fluctuation n_z; a supplementary variable, the electric field fluctuation E_x, will appear in the analysis.

We first write the conservation equation for the charge density

$$\partial_t \rho_e + \nabla \cdot \mathbf{J}_e = 0, \tag{19}$$

which defines the electric current J_e. Ohm's law, which relates the electric current to the electric field $\mathbf{E}$, has here the same structure as the thermal conduction law given by (16):

$$\mathbf{J}_e = \sigma_\perp \mathbf{E} + \sigma_a (\mathbf{n}\cdot\mathbf{E})\mathbf{n}. \tag{20}$$

At lowest order, we have

$$\mathbf{E} = E_x \hat{\mathbf{x}} + E_0 \hat{\mathbf{z}}.$$

In sufficiently purified nematics, the conductivity is rather low: $\sigma_\parallel \sim 10^{-9} \Omega^{-1} \text{cm}^{-1}$, and one has typically $\sigma_\parallel/\sigma_\perp \sim 1.5$ so that $\sigma_a = \sigma_\parallel - \sigma_\perp > 0$.

The x component of the current then reads

$$J_{e,x} = \sigma_\| E_x + \sigma_a E_0 n_z,$$

which we insert in the charge conservation equation (19):

$$\partial_t \rho_e + \sigma_\| \partial_x E_x + \sigma_a E_0 \partial_x n_z = 0. \tag{21}$$

Fluctuation E_x remains to be determined. In dielectrics, the electric induction $\mathbf{D}$ is linked to the field $\mathbf{E}$ by the dielectric tensor:

$$\mathbf{D} = \epsilon_\perp \mathbf{E} + \epsilon_a (\mathbf{n} \cdot \mathbf{E}) \mathbf{n} \tag{22}$$

(we consider here mostly the case of a negative dielectric anisotropy: $\epsilon_\| < \epsilon_\perp$) and the induction to the charge density by the Poisson equation

$$\nabla \cdot \mathbf{D} = 4\pi \rho_e. \tag{23}$$

At lowest order, (22) yields

$$D_x = \epsilon_\| E_x + \epsilon_a E_0 n_z$$

and, from (23),

$$\partial_x D_x = \epsilon_\| \partial_x E_x + \epsilon_a E_0 \, \partial_x n_z = 4\pi \rho_e \tag{23'}$$

from which we obtain $\partial_x E_x$, which is further inserted in (21) to give

$$\partial_t \rho_e + \frac{\sigma_\|}{\epsilon_\|}(4\pi \rho_e - \epsilon_a E_0 \, \partial_x n_z) + \sigma_a E_0 \, \partial_x n_z = 0. \tag{24}$$

Defining the charge relaxation rate $1/\tau_e$ and the Helfrich constant σ_H as

$$\frac{1}{\tau_e} = 4\pi \frac{\sigma_\|}{\epsilon_\|}, \quad \sigma_H = \sigma_\| \left(\frac{\epsilon_\perp}{\epsilon_\|} - \frac{\sigma_\perp}{\sigma_\|} \right)$$

we can write (24) in the form

$$\partial_t \rho_e + \frac{1}{\tau_e} \rho_e + \sigma_H E_0 \partial_x n_z = 0. \tag{25a}$$

3. Instability Mechanisms

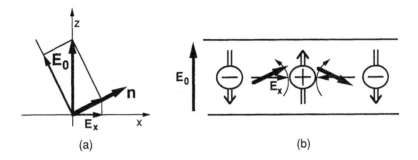

Fig. 7. Mechanism of the Carr–Helfrich instability: a) generation of a horizontal component of the electric field; b) sketch of the one-dimensional approximation to the complete mechanism.

The evolution equation for v_z is easily guessed from the analogy between buoyancy force due to thermal expansion, $\alpha \rho g \theta$, and the electric force, $\rho_e E_0$. We obtain

$$\rho \partial_t v_z = \eta_2 \partial_{x^2} v_z + \alpha_2 \partial_{tx} n_z + \rho_e E_0 \,. \tag{25b}$$

To the torque equation derived in the thermal case (18), we must add the direct contribution of the electric torque calculated as follows: firstly, the electric polarization $\mathbf{P}$ is defined from the induction by $\mathbf{D} = \mathbf{E} + 4\pi \mathbf{P}$, that is to say

$$4\pi \mathbf{P} = (\epsilon_\perp - 1)\mathbf{E} + \epsilon_a (\mathbf{n} \cdot \mathbf{E})\mathbf{n} \,,$$

and secondly, the electric torque is given by $\mathbf{\Gamma}_e = \mathbf{P} \times \mathbf{E}$ so that, at lowest order, we have

$$\Gamma_{e,y} = -\frac{\epsilon_a}{4\pi}\left(E_0^2 n_z + E_0 E_x\right)$$

and, adding this contribution to (17c) we get

$$\Gamma_{\text{total},y} = \alpha_2 \, \partial_x v_z + \gamma_1 \, \partial_t n_z - K_3 \, \partial_{x^2} n_z - \frac{\epsilon_a}{4\pi} E_0^2 n_z - \frac{\epsilon_a}{4\pi} E_0 E_x = 0$$

which we differentiate with respect to x to be able to eliminate $\partial_x E_x$ using (23'). Denoting the curvature $\partial_x n_z$ as ψ we obtain

$$\alpha_2 \, \partial_{x^2} v_z + \gamma_1 \, \partial_t \psi - K_3 \, \partial_{x^2} \psi - \frac{\epsilon_a}{4\pi} \frac{\epsilon_\perp}{\epsilon_\parallel} E_0^2 \psi - \frac{\epsilon_a}{\epsilon_\parallel} E_0 \rho_e = 0 \,. \tag{25c}$$

The couplings described by System (25) are sketched in Fig. 7.

4.2. Two Instability Regimes

In practice, it is sufficient to consider the slowest variables and an order of magnitude estimate immediately shows that the inertial term $\rho \partial_t v_z$ can be neglected so that v_z can be eliminated adiabatically, which leads to two equations (charge conservation and torque) for two unknowns (charge and curvature):

$$\partial_t \rho_e + \frac{1}{\tau_e}\rho_e + \sigma_H E_0 \psi = 0,$$

$$\left(\gamma_1 - \frac{\alpha_2^2}{\eta_2}\right)\partial_t \psi - \left(K_3 \partial_{x^2} - \frac{\epsilon_a}{4\pi}\frac{\epsilon_\perp}{\epsilon_\parallel}E_0^2\right)\psi - \left(\frac{\epsilon_a}{\epsilon_\parallel} + \frac{\alpha_2}{\eta_2}\right)\rho_e E_0 = 0. \quad (26)$$

When E_0 is independent of time, the instability described by these equations cannot be observed since it is usually preempted by the charge injection controlled instability, even though care has been taken to avoid it. However, this spurious effect can be ruled out by using an alternating field $E_0 = E_0(t)$.

System (26) is of the form

$$\partial_t \rho_e = -\frac{1}{\tau_e}\rho_e - \sigma_H E_0(t)\psi, \quad (27a)$$

$$\partial_t \psi = -\lambda(F^2 + E_0(t)^2)\psi - \gamma_{\text{eff}}^{-1} E_0(t)\rho_e, \quad (27b)$$

where F^2 stands for the elastic contribution to the relaxation of the curvature calculated for a normal mode of the form $\cos(kx)$. The expressions of λ and γ are easily found by identification.

As explained in Chapter 2, when the forcing is periodic the main part of the response is expected to be periodic so that amplification or damping must be appreciated after subtracting this trivial time dependence. The solution is thus sought in the form $(\rho_e, \psi) = (\widetilde{\rho}_e(t), \widetilde{\psi}(t))\exp(st)$, where $\widetilde{\rho}_e$ and $\widetilde{\psi}$ are unknown periodic functions of time with period $2\pi/\omega$, and $\exp(st)$ accounts for the nontrivial part of the evolution, stability or instability depending on the sign of the real part of s.

3. Instability Mechanisms

Functions $\tilde{\rho}_e$ and $\tilde{\psi}$ are usually expanded in Fourier series:

$$\left(\tilde{\rho}_e, \tilde{\psi}\right) = \frac{1}{2}\left(A_e^{(0)}, A_\psi^{(0)}\right)$$
$$+ \sum_{n=1}^{\infty}\left(\left(A_e^{(n)}, A_\psi^{(n)}\right)\cos(n\omega t) + \left(B_e^{(n)}, B_\psi^{(n)}\right)\sin(n\omega t)\right),$$

which are further inserted into Equation (27) to yield an eigenvalue problem for the the growth rate s of the perturbation.

Here, we assume a sinusoidal excitation $E_0(t) = \langle E_0\rangle\sqrt{2}\cos(\omega t)$ where $\langle E_0\rangle$ is the r.m.s. value of the field and we make simplifying supplementary assumptions sufficient to account for the two simplest regimes observed experimentally. First we truncate the expansion and keep only harmonics (0) and (1), which gives two uncoupled sets of three equations:

$$\left(s + \lambda\left(F^2 + \langle E_0\rangle^2\right)\right) A_\psi^{(0)} + \frac{1}{\gamma}\langle E_0\rangle\sqrt{2}A_e^{(1)} = 0,$$

$$\frac{\sqrt{2}}{2}\sigma_H\langle E_0\rangle A_\psi^{(0)} + \left(s + \frac{1}{\tau_e}\right) A_e^{(1)} + \omega B_e^{(1)} = 0,$$

$$-\omega A_e^{(1)} + \left(s + \frac{1}{\tau_e}\right) B_e^{(1)} = 0$$

and

$$\left(s + \frac{1}{\tau_e}\right) A_e^{(0)} + \sigma_H\langle E_0\rangle\sqrt{2}A_\psi^{(1)} = 0,$$

$$\frac{\sqrt{2}}{2}\frac{1}{\gamma}\langle E_0\rangle A_e^{(0)} + \left(s + \lambda\left(F^2 + \frac{3}{2}\langle E_0\rangle^2\right)\right) A_\psi^{(1)} + \omega B_\psi^{(1)} = 0,$$

$$-\omega A_\psi^{(1)} + \left(s + \lambda\left(F^2 + \langle E_0\rangle^2\right)\right) B_\psi^{(1)} = 0.$$

This decoupling implies the existence of two families of modes. In the first family the charge density oscillates with the field while the curvature is essentially frozen; this is called the *conduction regime*. On the contrary, in the second family the curvature oscillates while the charges remain at rest; this mode is called the *dielectric regime*.

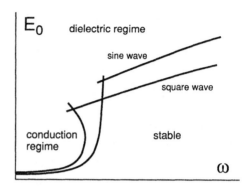

Fig. 8. Stability diagram of the EHD instability in planar nematics as a function of the frequency ω and intensity E_0 of the applied electric field (after Galerne, 1973).

We could have derived this classification from the symmetries of the equations, specifically from the fact that the electrical field is reversed every half-period ($E_0(t + \pi/\omega) = -E_0(t)$). This leads to:

Conduction regime: $\rho_e(t + \pi/\omega) = -\rho_e(t)$,
$\psi(t + \pi/\omega) = +\psi(t)$,
Dielectric regime: $\rho_e(t + \pi/\omega) = +\rho_e(t)$,
$\psi(t + \pi/\omega) = -\psi(t)$.

The conduction regime is expected when both the charge can relax rapidly, i.e., $\omega \ll 1/\tau_e$, and the curvature cannot respond, i.e., $\omega \gg \lambda(F^2 + \langle E_0 \rangle^2)$, that is to say at low frequency and low field; of course the reverse is expected to hold for the dielectric regime. Detailed calculations leads to the stability diagram displayed in Fig. 8 (see Bibliographic Notes).

5. Taylor–Couette Instability

5.1. Centrifugal Instabilities

We now return to isotropic liquids and consider a case where the basic state is no longer the rest state but rather a shear flow. More specifically, we study the stability of the flow between two coaxial cylinders rotating at different speeds (Fig. 9). We assume that the inner (outer) cylinder with radius r_1 (r_2) is rotating at angular ve-

3. Instability Mechanisms

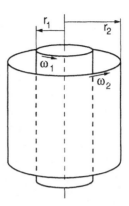

Fig. 9. Geometry and parameters of the Taylor–Couette experiment.

locity ω_1 (ω_2). Except for end effects that will be neglected, the unperturbed solution is a purely azimuthal flow with angular velocity at distance r from the axis given by the Couette profile

$$\omega_0(r) = a + b/r^2 \,, \tag{28a}$$

where a and b are determined from the no-slip boundary conditions ($\omega(r_i) = \omega_i$ for $i = 1, 2$)

$$a = \frac{\omega_2 r_2^2 - \omega_1 r_1^2}{r_2^2 - r_1^2}, \qquad b = \frac{(\omega_1 - \omega_2) r_1^2 r_2^2}{r_2^2 - r_1^2} \,. \tag{28b}$$

The angular momentum stratification associated with this velocity distribution may be stable or unstable. Centrifugal forces playing a role similar to that of gravity for the Rayleigh–Bénard instability, the resulting motion will be understood as an attempt to redistribute the angular momentum by convection. The instability mechanism has been analyzed first by Rayleigh (1916) in the inviscid case whereas Taylor (1923) performed a thorough theoretical and experimental study of the viscous case for the configuration considered here (Fig. 10a). The same mechanism is expected to work in different cases where the streamlines are curved, e.g., in the boundary layer at a concave wall (Görtler instability, Fig. 10b) or in the external part of a flow in a curved channel (Dean instability, Fig. 10c).

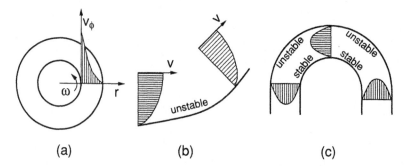

Fig. 10. Examples of centrifugal intabilities: a) Taylor instability between two co-axial cylinders; b) Görtler instability of the boundary layer at a concave wall; c) Dean instability of the Poiseuille flow in a curved channel.

5.2. Rayleigh Instability Mechanism

In the basic flow, at steady state, the centrifugal force at distance r from the axis $\rho r \omega^2(r)$ is compensated by a centripetal pressure gradient. Let us consider a fluid particle rotating displaced to a distance $r + \delta r > r$. In the absence of viscous losses, it keeps its angular momentum. The corresponding angular velocity fluctuation can therefore be obtained from

$$\delta(r^2 \omega) = 2\omega r\, \delta r + r^2\, \delta\omega = 0$$

which gives

$$\delta\omega = -2\frac{\omega}{r}\delta r.$$

At the same time, the angular velocity of fluid particles at distance $r + \delta r$ is given by

$$\omega(r + \delta r) = \omega(r) + \frac{d\omega}{dr}\delta r.$$

If the velocity of the displaced particle is larger than that of the surrounding fluid, the centrifugal pressure gradient it experiences is insufficient to bring it back to its initial position, and the purely azimuthal flow is unstable. On the other hand if the surrounding fluid rotates faster, the displaced particle is brought back to its initial position, and the flow is stable.

3. Instability Mechanisms

Explicitly, this stability condition reads

$$\frac{d\omega}{dr}\delta r \geq -2\frac{\omega}{r}\delta r \quad \Rightarrow \quad \left(2\omega + r\frac{d\omega}{dr}\right)\delta r \geq 0$$

or, upon multiplication by r,

$$\left(2r\omega + r^2\frac{d\omega}{dr}\right)\delta r = \delta(r^2\omega) \geq 0,$$

that is to say when the angular momentum $r^2\omega$ increases with r. This is precisely *Rayleigh's criterion*.

Let us apply this criterion to the angular velocity distribution given by Equation (28) and first consider the case when one of the cylinders is at rest. It is easily seen that the flow is stable (unstable) when the outer (inner) cylinder rotates. However, to obtain the criterion we have assumed the conservation of angular momentum for the displaced fluid particle. In the thermal case, Rayleigh's mechanism leads to convection at vanishing temperature gradient in the absence of dissipation, but in a finite height layer of real fluid, convective motions appear only above some threshold. In the same way, the centrifugal instability is expected to be delayed by viscous friction; the outer cylinder being at rest, Taylor instability will develop only above some critical angular velocity of the inner cylinder.

5.3. One-Dimensional Model

In principle, we have to consider a complete, 3-dimensional perturbed flow

$$\mathbf{v} = \mathbf{v}_0 + \mathbf{v}' = (v'_r, r\omega_0 + v'_\phi, v'_z)$$

where $\mathbf{v}_0$ is the purely azimuthal basic flow (and ϕ is the corresponding coordinate). However, the z component of the fluctuation is not involved directly in the mechanism since it is parallel to the rotation axis. In a one-dimensional model, it can be neglected and only reintroduced qualitatively at the end of the argument by dimensional considerations involving the closing of flow lines. On the contrary, radial and azimuthal fluctuations, lying in the plane perpendicular to the axis of the cylinders, are directly affected by the rotation. Fluctuations of the tangential velocity induce a centrifugal force that plays the role of a source term for the radial velocity, and we have now to find out the force induced by the radial velocity fluctuation that closes the feedback loop. We suppose first that the two cylinders rotate in the same direction, e.g., $\omega(r) \geq 0$; the general case will be briefly examined later.

The total centrifugal force reads

$$\rho v_\phi^2 / r = \rho(r\omega_0 + v'_\phi)^2/r \simeq \rho(r\omega^2 + 2\omega v'_\phi)$$

(the zero-th order term is compensated for by the centripetal pressure gradient, and the term quadratic in v'_ϕ is negligible at the linear approximation).

To obtain the radial balance equation, we simply add viscous damping:

$$\partial_t v'_r = 2\omega v'_\phi + \nu \partial_{x^2} v'_r \qquad (29a)$$

(signs are correct since the response to a positive v'_ϕ is directed outwards).

We now turn to the balance equation in the azimuthal direction and note first that fluctuations v'_r and v'_ϕ are defined with respect to a rotating frame attached to the fluid particle at distance r from the axis. At that place, the local rotation rate of this frame is not simply given by ω_0 but must include also the rotation rate due to the curvature of the flow lines. As a matter of fact, by its very definition, the vertical component of the vorticity $\mathbf{\Omega} = \nabla \times \mathbf{v}$ of the basic flow $\mathbf{v}_0$ is exactly twice the sought local rotation rate. We have $\mathbf{\Omega}_0 = \Omega \hat{\mathbf{z}}$ with

$$\Omega = \frac{1}{2}\left(\frac{dv_0}{dr} + \frac{v_0}{r}\right) = \frac{1}{2}\left(\frac{d(r\omega_0)}{dr} + \omega_0\right) = a,$$

where a is the coefficient introduced in the expression the unperturbed Couette flow profile (28b).

3. Instability Mechanisms

The tangential component of the Coriolis force F_C then reads

$$F_{C\phi} = 2\rho(\mathbf{\Omega} \times \mathbf{v'})_\phi = -2\rho\Omega v'_r = -2\rho a v'_r.$$

The tangential motion equation is then obtained by adding the viscous damping:

$$\partial_t v'_\phi = -2a v'_r + \nu \partial_{x^2} v'_\phi, \qquad (29b)$$

The stability analysis of model (29) is straightforward if we replace $\omega(r)$ by its mean value $\langle\omega\rangle = (\omega_1 + \omega_2)/2$, which is reasonable since the two cylinders are supposed to rotate in the same direction. Taking normal modes of the form $\sin(kx)\exp(st)$ with $k \sim \pi/h$, where $h = r_2 - r_1$ is the gap, we get the compatibility condition:

$$\begin{vmatrix} (s + \nu k^2) & -2\langle\omega\rangle \\ 2\Omega & (s + \nu k^2) \end{vmatrix} = 0$$

which reads

$$(s + \nu k^2)^2 + 4\Omega\langle\omega\rangle = 0,$$

with roots

$$s_\pm = -\nu k^2 \pm 2\sqrt{-a\langle\omega\rangle}.$$

When $a\langle\omega\rangle$ is positive, the flow is stable. In the opposite case, a stationary instability takes place for $s_+ = 0$, i.e.,

$$4a_c\langle\omega\rangle + (\nu k^2)^2 = 0.$$

Usually, the Taylor number is defined as $T = -4a\omega_1 h^4/\nu^2$. Setting $\mu = \omega_2/\omega_1$ we obtain the estimate

$$T_c \sim \frac{2\pi^4}{1 + \mu}.$$

When $\mu \to 1$, the Taylor–Couette problem can be mapped onto the Rayleigh–Bénard problem with no-slip boundary conditions on the velocity at the plates. Since in the thermal case the simplified model yields $R_c = \pi^4$, the exact result for the Taylor instability will read $T_c = 2R_c/(1 + \mu)$.

Fluctuations are periodic in the z direction, thus, the secondary flow has the form of toroidal rolls called *Taylor vortices*. These are the equivalent of Rayleigh–Bénard rolls for convection.

When the two cylinders rotate in opposite directions, ω goes through zero at some distance r_* of the axis, so that the absolute value of the angular momentum decreases close to the inner cylinder, for $r_1 < r < r_*$, whereas it increases close to the outer cylinder, for $r_* < r < r_2$. According to Rayleigh's criterion, the flow can be decomposed into an unstable inner sub-layer and a stable outer sub-layer. Assuming that the rolls remain localized within in the unstable sub-layer of width $h_{\text{eff}} = h/(1-\mu)$ ($\mu < 0$) and taking $k = \pi/h_{\text{eff}} = \pi(1-\mu)/h$, by the same argument as before we obtain $T_c \sim 2\pi^4(h_{\text{eff}}/h)^4 \propto (1-\mu)^4$, which nicely accounts for experimental observations at the limit $|\mu| \to \infty$.

6. Bénard–Marangoni Convection

6.1. Mechanism and Simplified Model

Bénard's experiments were performed with fluid layers open to ambient air. In spite of an apparent agreement with Rayleigh's theory, the buoyancy driven mechanism was not responsible for the regular tessellation of hexagonal cells observed, with the fluid rising at the centers of the hexagons and sinking along their sides. In fact, the presence of a free surface implied surface tension effects clearly recognized half a century later only (Pearson, 1958), and introduced here at a qualitative level.

Large scale surface tension inhomogeneities (see Appendix 1) induce surface stresses since the system always tends to minimize the area of regions with a high surface tension at the expense of low surface tension regions. This so-called *Marangoni effect* can be due to temperature fluctuations, the case considered here, but also to chemical variations, e.g., of the concentration of tensio-active substances, another case of great practical importance.

Surface tension, here denoted as A, usually decreases when the

3. Instability Mechanisms

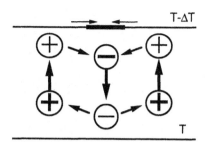

Fig. 11. Bénard–Marangoni instability mechanism.

temperature increases, i.e., $B = -dA/dT > 0$. Now, let us assume that the surface of the fluid becomes locally colder at some place. The corresponding negative temperature fluctuation θ induces a positive fluctuation of the surface tension $\delta A = -B\theta > 0$ so that the surface tend to contract around that place, therefore dragging some fluid parallel to the surface. In order to avoid fluid accumulation, a sinking motion must take place and, in compensation, some fluid must approach the surface farther in order to close of flow lines. Two cases are possible, either the interior is warmer or it is colder. When the uprising fluid is warm (cold), the temperature contrast at the surface is increased (decreased) and the layer at rest is unstable (stable). The mechanism is therefore destabilizing when the layer is heated from the inside, or equivalently, cooled from the outside (which is specially important for volatile liquids allowed to evaporate rapidly). Notice that there is no reference to the gravity (neither its direction nor its intensity: the mechanism can work on a vertical surface, e.g., drying of a paint, or in a spacecraft).

The difficulty in developing a simplified model stems from the fact that the surface force induces a flow component that is not directly coupled to the temperature fluctuation so that we need a conversion process involving the continuity condition.

The boundary condition directly involved in the mechanism expresses the continuity of the tangential stress at the interface (see appendix 1). Neglecting the viscosity of the gas above the fluid, we simply get:

$$\eta \partial_z v_x = -B \partial_x \theta \qquad (30)$$

with $B = -dA/dT$. On the other hand, the evolution of the temperature field is governed by

$$\partial_t \theta + (-\beta)v_z = \kappa \partial_{x^2}\theta, \qquad (31)$$

which indeed requires the determination of v_z in terms of v_x and, in fact, a complete account of the velocity field. To do this properly we have to return to the general hydrodynamic equations, but we can simplify them by considering only the linear and steady response to a perturbation at the boundary. Therefore, we have

$$\partial_x p = \eta(\partial_{x^2} + \partial_{z^2})v_x, \qquad (32a)$$
$$\partial_z p = \eta(\partial_{x^2} + \partial_{z^2})v_z, \qquad (32b)$$
$$\partial_x v_x + \partial_z v_z = 0. \qquad (32c)$$

Elimination of the pressure by cross-differentiation of Eqs. (32a,b) leads to

$$(\partial_{x^2} + \partial_{z^2})v_z = 0.$$

Analyzing the perturbation in Fourier modes $v_z \sim \sin(kx)$ we get

$$(-k^2 + \partial_{z^2})v_z = 0,$$

whose general solution reads

$$v_z = \sin(kx)\left(\sum_{\pm}(a_\pm + b_\pm z)\exp(\pm kz)\right).$$

The flow component v_x is then derived from the continuity condition (32c), which yields

$$v_x = \cos(kx)\left(\sum_{\pm}\left(\frac{b_\pm}{k} \pm (a_\pm + b_\pm z)\right)\exp(\pm kz)\right).$$

We have four unknowns $(a_\pm, b_\pm)$ and four boundary conditions: Equation (30) plus condition $v_z = 0$ at the interface $(z = 0)$ and no-slip conditions $v_z = v_z = 0$ at the bottom of the vessel $(z = -h)$.

3. Instability Mechanisms

We suppose first that the wavelength is short when compared with the height h (limit of a semi-infinite medium). Then, it is easily checked that the solution that fulfills the boundary conditions simply reads

$$v_z = b_+ z \exp(kz) \sin(kx), \qquad v_x = \frac{b_+}{k}(1+kz)\exp(kz)$$

with $b_+/k = -B\Theta/2\eta$, where Θ is the amplitude of the temperature fluctuation at the surface $(\theta|_{z=0} = \Theta\sin(kx))$. We notice immediately that in this limit, only a thin sub-layer of fluid of thickness $1/k$ is involved. The temperature fluctuation being given, though the x component of the velocity induced at the surface remains the same, the mechanism is expected to be inefficient at large k since the fluid that takes part in the instability can come only from a close neighborhood of the surface, i.e., without any effect on the temperature fluctuation at the interface.

When the wavelength of the perturbation is of the order of the height h, the whole fluid layer takes part in the mechanism. In principle the complete expression of the velocity field should be used. We assume instead that $k \simeq \pi/h \sim \partial_z$ in the continuity equation, so that $v_z \sim v_x$ and, from the boundary condition $\eta v_x \sim B\theta$, therefore $v_z \sim B\theta/\eta$, which we introduce in Eq. (31) to get the effective equation

$$\partial_t \theta = \left(-\kappa\frac{\pi^2}{h^2} + \frac{\beta B}{\eta}\right)\theta.$$

Therefore, the fluid layer is unstable when $\beta > \beta_c = \eta\kappa\pi^2/Bh^2$ or, in terms of the Marangoni number

$$M = \frac{\beta B h^2}{\kappa\eta}$$

for $M > M_c = \pi^2 \sim 10$ again underestimating the actual value $M_c \simeq 80$ for $k_c \simeq 2/h$.

6.2. Marangoni versus Buoyancy-Driven Convection

The distinction should now be well apparent between *Bénard–Marangoni convection* driven by surface tension effects and *Rayleigh–Bénard convection* resulting from buoyancy. But in a fluid layer open to air, the two mechanisms can operate simultaneously and we need an estimate of their relative strength, which can be obtained by a comparison of the induced velocities.

The typical velocity of the Marangoni-driven flow is given by $\partial_x A = \partial_T A \, \partial_x T \sim B \Delta T / h \sim \eta \partial_z v_x \sim \eta V_M / h$ ("M" for Marangoni), so that $V_M \sim B \Delta T / \eta$. On the other hand, the buoyancy-driven velocity can be derived from $\delta \rho g = \rho \alpha \, \Delta T \, g \sim \eta \, \partial_{x^2} v_z \sim \eta V_R / h^2$ ("R" for Rayleigh), so that $V_R \sim \rho \alpha \, \Delta T \, g h^2 / \eta$.

Accordingly, we get

$$\frac{V_R}{V_M} = \frac{\alpha g \rho}{B} h^2$$

(notice that this ratio is independent of the viscosity $\eta = \rho \nu$ and the heat diffusivity κ, which characterize the dissipative processes). Assuming $\alpha \sim 10^{-3}$, $\rho \sim 1\text{g/cm}^3$, $g \sim 10^3 \text{cm/s}^2$, and $B \sim 0.1$ CGS, we obtain $V_R / V_M \sim 10 h^2$ with h in centimeters. The two velocities would be of the same order of magnitude for $h \sim 0.3$ cm; this means that surface tension effects dominate in thin layers while buoyancy controls the onset of convection in thick layers. We can arrive at the same conclusion by noticing that this ratio is also equal to that of the Rayleigh number R to the Marangoni number M which vary as $\beta h^4 = \Delta T \, h^3$ and $\beta h^2 = \Delta T \, h$, respectively.

It turns out that, at threshold, convection cells are hexagonal in both cases for layers with a free surface. Thus, the shape of the cells is not a signature of the driving mechanism (for comparison, rolls occur usually in Rayleigh–Bénard convection between rigid plates). By contrast, the shape of the interface depends on the nature of the destabilizing forces. This can be anticipated from the fact that in buoyancy-driven convection the rising fluid is pushed upwards by a bulk process, whereas Marangoni effect induces a vertical motion more indirectly by dragging the fluid parallel to the surface.

3. Instability Mechanisms

To analyze the response of the interface we need only suppose that the horizontal temperature modulation is given from the outside rather than determined self-consistently. In projection along the normal to the interface, the equilibrium condition reads at lowest order

$$(p - 2\eta\, \partial_z v_z)\big|_{z=\xi} = p_{\text{ext}} - A\, \partial_{x^2}\xi, \tag{33}$$

where $\xi(x)$ denotes the position of the interface. On the r.h.s., the first term is the external pressure and the second term is the Laplace correction associated with the curvature of the interface (for a one-dimensional infinitesimal perturbation). The hydrodynamic pressure on the l.h.s. can be derived for example by integration of (32a), once v_x has been determined. This integration with respect to x introduces a function of z which is further obtained by identification of x-independent contributions.

The pressure and flow fields induced by the two mechanisms remain to be determined. Though this is certainly not the case in experiments for which wavelengths of the order of $2h$ are selected at threshold, here we consider the long wavelength limit (small k's). Results will not be inconsistent because, as we know from the previous analysis, both mechanisms are most efficient in this limit.

Let us consider first the surface tension-induced force and assume a temperature perturbation of the form $\theta = \Theta \sin(kx)$ so that the Marangoni term reads $\partial_x A = -Bk\Theta \cos(kx)$. Thanks to the assumption about k, the complete solution is easily obtained. Taking boundary conditions $v_z(z = 0, -h) = 0$ and $v_x(z = 0)$ into account we get

$$v_z = az(z+h)^2 \sin(kx)$$

from which we derive

$$v_x = \frac{1}{k} a(z+h)(3z+h) \cos(kx)$$

and, adding the buoyancy term $-\rho g$ in (32b) we obtain

$$p = \frac{6a\eta}{k^2} \sin(kx) + p_{\text{ext}} - \rho g z\,.$$

In these expressions, a is a yet unknown constant to be derived from the last boundary condition (Marangoni effect): $a = -Bk^2\Theta/4\eta h$. Introducing these results in Equation (33) for the interface elevation, $\xi(x) = \Xi \sin(kx)$, we get

$$\Xi_{Ma} = -\frac{1}{1+\lambda^2 k^2}\frac{3Bh\Theta}{2\rho g},$$

where $\lambda = \sqrt{A/\rho g}$ is the capillary length.

Let us now turn to the case of buoyancy-driven convection. We first write the density as $\rho(1-\alpha\theta) = \rho + \delta\rho(x)$ with $\delta\rho = -\alpha\rho\Theta\sin(kx)$ so that

$$\partial_z p = -\rho g - \delta\rho g + \eta(\partial_{x^2} + \partial_{z^2})v_z. \qquad (34)$$

The equation for v_z,

$$\eta(\partial_{x^2} + \partial_{z^2})^2 v_z = g\partial_{x^2}\delta\rho,$$

is easily solved in the small-k limit using boundary conditions $v_z(z=0,-h) = 0$, $v_x(z=0)$, and $\partial_z v_x(z=0) = 0$ since the surface is free with negligible surface tension variations. We obtain

$$v_z = cz(z+h)^2(2z-h)\sin(kx),$$

where $c = k^2 g\alpha\rho\Theta/48\eta$ takes the buoyancy force into account. v_x can be obtained from the continuity equation

$$v_x = \frac{c}{k}(z+h)\left(8z^2 + hz - h^2\right)\cos(kx)$$

and from (34):

$$\begin{aligned}p &= \frac{6\eta c}{k^2}(8z+3h)\sin(kx) + p_{ext} - (\rho+\delta\rho)gz \\ &= \frac{3}{8}\alpha\rho g\Theta h\sin(kx) + p_{ext} - \rho gz.\end{aligned}$$

Inserting these results in the equilibrium condition (33), we get

$$\Xi_R = \frac{1}{1+\lambda k^2}\frac{3}{8}\alpha h\Theta.$$

3. Instability Mechanisms

Sign considerations immediately show that the position of the rising fluid is marked by a bump when convection is driven by buoyancy and a trough in the Marangoni case. Since both mechanisms generally operate simultaneously when the layer is open to air, at the linear stage, we expect an algebraic addition of the two deformations $\Xi_{tot} = \Xi_M + \Xi_{Ra}$ and, hence, the existence of a compensation height when $\Xi_{tot} = 0$, that is to say when $h = h_{comp} = \sqrt{4B/\alpha\rho g}$, in agreement with the previous comparison of the strengths of the mechanisms from a dimensional view point.

In the following chapter, we will study plain Rayleigh–Bénard convection in pure isotropic liquids between plates, therefore avoiding complications associated with concentration and orientation fluctuations or surface tension variations.

7. Bibliographical Notes

In this chapter, we discussed several instability mechanisms taking place in closed systems. Developments concerning plain Rayleigh–Bénard convection will be quoted in the notes at the end of the following chapter. A more detailed presentation of convection in mixtures can be found in:

[1] J.K. Platten and J.C. Legros, *Convection in liquids* (Springer, Berlin, 1984).

An introduction to instabilities in nematic liquid crystals discussing mechanisms in much the same spirit as here is given by:

[2] E. Dubois-Violette, G. Durand, E. Guyon, P. Manneville, and P. Pieranski, "Instabilities in nematic liquid crystals" in *Liquid crystals*, L. Liebert, ed., Solid State Physics, Supplement **14**, p. 147 (Academic Press, New York, 1978).

Rayleigh–Bénard and Taylor–Couette instabilities are studied in detail by

[3] S. Chandrasekhar, *Hydrodynamic and hydromagnetic stability* (Clarendon Press, Oxford, 1961) .

who also considers the stratification of two immiscible liquids, heavy above light (Rayleigh–Taylor instability) and several cases involving electrical currents and magnetic fields in conducting fluid.

Of great practical importance is another class of instabilities involving shear in open flows, see Tritton BN 1 [7], but also

[4] S. A. Maslowe, "Shear flow instabilities and transition", Chapter 7 of Swinney and Gollub BN 1 [9].

and for a detailed analytic approach:

[5] P. G. Drazin and W. H. Reid, *Hydrodynamic stability* (Cambridge University Press, Cambridge, 1981).

Interface instabilities have been much studied recently, notably the instability of solidification fronts, see the reviews by, e.g.,

[6] J. S. Langer, "Instabilities and pattern formation in crystal growth", Rev. Mod. Phys. **52**, 1 (1981).

and the instability of the interface between two fluids, a low viscosity fluid pushing a more viscous fluid studied first by Saffman and Taylor (1958). In porous media this leads to the formation of *fingers*, hence the name of Saffman–Taylor fingering, see e.g.,

[7] G.M. Homsy, "Viscous fingering in porous media", Ann. Rev. Fluid Mech. **19**, 271 (1987).

Chapter 4

Thermal Convection

This chapter is devoted to a thorough presentation of theory and experiments on plain Rayleigh–Bénard convection. We begin with setting the theoretical frame, Section 1, and deriving the analytical solution of the marginal stability problem, Section 2. Then we turn to nonlinear aspects, discussing first the universal/specific features of secondary instabilities, according to the distance to the threshold and to the value of the Prandtl number that controls the thermal/hydrodynamic nature of the modes (Section 3). Finally, we review experimental findings on the transition to turbulence in convection (Section 4), stressing the fundamental role of the aspect ratio, which controls the nature of weak turbulence and illustrating the different scenarios.

1. Boussinesq Equations and Boundary Conditions

The Boussinesq approximation is basically an assumption of moderate heating reasonably valid in usual experimental situations. Thermodynamic properties of the fluid are contained in a state equation that simply reads:

$$\rho = \rho_{\text{ref}} \left(1 - \alpha \left(T - T_{\text{ref}}\right)\right) \qquad (1)$$

where T_{ref} is a reference temperature, ρ_{ref} the density at that temperature, and α the coefficient of thermal expansion [most of the time we will take T_{ref} at the bottom plate: $z_{\text{ref}} = z_{\text{b}}(= 0)$]. Mechanically, the fluid can be treated as incompressible and all density fluctuations neglected except in the buoyancy term.

1.1. Evolution Equations

The incompressibility condition (continuity of matter) reads

$$\mathbf{v}\cdot\nabla\mathbf{v} = \nabla_h\cdot\mathbf{v}_h + \partial_z v_z = 0, \tag{2a}$$

where $\mathbf{v}_h$ and ∇_h are the horizontal components of $\mathbf{v}$ and ∇.

According to our qualitative understanding of the instability mechanism we know that the vertical direction is singled out. Therefore, we separate Navier–Stokes equations into a horizontal component

$$\rho_{\text{ref}}(\partial_t + \mathbf{v}\cdot\nabla)\mathbf{v}_h = -\nabla_h p + \eta\nabla^2\mathbf{v}_h \tag{2b}$$

and a vertical component

$$\rho_{\text{ref}}(\partial_t + \mathbf{v}\cdot\nabla)v_z = -\partial_z p + \eta\nabla^2 v_z - \rho(z)\,g \tag{2c}$$

containing the buoyancy term (the dynamical viscosity $\eta = \rho_{\text{ref}}\nu$ is assumed independent of the local temperature).

The heat equation reads

$$C(\partial_t + \mathbf{v}\cdot\nabla)T = \chi\nabla^2 T \tag{2d}$$

where the heat capacity *per* unit volume C and the thermal conductivity χ are assumed constant (heating due to viscous dissipation is also neglected).

The validity of these approximations will not be discussed here. We just note that simplifications introduced above are generally supported by order of magnitude estimates but may be insufficient in certain cases, e.g., for water around 4° where it presents a density maximum, thus calling for "non-Boussinesq corrections".

The unperturbed problem corresponds to a fluid at rest with a temperature increasing from bottom to top according to the Fourier law

$$T_0(z) = T_b - \beta z \quad \text{with } \beta = (T_b - T_t)/h,$$

4. Thermal Convection

T_b and T_t being the temperatures of the bottom and top plates, and h height of the fluid layer. The hydrostatic pressure inside the fluid is given by
$$P_0(z) = P_b - \int_{z_b}^{z} \rho(z) g \, dz$$
with $\rho(z) = \rho(T_0(z))$.

Fluctuations are defined by $T = T_0(z) + \theta$, $P = P_0(z) + p$, and $\mathbf{v}$ itself since the basic state is the rest state. Turning to dimensionless variables, we take h and $\tau_\theta = h^2/\kappa$ as length and time units, $h/\tau_\theta = \kappa/h$ for the velocity, $\kappa\nu/\alpha g h^3$ for the temperature, and κ^2/h^2 for p/ρ_{ref}. Adding $P = \nu/\kappa$, the Prandtl number, and $R = \alpha g \beta h^4/\kappa\nu$, the Rayleigh number (the usual control parameter), we obtain the full set of nonlinear equations in dimensionless form:

$$\nabla_h \cdot \mathbf{v}_h + \partial_z v_z = 0, \tag{3a}$$

$$\frac{1}{P}[(\partial_t + \mathbf{v}\cdot\nabla)\mathbf{v}_h + \nabla_h p] = \nabla^2 \mathbf{v}_h, \tag{3b}$$

$$\frac{1}{P}[(\partial_t + \mathbf{v}\cdot\nabla)v_z + \partial_z p] = (\nabla^2 v_z + \theta), \tag{3c}$$

$$(\partial_t + \mathbf{v}\cdot\nabla)\theta = \nabla^2 \theta + R v_z. \tag{3d}$$

At this stage, it is customary to eliminate the pressure field by taking the curl of Equations (3b) and (3c). Let us first consider the equation for the vertical component of the curl $(\partial_x(3b)_y - \partial_y(3b)_x)$. A straightforward calculation leads to

$$(\partial_t - P\nabla^2)(\partial_x v_y - \partial_y v_x) = \partial_y(\mathbf{v}\cdot\nabla v_x) - \partial_x(\mathbf{v}\cdot\nabla v_y), \tag{4}$$

where terms have been reordered to present the r.h.s. as a quadratic source term for the evolution of the linearly relaxing vertical component ζ_z of the vorticity $\nabla \times \mathbf{v}$. The *linearized equation* then simply reads

$$(\partial_t - P\nabla^2)\zeta_z = 0. \tag{4'}$$

The equations for the horizontal components of the curl are obtained in the same way. We get

$$(\partial_t - P\nabla^2)(\partial_y v_z - \partial_z v_y) - P\partial_y\theta = \partial_z(\mathbf{v}\cdot\nabla v_y) - \partial_y(\mathbf{v}\cdot\nabla v_z), \tag{5a}$$
$$(\partial_t - P\nabla^2)(\partial_z v_x - \partial_x v_z) + P\partial_x\theta = \partial_x(\mathbf{v}\cdot\nabla v_z) - \partial_z(\mathbf{v}\cdot\nabla v_x). \tag{5b}$$

Making use of the continuity condition, we can suppress the reference to the horizontal velocity field on the l.h.s. Indeed, noticing that

$$\partial_y(\partial_y v_z - \partial_z v_y) - \partial_x(\partial_z v_x - \partial_x v_z)$$
$$= (\partial_{x^2} + \partial_{y^2})v_z - \partial_z(\partial_x v_x + \partial_y v_y)$$
$$= (\partial_{x^2} + \partial_{y^2})v_z - \partial_z(-\partial_z v_z)$$
$$= (\partial_{x^2} + \partial_{y^2} + \partial_{z^2})v_z = \nabla^2 v_z$$

we differentiate (5a) and (5b) with respect to y and x. Then, subtracting the second equation from the first, we get

$$\partial_t \nabla^2 v_z - P(\nabla^4 v_z + \nabla_h^2 \theta)$$
$$= \partial_z \Big(\partial_x(\mathbf{v}\cdot\nabla v_x) + \partial_y(\mathbf{v}\cdot\nabla v_y) \Big) - \nabla_h^2(\mathbf{v}\cdot\nabla v_z). \quad (6)$$

We need only drop the r.h.s. of (6) and the $\mathbf{v}\cdot\nabla\theta$ term in (3d) above to obtain the complete set of exact linearized equations for fluctuations v_z and θ.

However, the horizontal velocity field cannot be eliminated from the nonlinear problem, which makes things more delicate. It may be practical to split the horizontal component into an irrotational part deriving from a potential ϕ and a rotational part given by a stream function ψ:

$$\mathbf{v}_\text{h} = \mathbf{v}_\text{h}^\text{irr} + \mathbf{v}_\text{h}^\text{rot} = \nabla_\text{h}\phi + \nabla \times (\psi\hat{\mathbf{z}})$$
$$= (\partial_x\phi, \partial_y\phi) + (\partial_y\psi, -\partial_x\psi) = (\partial_x\phi + \partial_y\psi, \partial_y\phi - \partial_x\psi).$$

Inserting this expression into the continuity equation yields the potential part as a function of the vertical velocity component

$$\nabla^2 \phi = -\partial_z v_z,$$

whereas ψ is obtained from the equation for the vertical component of the vorticity since $\nabla^2\psi = \zeta_z$. When needed, the pressure can be obtained from

$$\nabla^2 p = \partial_z \theta - \nabla_\text{h}(\mathbf{v}\cdot\nabla\mathbf{v}_\text{h}) - \partial_z(\mathbf{v}\cdot\nabla v_z)$$

4. Thermal Convection   99

which is obtained by taking the divergence of the velocity equations.

1.2. Boundary Conditions

The complete specification of a convection experiment requires information on the geometrical and physical properties of the 'walls' enclosing the fluid. Here, we are mainly interested in laterally unbounded layers. Therefore, we neglect lateral boundary and consider only the horizontal boundaries (usually plates), denoting their position by $z = z_p$, that is to say, z_b and z_t for bottom and top respectively.

Let us consider first the conditions on the temperature field. In the introductory presentations, it was implicitly assumed that the temperature was fixed at the boundary further viewed as a geometrical plane. In fact the temperature field inside the boundary, which has some thickness, is always a part of the complete problem and the solution of the Fourier equation in the boundary has to be found, which can bring major complications especially at the nonlinear stage.

For specificity, we consider a situation where the fluid layer is bounded by solid plates as depicted in Fig. 1. Usually, either the temperature outside the top and bottom plates is regulated by thermal baths (Fig. 1a), or only the top one is regulated and some electrical power is dissipated in a resistor at the bottom plate (Fig. 1b). In the first case, the control parameter is the temperature difference between the two thermal baths, and the temperature gradient applied to the fluid layer results from a (possibly trivial) subsidiary calculation. In the second case, the control parameter is either the temperature difference or the imposed thermal flux according to whether the electric power dissipated in the resistor is slaved or not to a temperature measurement at the interface between the plate and the fluid. As long as convection remains steady this makes no difference but some care is required when convection becomes time-dependent.

At the interface between the fluid and the plate, the temperature and the heat flux are usually assumed to be continuous. Using primes for quantities related to the plate, i.e., h' the thickness of the plate

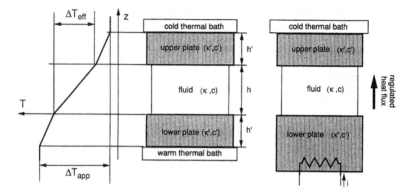

Fig. 1. Concrete models of thermal boundary conditions. Left: the fluid layer is intercalated between two solid plates and, outside the plates, the temperature is held constant by two regulated thermal baths. Right: electric power is dissipated in the bottom plate and the temperature of the top plate is held constant.

(in units of h), χ' its thermal conductivity, and θ' the temperature fluctuation around the basic temperature profile in the plate, we have

$$\theta'\big|_{z_p} = \theta\big|_{z_p}, \qquad (7a)$$

$$\chi' \partial_z \theta'\big|_{z_p} = \chi \partial_z \theta\big|_{z_p}. \qquad (7b)$$

In principle, θ' can be eliminated by solving the Fourier equation in the plate:

$$\partial_t \theta' = \kappa' \nabla^2 \theta' \qquad (8)$$

($\kappa' = \chi'/C'$ is the thermal diffusivity in the plate and C' its specific heat). Boundary conditions relevant to this equation are given by $(7a, b)$ at $z = z_p$, plus $\theta' = 0$ at the interface between the plate and the thermal bath. From this viewpoint, the presence of a temperature fluctuation θ in the fluid implies a forcing at the corresponding boundary. Recalling that the superposition principle holds for Equation (8), we assume that fluctuation θ is reduced to a single, time independent, trigonometric mode with wavevector k in the x-direction: $\theta = \theta(z)\cos(kx)$. The extension to the time dependent

4. Thermal Convection

case is trivial as long as fluctuation θ is slow when compared to a fluctuation θ' with the same horizontal space dependence. When this is no longer the case, the interaction between the plate and the fluid layer becomes more complicated owing to delays implied by a temporary storage of heat in the plate. However, such a difficulty is not expected at, and close to, the primary threshold of stationary convection.

The time-independent solution for fluctuation θ' with wavevector k fulfills $(\partial_{x^2} + \partial_{z^2})\theta' = 0$, that is to say $\partial_{z^2}\theta' = k^2\theta'$. The solution then reads

$$\theta'(x, z) = \left(\sum_{\pm} a_{(\pm)} \exp\left(\pm k\left(z - z_\mathrm{p}\right)\right)\right) \cos(kx).$$

Let us consider the top plate. Then, from the condition $\theta' = 0$ at $z = z_\mathrm{t} + h'$, we get $a_{(-)} = -a_{(+)}\exp(2kh')$, $a_{(+)}$ being determined from condition (7a): $a_{(+)} = \left.\theta\right|_{z_\mathrm{t}}/(1 - \exp(2kh'))$. The fulfillment of condition (7b) thus implies a relation between $\left.\theta\right|_{z_\mathrm{t}}$ and $\left.\partial_z\theta\right|_{z_\mathrm{t}}$:

$$\chi'\left.\partial_z\theta'\right|_{z_\mathrm{t}} = \chi' k\left(a_{(+)} - a_{(-)}\right) = \left.\theta\right|_{z_\mathrm{t}} \chi' k \frac{1 + \exp(2kh')}{1 - \exp(2kh')}$$

$$= \frac{\chi' k}{\tanh(kh')}\left.\theta\right|_{z_\mathrm{t}} = \chi\left.\partial_z\theta\right|_{z_\mathrm{t}}.$$

Thus we see that the presence of the solid plate can be taken into account through an effective boundary condition of the form

$$\left.\theta\right|_{z_\mathrm{t}} = B(k)\left.\partial_z\theta\right|_{z_\mathrm{t}} \tag{7c}$$

(it is easily seen that at the bottom plate, a similar relation holds with a minus sign). The parameter $B(k) = \chi\tanh(kh')/(\chi' k)$ is called a *Biot number*. It depends on the ratio of the thermal conductivity, the thickness of the plate and the wavevector of the instability mode. Dimensionally, it is homogeneous to a length, and can serve as a measure of the penetration of temperature fluctuations in the wall. When the wall is made of a material with a thermal conductivity much larger than that of the fluid, B remains small, so that one can consider the temperature in the fluid as practically imposed at the wall. At the *perfectly conducting* limit, we have

$$\theta\big|_{z_p} = 0. \qquad (9a)$$

Experimentally, this limit can be approached using fluids such as silicon oil, water, etc., and plates made of copper, sapphire, etc.

The opposite case, $B \to \infty$, is obtained when $K' \ll K$; it is said to be *perfectly isolating*. It corresponds to a *fixed thermal flux* condition. In this limit, we had better rewrite (7c) as $B^{-1}\theta\big|_{z_p} = \pm\partial_z\theta\big|_{z_p}$, which leads to the ideal boundary condition

$$\partial_z\theta\big|_{z_p} = 0. \qquad (9b)$$

However, since the extrapolation length tends to infinity, the characteristic length of the system 'plates + fluid' is much larger than h and a singular behavior can be expected when k tends to zero. Though appealing from a theoretical viewpoint, this limit is not realistic and experimental situations with *poorly conducting* plates instead require returning to condition (7c).

Boundary conditions on the velocity are treated in the same way. Usually, the fluid layer is confined by horizontal impervious (i.e., not porous) solid plates at which the first layers of fluid molecules are thought to adhere. Consequently, the fluid velocity has to vanish: $\mathbf{v}(z_p) = 0$, that is to say

$$v_z\big|_{z_p} = 0, \qquad (10a)$$

$$\mathbf{v}_h\big|_{z_p} = 0 \qquad (10b)$$

(*no-slip* condition at a *rigid* plate). Making use of the continuity condition, we can convert the boundary condition on $\mathbf{v}_h$ into a boundary condition on v_z. Indeed since $\mathbf{v}$ must vanish identically at $z = z_p$, all horizontal derivatives also vanish, especially $\partial_x v_x$ and $\partial_y v_y$. Inserting these conditions into the continuity equation implies

$$\partial_z v_z\big|_{z_p} = 0. \qquad (10b')$$

4. Thermal Convection

Experiments can also be performed with the fluid layer covered by another fluid, or else with a free surface open to air. At such an interface, boundary conditions have to account for the continuity of the velocity normal to the interface and that of the tangential stress in the plane of the interface. Assuming that surface tension effects are not involved and that the interface remains flat, we see that condition (10a) remains valid. The tangential stress components in the (x, y)-plane are given by $\sigma_{uz}(z_p) = \eta(\partial_z v_u + \partial_u v_z)$, where $u \equiv x$ or y, and η is the dynamical viscosity. The continuity condition for the tangential stress reads

$$\eta' \partial_z v'_u \big|_{z_p} = \eta \partial_z v_u \big|_{z_p},$$

where primes denote quantities related to the fluid above the convecting layer (since $v_z\big|_{z_p} \equiv 0 \Rightarrow \partial_u v_z\big|_{z_p} \equiv 0$). In principle, we should have to solve for $\mathbf{v}'$ in the same way as for θ' in the thermal case but here we simply assume a large viscosity contrast $(\eta'/\eta \to 0)$. Then, neglecting the l.h.s. of the equation above, we get the so-called *stress-free* condition

$$\partial_z \mathbf{v}_h \big|_{z_p} = 0. \tag{10c}$$

As above, this condition on the horizontal velocity components can be transformed into a condition on the vertical component; differentiating Equation (3a) with respect to z first, we get $\nabla_h \partial_z \mathbf{v}_h + \partial_{z^2} v_z = 0$. Evaluating this expression at $z = z_p$ using condition (10c), we get

$$\partial_{z^2} v_z \big|_{z_p} = 0. \tag{10c'}$$

Finally, one usually considers velocity boundary conditions either no-slip —Equations $(10a, 10b')$— or stress-free —Equations $(10a, 10c')$. These conditions can apply symmetrically at top and bottom, e.g., rigid, or asymmetrically: in practice, free at top and rigid at bottom (which requires some caution owing to the Marangoni effect). The symmetrical stress-free case, though unrealistic, is often considered since it is can be handled most easily form an analytical viewpoint.

2. Normal Mode Analysis

Neglecting nonlinear terms in Equations (6) and (3d), we are left with a set of two equations for two unknowns, v_z and θ,

$$\partial_t \nabla^2 v_z = P\left(\nabla^4 v_z + \nabla_h^2 \theta\right), \qquad (11a)$$
$$\partial_t \theta = \nabla^2 \theta + R v_z. \qquad (11b)$$

Since the fluid layer is supposed to be laterally unbounded, the horizontal dependence can be analyzed using Fourier modes:

$$\{v_z, \theta\} = \{V(z), \Theta(z)\} \exp(i\mathbf{k}_h \cdot \mathbf{x}_h) \exp(st), \qquad (12)$$

the general solution being obtained by *superposition* of individual solutions associated with every wavevector of arbitrary length and orientation. In the following, we assume symmetrical boundary conditions at top and bottom, good conducting and either stress-free (easier analytically) or no-slip (more realistic).

2.1. Stress-Free Solution, Rayleigh (1916)

Recalling the boundary conditions $\theta = v_z = \partial_{z^2} v_z = 0$ at $z_b = 0$, and $z_t = 1$, we see that the solution can be sought as a superposition of individual modes of the form

$$\{V_n, \Theta_n\} \sin(n\pi z) \exp(i\mathbf{k}_h \cdot \mathbf{x}_h) \exp(s_n t),$$

which fulfill all boundary conditions identically. Upon insertion in system (11), with $k = |\mathbf{k}_h|$, we get

$$-(n^2\pi^2 + k^2)s_n V = P\left((n^2\pi^2 + k^2)^2 V - k^2 \Theta\right),$$
$$s_n \Theta = -(n^2\pi^2 + k^2)\Theta + RV$$

which have the same form as the model equations of Chapter 3 but now take correctly into account the effects of the horizontal flow.

4. Thermal Convection

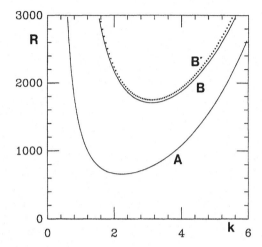

Fig. 2. Marginal stability curves: stress-free case, Equation (13'), curve A; no-slip case, exact numerical solution of (15), curve B; approximate result, Equation (18), curve B' (dotted).

The dispersion equation reads

$$\left(s_n + \left(k^2 + n^2\pi^2\right)\right) \\ \times \left(s_n \left(k^2 + n^2\pi^2\right) + P\left(k^2 + n^2\pi^2\right)^2\right) - RPk^2 = 0. \tag{13}$$

Expanding this quadratic equation, we can show easily that the instability is stationary. The marginal stability condition for mode n, $s_n = 0$, yields the threshold condition for the nth mode,

$$R_n(k) = \frac{(n^2\pi^2 + k^2)^3}{k^2}. \tag{13'}$$

The instability threshold is given by the minimum of the lowest curve corresponding to $n = 1$ (see Fig. 2, curve A). We get

$$R_c = \frac{27\pi^4}{4} \quad \text{for} \quad k_c = \frac{\pi}{\sqrt{2}}.$$

When k tends to zero, R_n is seen to diverge as k^{-2}, which expresses the stabilizing effect of the viscous dissipation associated with the horizontal shear not included in the simplified model and supports the argument given in Chapter 3, Section 1.3. By contrast, the k^4 divergence as k tends to infinity was already correctly taken into account by the semi-qualitative analysis.

2.2. No-Slip Solution, Pellew and Southwell (1940)

Boundary conditions (10a) and (10b) for the vertical velocity component v_z now differ by only one differentiation order. This means that simple trigonometric lines will not fulfill both conditions and that we must return to the initial guess (12) without making any assumption on the analytic expression of $V(z)$ and $\Theta(z)$.

Inserting (12) into system (11) and denoting the differentiation with respect to z simply as ∂, we are lead to

$$(\partial^2 - k^2)sV = P\left((\partial^2 - k^2)^2 V - k^2\Theta\right), \qquad (14a)$$

$$s\Theta = (\partial^2 - k^2)\Theta + RV. \qquad (14b)$$

Though this can be proven by a rigorous argument, we simplify the present calculation by supposing from the start that the instability is stationary, i.e., $s = 0$ at threshold. The elimination of one of the two variables, either V or Θ is then a simple matter. For example, keeping V, we obtain a single sixth-order differential equation, cubic in ∂^2,

$$\left((\partial^2 - k^2)^3 + k^2 R\right) V = 0.$$

The general solution is thus a superposition of six complex exponentials,

$$V(z) = \sum_{m=1}^{6} V_m \exp(iq_m z),$$

where the V_ms are now pure numbers and the q_ms are the roots of

$$(k^2 + q_m^2)^3 - k^2 R = 0,$$

that is to say, three pairs $\pm q_m$ given by

$$q_m^2 = -k^2 + \sqrt[3]{Rk^2}\, j^m, \qquad m = 0, 1, 2,$$

where $j = -1/2 + i\sqrt{3}/2$.

Owing to the symmetry of the boundary conditions, it is more practical to set the origin of coordinates at mid-height ($z_p = \pm 1/2$),

4. Thermal Convection

which allows a simpler classification of solutions according to their parity. Complex exponentials can then be recombined into sines and cosines. For simplicity, we consider only the lowest lying mode. In order to minimize the horizontal shear, this mode should correspond to a solution with an even vertical velocity component resembling a cosine arch so that we choose

$$V(z) = \sum_{m=0}^{2} V_m \cos(q_m z).$$

The corresponding temperature fluctuation is given by Equation (14a) or (14b); for example, from (14a),

$$\Theta(z) = \sum_{m=0}^{2} \frac{(q_m^2 + k^2)^2}{k^2} V_m \cos(q_m z).$$

Boundary conditions $V(\pm 1/2) = 0 = \partial V(\pm 1/2) = 0 = \Theta(\pm 1/2)$ then read

$$0 = \sum_{m=0}^{2} V_m \cos(q_m/2),$$

$$0 = \sum_{m=0}^{2} V_m q_m \sin(q_m/2), \qquad (15)$$

$$0 = \sum_{m=0}^{2} V_m (q_m^2 + k^2)^2 \cos(q_m/2).$$

System (15) is a homogeneous algebraic system of three equations with three unknowns V_m. It contains k and R as parameters through the q_ms. It will have a trivial solution $V_m = 0$ except when its determinant vanishes. The fulfillment of this *compatibility condition* implies a relation between R and k defining the *marginal stability curve* in implicit form. This transcendent equation has to be solved numerically. The result is displayed as curve B in Fig. 2. The threshold is reached for

$$k = k_c \simeq 3.11632, \qquad R = R_c \simeq 1707.76.$$

Notice that the critical wavelength $\lambda_c = 2\pi/k_c$ is close to 2 ($2h$ in dimensional units), i.e., the diameter of a roll is roughly equal to the height of the cell, which supports the assumptions made to conclude the qualitative analysis.

Once a horizontal pattern has been selected (i.e., a distribution of $\mathbf{k}_h$), the horizontal velocity can be obtained from the continuity equation. The z-dependence associated with a given wavevector of length k is then given by

$$U(z) = -i \sum_{m=0}^{2} \frac{q_m V_m}{k} \sin(q_m z).$$

Note that the analysis just developed is both nontrivial and remarkably simple; the differential system has constant coefficients, and the order is sufficiently low to make it tractable by hand (although a numerical calculation is required at the last step). This analysis can easily be extended to more complicated cases, e.g., imperfect thermal boundary conditions (Equation (7c)). In the absence of symmetry, the odd/even classification of solutions no longer holds and we are obliged to return to the sixth-order problem. Otherwise, the same approach can be developed for systems with a larger number of coupled variables (binary mixtures,...) with the possibility of oscillating neutral modes. Problems can be much more serious analytically when the resulting differential system has variable coefficients (non-Boussinesq effects, Taylor instability,...). This calls for approximation methods to be introduced in Section 2.4 just after the study of the neighborhood of the critical point.

2.3. Vicinity of the Threshold (Linear Stage)

A given Fourier mode k becomes marginal when the Rayleigh number R reaches the value corresponding to the marginal stability condition for the lowest lying normal mode, $R = R_1(k)$. Here we are interested in the behavior of such a mode at slightly different Rayleigh numbers but for k sufficiently close to k_c.

Let us first consider the stress-free case, the dispersion relation of which is known analytically. For $R \sim R_c$ and $k \sim k_c$, Equation (13)

4. Thermal Convection

with $n = 1$ has two roots, one of which is strongly negative. The other root is close to zero and can be obtained by an expansion in powers of $\delta k = (k - k_c)$ and $r = (R - R_c)/R_c$ which leads to

$$\tau_0 s = r - \xi_0^2 \delta k^2 \tag{16}$$

with

$$\tau_0 = \frac{3}{2\pi^2} \frac{1+P}{P} \quad \text{and} \quad \xi_0^2 = \frac{8}{3\pi^2},$$

where τ_0 is called the *natural relaxation time* and ξ_0 the *coherence length*.

The no-slip case is more delicate. The basic reason is that for stress-free boundary conditions, the simple projection onto the sine basis is adequate in all circumstances. In contrast, the analytical structure of no-slip eigenmodes evolves continuously as (k, R) are shifted away from (k_c, R_c). However, for k and R sufficiently close to k_c and R_c, a relation of the form (16) with adapted coefficients can be obtained by a perturbation expansion in powers of δk and r. This calculation is an interesting example of the use of algebraic properties of linear differential boundary value problems; however it will not be reproduced here owing to its technical character. The result reads

$$\tau_0 \simeq \frac{1 + 1.9544\, P}{38.4429\, P} \quad \text{and} \quad \xi_0^2 \simeq 0.1479.$$

A comparison between the stress-free and no-slip results is best obtained by calculating

$$(k_c \xi_0)_{\text{stress-free}} = \frac{2}{\sqrt{3}} \simeq 1.155 \approx (k_c \xi_0)_{\text{no-slip}} = 1.199,$$

which shows that the coherence lengths are both close to $\lambda_c/2\pi$ though the values of λ_c are different in each case.

2.4. Approximate Determination of Threshold Curves

The exact analysis just presented may be difficult to adapt to more complicated physical situations so that it may be useful to master methods leading to accurate approximations of the solution of the linear stability problem. Postponing the presentation of purely numerical methods to Appendix 3, we restrict ourselves here to the consideration of a method based on the truncated projection of the solution on a set of well chosen basis functions, the so-called *Galerkin method*.

Basic steps are (i) the choice of a family of basis functions that fulfill the boundary conditions, and (ii) the definition of projection rule on the space spanned by the basis functions. The basis chosen need not be orthogonal according to the projection rule chosen but should be complete. Low order truncations are often easily tractable by hand, which gives a practical way of getting simplified but near-realistic models of instability as will be shown below. However the numerical solution of high order truncations is required to obtain reliable quantitative results.

Here, owing to the simplicity of the calculation, we consider polynomials which are easily seen to form a complete basis. The boundary condition on the temperature field, $\Theta(z = \pm 1/2) = 0$, implies

$$\Theta(z) = (1/4 - z^2) P_\theta(z) = (1/4 - z^2) \sum_{n=0}^{\infty} \Theta_n z^n. \qquad (17a)$$

In the same way, for the vertical velocity component we have $V_z(\pm 1/2) = \partial V_z(\pm 1/2) = 0$, so that the relevant prefactor is now $(1/4 - z^2)^2$:

$$V(z) = (1/4 - z^2)^2 P_v(z) = (1/4 - z^2)^2 \sum_{n=0}^{\infty} V_n z^n. \qquad (17b)$$

Two series of unknown coefficients $\{\Theta_n\}$, and $\{V_n\}$ are thus introduced and, in fact, truncated at some order $n_{\max}$.

We now turn to the definition of the projection rule that converts the differential problem into an algebraic problem. A general approximation technique, called the *method of weighted residuals*, consists in expanding the equations on the chosen basis and projecting the

4. Thermal Convection

equations on a second basis of so-called weight functions. For example, we may ask that the equations be satisfied at n_{max} given points on the interval on which the problem is defined, therefore choosing delta functions centered at these points as weight functions. In the Galerkin method, the weight functions are the basis functions themselves so that Equation (14a) for the velocity is projected on the basis functions for the velocity (17b) and Equation (14b) for the temperature on functions for the temperature (17a). We have then to compute integrals involving terms of the form $z^m(1/4 - z^2)^n$ on the interval $(-1/2, 1/2)$. For convection, this choice turns out to be optimal, owing to the variational structure of the underlying differential problem.

This projection procedure converts the differential problem into a linear homogeneous algebraic problem containing k and R as parameters. For simplicity, we consider the lowest order nontrivial case and thus insert only one term for each fluctuation

$$\Theta \to \widetilde{\Theta}(1/4 - z^2), \qquad V \to \widetilde{V}(1/4 - z^2)^2$$

in the steady state equations (14a, b) with $s = 0$. Dropping all unnecessary subscripts, we multiply (14a) by $(1/4 - z^2)^2$ and (14b) by $((1/4 - z^2)$, and by integration over $(-1/2, 1/2)$, we obtain:

$$\int_{-1/2}^{1/2} (1/4 - z^2)^2$$
$$\times \left((\partial^4 - 2k^2 \partial^2 + k^4) \widetilde{V} (1/4 - z^2)^2 - k^2 \widetilde{\Theta} (1/4 - z^2) \right) dz = 0,$$

$$\int_{-1/2}^{1/2} (1/4 - z^2)$$
$$\times \left(R\widetilde{V} (1/4 - z^2)^2 + (\partial^2 - k^2) \widetilde{\Theta} (1/4 - z^2) \right) dz = 0$$

which yields

$$(k^4 + 24k^2 + 504)\widetilde{V} - (9/2)k^2 \widetilde{\Theta} = 0,$$
$$(3/14)R\widetilde{V} - (k^2 + 10)\widetilde{\Theta} = 0.$$

As usual, the vanishing of the determinant of this system gives us
the marginal stability condition, here
$$R = f(k) = \frac{28}{27} \frac{(k^4 + 24k^2 + 504)(k^2 + 10)}{k^2}. \tag{18}$$
In Fig. 2, the corresponding curve is displayed as B'. The minimum takes place at $R_c \simeq 1750$ and $k_c = 3.1165$, in excellent agreement with the exact values. The curvature is also well approximated ($\xi_0^2 \simeq 0.1497$) and, once extended to the time-dependent problem, the calculation gives a good estimate of the relaxation time ($\tau_0 \simeq (1 + 1.9425\,P)/38.2927\,P$).

Clearly, this method can be adapted to cases more difficult to handle analytically, e.g., linear differential systems with variable coefficients (non-Boussinesq convection, Taylor instability,...), whereas expansions of the full nonlinear evolution problem are at the heart of numerical methods using 'spectral methods'.

2.5. Treatment of the Vertical Vorticity

At the linear stage, the vertical vorticity is simply governed by
$$\partial_t \zeta_z = P \nabla^2 \zeta_z. \tag{4'}$$

In the no-slip case, condition (10b) implies $\zeta_z(z_p) = 0$. Assuming $z_b = 0$ and $z_t = 1$, we search for the solution as a sine series in z:
$$\zeta_z = \sum_{n=1}^{\infty} Z_n \sin(n\pi z) \exp(i\mathbf{k}_h \cdot \mathbf{x}_h) \exp(s_n t),$$
which leads to
$$s_n Z_n = -P\,(n^2\pi^2 + k^2) Z_n \tag{19}$$
so that s_n remains negative for all n; only modes without vertical vorticity are retained at the linear level.

In the stress-free case, the situation is slightly different. The boundary condition is now $\partial_z \zeta_z = 0$. Keeping the boundaries at $z_p = 0$ and 1, we look for the solution as a cosine series:
$$\zeta_z = \sum_{n=0}^{\infty} Z_n \cos(n\pi z) \exp(i\mathbf{k} \cdot \mathbf{x}_h) \exp(s_n t).$$

4. Thermal Convection 113

The evolution is still given by Equation (19) so that all modes are damped except the mode $n = 0$ for $k = 0$ which is neutral ($s_0(k = 0) = 0$). Since the vertical vorticity is not coupled to (v_z, θ) at the linear stage, the determination of the threshold is not affected by the existence of this mode which, in turn, will strongly influence the nonlinear dynamics.

3. Phenomenology of Nonlinear Convection

3.1. Introduction

In the first part of this chapter, Rayleigh–Bénard convection has served as a pretext for a detailed presentation of some technicalities involved in linear stability analysis. It will now serve us to introduce the nonlinear dynamics of dissipative structures, at a phenomenological level.

From the form of the general nonlinear equations (3) and the results of linear theory, we guess that, at the limit of the laterally unbounded layer, relevant parameters controlling the behavior above threshold will be the Rayleigh number R, the Prandtl number P, and the wavevector k of the disturbance. From an experimental viewpoint, R is the usual control parameter, whereas P, which characterizes the fluid properties can be varied only by changing the fluid and to a certain extent by keeping the same fluid but changing the average temperature of the layer.

In fact, though its role is important at the nonlinear stage, the wavevector k can hardly be considered a parameter. Indeed, in the absence of forcing, a disordered texture is usually observed with *grains* made of locally periodic rolls with wavelengths close to λ_c (see Chapter 1, Fig. 3). However, a detailed understanding of the elementary processes involved in the transition to turbulence requires some control on the structure. A first popular method takes advantage of lateral boundary effects to monitor the orientation of the convection rolls. A second, more efficient, method called *thermal printing* consists in heating locally the fluid below the convection threshold. This induces slight density differences and consequently a slow "sub-critical" motion further amplified by the instability mechanism that freezes the desired pattern. The stability of large, well-oriented, single domains with given wavevector far from k_c can be studied in this way but so can the dynamics of more exotic patterns containing specific defects (see Bibliographic Notes).

The linear stability theory has left us with the possibility of arbitrary superpositions of noninteracting modes. The first effect of nonlinear couplings is to select specific wavevector combinations and to discard others. For Rayleigh–Bénard convection between rigid, good-conducting plates, only rolls are observed, i.e., superpositions made of a single pair of opposite wavevectors. In this regime, the perturbations to the uniform basic state of rest depend locally on two space coordinates, hence the name of *two-dimensional convection*. However, selected patterns may be different. For example, convection between poorly conducting plates takes the form of squares (superpositions of two pairs of wavevectors at right angles $\pm \mathbf{k}_{1,2}$), whereas convection with a free surface usually yields hexagons (3 pairs of wavevectors at 120°). For specificity, we concentrate our attention on Rayleigh–Bénard rolls.

The stability of convection rolls has been studied in detail, both theoretically and experimentally by Busse and his co-workers (since 1969). In the three-dimensional parameter space (R, P, k), the stability condition of straight rolls define a surface called the *Busse balloon*, of which we will consider specific portions only. Basically, two kinds of secondary instabilities can take place: universal or specific. The presence of universal secondary modes depends only on the symmetry properties of the structure whereas specific modes involve physical couplings depending on the value of the Prandtl number. In practice, the situation is a little less clear cut since the threshold of universal modes can depend quantitatively on P and some secondary modes mix universal and specific features.

4. Thermal Convection

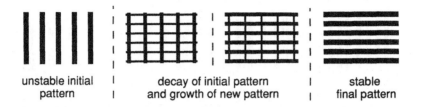

Fig. 3. Cross roll instability: weak rolls with too large a wavelength (left) are unstable against of rolls at right angles (middle) which finally replace the original pattern (right).

3.2. Universal Secondary Modes

From a general viewpoint, one can say that a roll pattern can be characterized by the intensity of the convective motion, the length and orientation of the wavevector, and the absolute position of the pattern in the laboratory frame. As will be discussed in detail in the final chapters, the structure can be described locally by a complex envelope with a modulus and a phase. The first instability mode to be presented is linked to the local intensity of convection as measured by the modulus of the envelope. The two other modes are *phase instabilities* associated with the position of the rolls.

3.2.1. Cross-Roll Instability

At distance $(R-R_c)$ of the threshold, a wavevector bandwidth of the order of $\sqrt{R-R_c}$, centered around k_c is linearly unstable. At the same time, the intensity of convection is of the order of $\sqrt{R-R_c}$ at the center of the band and decreases to zero at the band edge, i.e., for $k - k_c \sim \mathcal{O}(\sqrt{R-R_c})$. Rolls at the band edge (induced by thermal printing) are thus weak and the fluid layer is likely to "prefer" a state of more intense convection with a wavevector closer to k_c. The initial pattern is merely replaced by another one that result from the growth of the fastest growing unstable mode with k closer to k_c (Fig. 3). On general grounds, an *oblique-roll* instability would be expected. For convection, it turns out that this new pattern is at right angles with the old one, hence the name of *cross-roll instability*

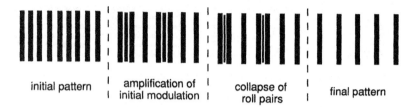

Fig. 4. Evolution of a wavelength modulation for rolls compressed on average beyond the threshold of Eckhaus instability.

(subscript "CR" in the following). Close to R_c, in the $(\delta k, r)$-plane (with $\delta k = k - k_c$ and $r = (R - R_c)/R_c$ as usual), the domain where this instability develops is the region between the marginal stability curve (subscript "MS") given by $r_{\mathrm{MS}} = \xi_0^2 \delta k^2$ and the parabola $r_{\mathrm{CR}} = \xi_{\mathrm{CR}}^2 \delta k^2$.

3.2.2. Compression/Dilatation Instability

Translation invariance in the direction of the wavevector is associated with a long wavelength, compression/dilatation, instability —the *Eckhaus instability* (subscript "EI"). This instability develops between the marginal stability curve and the parabola $r_{\mathrm{EI}} = 3\xi_0^2 \delta k^2$.

The Eckhaus instability takes place for rolls with both $k < k_c$ and $k > k_c$ but sufficiently far from k_c, again in a k-region where convection is weak. The growing modulation is not expected to saturate. Indeed, let us assume for example basic rolls with k larger than k_c, i.e., a structure compressed on average; the modulation generates under-compressed and over-compressed regions but over-compressed regions are even more unstable, which is suitable for the collapse of a pair of rolls. After this collapse, the average wavelength has increased or, equivalently, k has got closer to k_c so that the roll pattern is "more optimal", as illustrated in Fig. 4. In the same way, for a pattern dilated on average $(k < k_c)$, regions over-dilated by the modulation witness the birth of a pair of rolls, which increases the average wavevector. These processes can take place because when the phase modulation becomes large, the local effective wavevector gets so close to the

4. Thermal Convection

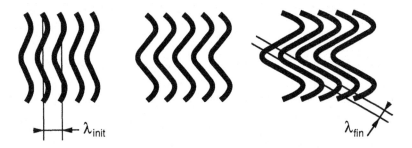

Fig. 5. Growth and saturation of the zigzag instability.

edge of the unstable band that the amplitude of primary convection decreases, leaving room for a localized secondary convection with a more optimal wavevector as for the cross-roll instability.

3.2.3. Torsion Instability

The isotropy of the Rayleigh mechanism in the horizontal plane makes rolls neutral with respect to a local rotation. The second phase instability is associated with this invariance property. Leading to an undulation of the rolls along their axis (i.e., in a direction perpendicular to that of the wavevector) it is called the *zigzag instability* (subscript "ZZ"). In the $(\delta k, r)$-plane, the domain of the zigzag instability is the region on the left of a line simply given at lowest order by $\gamma r_{ZZ} + \delta k = 0$ in which the coefficient γ depends on the Prandtl number, down to the marginal stability curve.

Associated perturbations correspond to a sinusoidal displacement of the rolls as pictured in Fig. 5, hence the name of the instability. Rolls unstable to the zigzags have wavelengths longer than the optimum. The modulation can still be understood as an attempt to get closer this optimum. Indeed, assuming a basic wavevector $k < k_c$ with rolls well-aligned parallel to the y-axis ($\mathbf{k}_0 = k_0 \hat{\mathbf{x}}$), adding the zigzag perturbation, i.e., a small component $k_y \hat{\mathbf{y}}$, we see that the length of the effective wavevector is $(k_0^2 + k_y^2)^{1/2} > k_0$, i.e., closer to k_c. However, in contrast with the previous case, the

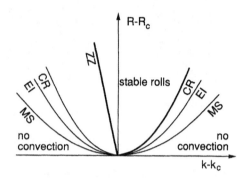

Fig. 6. Stability of rolls against universal modes close to the threshold: marginal stability condition → MS; Eckhaus instability → EI; zigzags → ZZ; cross-rolls → CR.

instability is expected to saturate at a finite amplitude depending on the shift from k_c, and this is actually observed in adequately prepared systems of rolls. The threshold curves of the different universal instabilities close to the onset of convection is depicted in Fig. 6.

3.3. Specific Secondary Instabilities

It has been seen that the threshold of Rayleigh–Bénard convection was independent of the Prandtl number $P = \nu/\kappa$ which compares the relaxation rate of temperature fluctuations $\tau_\theta = h^2/\kappa$ to that of velocity fluctuations $\tau_v = h^2/\nu$. However, as already discussed in Chapter 3, Section 1.4, the value of P controls the nature of the convective mode and the nonlinear behavior beyond the threshold. These effects originate from the inertia term $(\partial_t + \mathbf{v}\cdot\nabla)\mathbf{v}$ in the complete Boussinesq equations, system (3). At the linear stage, only the ∂_t-term is involved (τ_0 depends on P) but the second contribution appears to play the major role at the nonlinear stage. From a general viewpoint one can say that the slowest is the most important variable, since rapid fluctuations can be adiabatically eliminated to a large extent. Therefore, strongly different nonlinear behaviors can be expected according to whether the temperature, a scalar field, or the velocity, a three-dimensional vector field, is the more relevant variable.

4. Thermal Convection   119

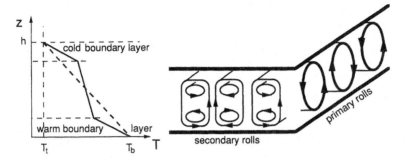

Fig. 7. When R increases, thermal boundary layers develop close to the top and bottom plates; they can become unstable against the formation of localized secondary rolls at right angle with primary rolls (bimodal instability, after Busse, 1982).

3.3.1. High Prandtl Number Instabilities

When P is "large," the temperature is the slowest variable and the vertical velocity component is directly slaved to it by the mechanism of the primary instability. In turn, the horizontal component of velocity field is slaved to its vertical component. Indeed, at the limit "$P \to \infty$" the term $(\partial_t + \mathbf{v}\cdot\nabla)\mathbf{v}$ completely disappears from the primitive equations, so that the vertical vorticity equation (4) remains trivial. The horizontal component of the velocity field is then entirely potential and can be derived from the continuity equation (3a) after v_z has been obtained from θ using Equation (6) which simply reads $\nabla^4 v_z + \nabla_h^2 \theta = 0$.

The feedback between θ and $\mathbf{v}$ which enters the problem through the term $\mathbf{v}\cdot\nabla\theta$ in Equation (3d) is purely passive. As a result, secondary instabilities develop in thermal boundary layers close to the horizontal plates and take place "far" from the onset of convection (R large).

Here we consider a secondary instability called *bimodal*. Beyond the convection threshold, the temperature profile, once averaged over a wavelength of the primary structure, displays two thermal boundary layers at top and bottom, as shown in Fig. 7 (left). Indeed, in the bulk of the layer, the local temperature gradient is decreased with respect to the applied gradient $\beta = \Delta T/h$ since part of the heat is transported by convection rather than by conduction; there should be a whole nearly isothermal sub-layer at mid-height in the fluid where the vertical velocity is maximum. On the other hand, close to the plates where the vertical motion is inhibited by their presence, the local gradient must be larger than β since half of the applied temperature difference ΔT is applied to the two complementary sub-layers where conduction is the dominant transport process.

These two *thermal boundary layers* can become unstable against Rayleigh's mechanism, so that a new set of rolls localized close to the plates can develop. The specific form of nonlinear interactions between the primary rolls and this secondary structure favors the growth of a system of rolls at right angles with the primary mode. This *bimodal instability*, depicted in Fig. 7 (right), gives rise to a rectangular pattern. Before this secondary instability, the structure was two-dimensional, now it depends on the three space coordinates. A transition to three-dimensionality has taken place. However the bimodal instability does not introduce time dependence.

Figure 8 displays the aspect of the section of the Busse balloon at high P, from which we see that the only "dangerous" instabilities are the zigzag instability for $k < k_c$ and R close to R_c and the cross-roll instability, which is progressively converted into the bimodal instability for $k > k_c$ when R is increased. This diagram is well corroborated by experimental results with prepared convective structures in extended geometry.

Time dependence appears at even higher R, e.g., via Howard's mechanism, presented here mainly because it involves an interesting dimensional argument involving the thermal boundary layer just introduced.

Let us consider again the spatially averaged temperature profile. After the bimodal instability has taken place, it still displays two thermal boundary layers. Let δ_{bl} be the "thickness" of the such a sub-layer and define the corresponding Rayleigh number by $R_{bl} = \alpha g (\Delta T/2) \delta_{bl}^3 / \kappa \nu$ where $\Delta T/2$ is the order of magnitude of the temperature difference applied to one of the two boundary layers.

4. Thermal Convection

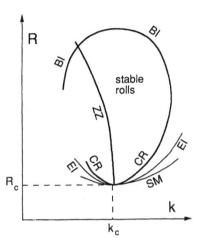

Fig. 8. Typical aspect of a section of the Busse balloon for P large, say $P \gg 10$, MS, EI, ZZ, CR, as before; bimodal instability → BI (after Busse, 1970).

When R_{bl} is larger than some critical value R_H, the boundary layers are expected to be unstable. Let us assume that the boundary layers are periodically destroyed and rebuilt so that this tertiary instability is oscillatory with frequency ω_H. The corresponding period $2\pi/\omega_H$ can be guessed to scale as the thermal diffusion time over the thickness of the boundary layers: $\tau_{bl} = \delta_{bl}^2/\kappa$. Now recalling the definition of the Rayleigh number for the full layer $R = \alpha g \, \Delta T \, h^3/\kappa\nu$ we have $R_{bl}/R \propto \delta_{bl}^3/h^3$ so that, at the threshold of the oscillatory instability $R_{bl} = R_H$, we have $\delta_{bl} \propto R^{-1/3}$. Inserting this into $\omega_H \sim \tau_{bl}^{-1}$, we obtain $\omega_H \propto R^{2/3}$, as actually observed in some experiments. Other processes also induce time-dependence, often both less intrinsic and less regular since they involve defects of the bimodal structure and lead to the formation of unsteady "thermal plumes".

3.3.2. Intermediate Prandtl Numbers Instabilities

When P gets "smaller," the velocity field takes on more and more importance since it becomes comparatively slower. The terms $\mathbf{v} \cdot \nabla \mathbf{v}$ cannot be ignored, especially as source terms for the vertical vorticity, which was damped to zero at the linear stage as well as at the limit $P \to \infty$. As a rule, secondary instabilities appear closer to R_c, at values for which the definition of thermal boundary layers would be meaningless. The case of moderate Prandtl numbers is the least universal and the most complicated. One must resort to numerical analysis to find the "zoo" of possible secondary modes for $5 < P < 50$ which can be observed in specially designed experiments.

3.3.3. Low Prandtl Number Instabilities

The situation at lower P, typically $P < 2$, becomes less confused. It can be seen from the nonlinear Boussinesq equations that straight rolls do not generate vertical vorticity. In contrast, except in special cases such as that of a strictly axisymmetric roll pattern, the horizontal velocity field associated with curved rolls is rotational. Since only deviations with respect to the ideal pattern are expected to take part in the forcing of the vertical vorticity, we may expect that its response will be a flow at a scale large when compared to the diameter of the rolls. The corresponding horizontal velocity field does not average to zero over the thickness of the fluid layer so that it can push or pull the roll pattern, hence the name *drift flow*. The net result is a feedback mechanism, extremely complicated when operating on typical disordered textures but slightly more accessible when acting on slightly distorted straight rolls.

The first consequence of the presence of the drift flow is the dependence of the slope of the zigzag instability threshold on the Prandtl number (the coefficient γ introduced earlier). The drift flow can be shown to counteract the curvature of the pattern (see Fig. 9a); its stabilizing effect leads to an enlargement of the domain of zigzag-stable rolls as P decreases. This instability is therefore expected to play a minor role at low P (see Fig. 10 below and compare it with Fig. 8).

The drift flow is a manifestation of the lowest order vorticity mode examined at the linear stage in Section 2.5. It was seen to be damped for no-slip boundary conditions, but simply neutral in the stress-free case. This latter property is linked to the Galilean invariance of Boussinesq equations with stress-free boundary conditions. Indeed, it can be checked that, replacing a solution $\mathbf{v}$ of the equations by $\mathbf{v} + \mathbf{U}_h$, where $\mathbf{U}_h$ is an arbitrary uniform horizontal velocity field,

4. Thermal Convection

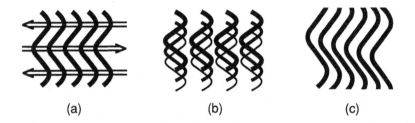

Fig. 9. a) Drift flow associated with the zigzag mode; b) Busse oscillations, i.e., zigzag undulation propagating along the axis of primary rolls; c) skewed-varicose instability.

and turning to a moving frame in uniform translation at velocity $\mathbf{U}_h$, i.e., $\mathbf{x}_h \to \mathbf{x}_h - \mathbf{U}_h t$, we still have an exact solution of the stress-free problem. A new secondary mode called *Busse oscillatory instability* is associated with this typically inertial mode. It reveals itself as a propagation of sinusoidal waves along the axis of the primary rolls (Fig. 9b). Though most easily found in the stress-free case (Busse, 1972), this mode is also present in the realistic no-slip case at sufficiently small P. The reason is that, in much the same way as thermal boundary layers build up when $P \to \infty$, for $P \to 0$ viscous boundary layers are present close to the solid plates, which relaxes the no-slip condition. Experimentally, this mode can be observed for $P < 1$.

The drift flow should not affect the Eckhaus instability directly since the wavevector of the modulation is strictly parallel to the wavevector of primary rolls and thus does not involve curvature. In fact, the existence a supplementary mode called the *skewed varicose instability* can indirectly be traced back to it. This new mode, sketched in Fig. 9c can be understood as the result of a buckling of the Eckhaus mode, i.e., a compression/dilatation instability at an oblique angle with the wavevector of the primary pattern. Indeed, at finite Prandtl number, the Eckhaus instability involves the building of a internal longitudinal pressure gradient that can be relaxed *via* the drift flow associated with an oblique modulation. The skewed varicose mode plays an essential role in the transition to turbulence for $0.5 < P < 5$.

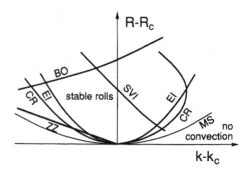

Fig. 10. Typical section of the Busse balloon at low Prandtl numbers: MS, EI, ZZ, as before; Busse oscillations → BO; skewed varicose instability → SV.

Figure 10 displays the typical aspect of the section of the Busse balloon for P small.

To conclude this section, let us remark that modes important for P large involve a strong three-dimensionality and require specific calculations even though some physical intuition of the mechanisms can be gained by qualitative and dimensional arguments. In contrast, modes that are relevant for P moderate to small introduce much weaker three-dimensional modulations, which leave the primary pattern easier to recognize. These secondary instabilities, EI, ZZ, BO, and SV, are all closely related to each other and to invariance properties of the equations. They all appear relatively close to R_c and involve modulations at scales somewhat larger than the diameter of the rolls (wavevectors smaller than k_c). All these features make them particularly appealing for an analytical study in terms of envelope and phase equations (see Chapter 9 and 10).

4. Phenomenology of the Transition to Turbulence

In order to better situate the change of perspective introduced by Ruelle and Takens, we first survey briefly experimental results obtained before 1974. Then we present more recent results obtained in confined geometry and fitting this new approch. Finally we conclude by discussing experimental evidence of its limitations for weakly confined systems.

4. Thermal Convection

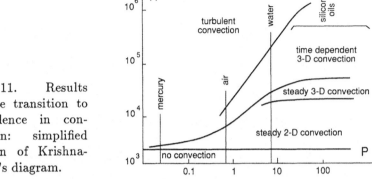

Fig. 11. Results on the transition to turbulence in convection: simplified version of Krishnamurti's diagram.

4.1. "Old" Results

A compilation of the experimental findings about the transition to turbulence in convection has been given by Krishnamurti (1973). Experiments were performed with containers of wide lateral extension. Fluids available were mainly mercury (liquid metal, low viscosity, high thermal conductivity, $P = 0.025$), air under normal conditions (perfect gas, $P = 0.7$), water, alcohol ($P \sim 5$–10), and silicon oils of different viscosities ($P > 50$, possibly very large).

At a qualitative level, the detection of instabilities was made by simple but efficient optical methods. More quantitatively, "hot wire anemometry" was used to probe locally the velocity field. On a global scale, heat flux measurements were performed and instabilities detected by small slope discontinuities in curves giving the Nusselt number, i.e., the ratio of the actual heat flux to the conductive heat flux determined from the applied temperature difference using Fourier law, as a function of the Rayleigh number.

Figure 11 is a simplified version of Krishnamurti's main diagram that displays the nature of the regime observed as a function of the reduced Rayleigh number R/R_c for different values of Prandtl number P (both in logarithmic scales).

At large P, the following sequence was observed: (1) conduction, (2) steady two-dimensional convection, (3) steady three-dimensional convection, (4) time periodic convection, and (5) turbulence. Well-defined steps were less easily identified at low Prandtl number and turbulence was seen to occur at much lower Rayleigh numbers.

The general impression that comes out from this picture is that experimental results are consistent both with results of the nonlinear stability analysis sketched above and with the Ruelle-Takens idea of a finite number of steps before turbulence. The actual situation is less idyllic for several reasons.

Firstly, most experiments were performed on disordered patterns and the values of the successive thresholds were not accurately determined. In fact, at large P, the transition to three-dimensionality and time-dependence is strongly affected by the imperfections of the pattern. The new structures develop primarily at structural defects, well below the nominal threshold for a perfect pattern. The dispersion of the corresponding thresholds could be considerable from one experiment to another (sometimes by a factor of 10, which may not be so striking when using logarithmic coordinates).

The second reason relates to the value of the wavevector underlying these disordered structures, which was known to decrease regularly when R was increased whereas it is not expected vary with R as long as no secondary instability mode is involved. This variation, shown in Fig. 12 adapted from Koschmieder's compilation (1973), is likely to play an important role since the nature of secondary instabilities is strongly dependent on the value of the wavevector especially at moderate to low P, where the transition to turbulence is ill resolved.

The third reservation comes from the fact that, except possibly at very large P, regimes reported as steady two- or three-dimensional convection, were most often slowly time dependent due to the motion of defects. The low frequency noise associated with this slow residual time-dependence, was considered as "extrinsic" but could well be more "intrinsic". Nowadays, we would probably rather consider such a residual time-dependence as a manifestation of *weak turbulence* as opposed to "turbulence" previously synonymous of strongly irregular time dependence that would be better called *strong* or *hard turbulence*.

4. Thermal Convection   127

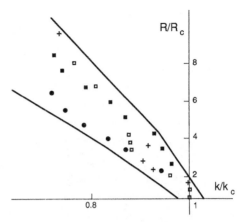

Fig. 12. Variation of the average wavevector as a function of the relative distance to threshold for disordered patterns at large Prandtl numbers (after Koschmieder, 1973).

4.2. "Recent" Results: Concrete Examples

The Ruelle-Takens picture stresses on general properties of nonlinear systems, especially low-dimensional systems, irrespective of their physical origin. A new tendency has then appeared to perform experiments which fits this framework *a priori*. As already explained, systems with loosely constrained spatial structure are expected to be described by a large number of degrees of freedom, so that confinement effects have been used to freeze the spatial structure. Among a very large set of experiments, we shall select some specific examples, intended to illustrate the emergence of *temporal chaos* and the role of the confinement effects on the transition to turbulence, or more strictly speaking, to *spatio-temporal chaos*.

4.2.1. Quasi-Periodicity

We begin with the description of a convection experiment developed by Bergé and Dubois (1981) using silicon oil (high Prandtl number, $P \simeq 130$) in a rectangular container with aspect ratios $\Gamma_x = L_x/h = 2$, $\Gamma_y = L_y/h = 1.2$. The situation considered here is precisely that given as an example of convection in confined geometry in Chapter 1 (see Fig. 2). The visualization of the convective structure is obtained by differential interferometry that probes the average vertical variation of the optical index along a horizontal path

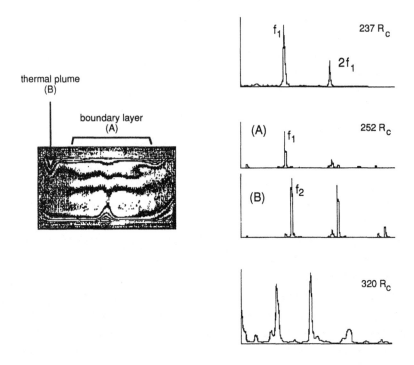

Fig. 13. Left: localization of the oscillators in the convection experiment developed by Bergé and Dubois (1981). Right: Fourier spectra of time series of the velocity; top: monoperiodic regime for $R = 237R_c$ observed in the thermal boundary layer (site A—first oscillator); middle: biperiodic regime for $R = 252R_c$, spectra of signals detected at site A and site B (location of the periodic thermal plume—second oscillator); bottom: weakly turbulent regime for $320R_c$.

through the experimental cell. Interference fringes observed lead to an interpretation of the structure as a superposition of two "rolls" parallel to the y-axis and one "roll" along the x-axis. In addition, "Laser Doppler anemometry" provided a nonintrusive determination of the fluid velocity at different places in the container (Fig. 13, left).

In this experiment, convection remains steady up to $215R_c$ where it becomes periodic owing to an oscillation of the cold thermal boundary layer that alternatively thickens and withdraws (Fig. 13, top-right). At $250R_c$, a second frequency appears, corresponding to the

4. Thermal Convection

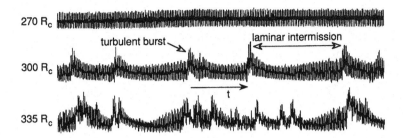

Fig. 14. Transition to temporal chaos *via* intermittency observed under the same experimental conditions as that in Fig. 13 but with a different convection structure (times series of fluctuations at a point in the cell).

periodic birth and separation of a "thermal plume" in a cold corner. Depending on the position of the probe, one or the other frequency can be isolated in the Fourier spectrum as shown in Fig. 13 (middle-right). For most values of the Rayleigh number, the two frequencies are incommensurate and the regime is doubly-periodic. However, the two modes are undoubtedly coupled by the general flow in the box and several high order frequency lockings are observed. Finally, for $r > 305$, the time-dependence becomes chaotic, with a widening of spectral lines and a continuous rise of the noise at frequencies close to zero (Fig. 13, bottom-right).

4.2.2. Intermittency

Other transition scenarios can be observed in confined geometry, using the same box and under apparently the same experimental conditions. As an example, a transition *via intermittency* can be observed when the spatial structure of the steady convective state is made of two "rolls" along the y-axis instead of one. Time dependence then enters the system at $250R_c$ (hot droplet growing periodically in one corner and further advected by the main flow). The regime becomes chaotic at $290R_c$. The chaotic state is characterized by long intermissions of regular nearly periodic behavior interrupted more or less randomly by short bursts, the frequency of which increases with the Rayleigh number (Fig. 14).

As already noticed, in large Prandtl number fluids, convecting structures are formed with rolls, boundary layers, plumes, hot or cold streams, etc., and results presented above suggest that several configurations can coexist in a given set-up, each following a specific scenario. When the aspect ratios increase, the *number of degrees of freedom*, i.e., the number of available structures, becomes large. At the same time, the threshold for chaotic dynamics is seen to lower. However, the situation is extremely complicated and, for some values of the aspect ratios, a well-defined permanent structure can exist and evolve chaotically as described above, whereas, for other values, one can observe a chaos due to *pattern competition*. Still in other cases, a permanent wandering of hot and cold streams takes place so that no definite long-lived average structure exists.

4.2.3. Sub-harmonic Cascade

When P is small, secondary instabilities and the transition to chaos take place closer to the convection threshold, even in confined geometry. Let us first consider an experiment developed by Libchaber and Maurer (1980) using liquid helium around $4K$. The geometry is rectangular with aspect ratios similar to those of the first two examples but, now, P is of the order of 0.7. At such low temperatures, visualization is not possible but convection is detected by precise temperature measurements. The following sequence is observed:

- steady convection for $R_c < R < 30R_c$;
- in the range $30R_c < R < 39R_c$, periodic oscillations at frequency $\omega_1 = 2\pi/T_1$ (likely due to Busse oscillatory instability modified by lateral confinement effects);
- at $R \simeq 39.5R_c$, onset of a second frequency ω_2 lower than and incommensurate with ω_1, hence a doubly-periodic regime;
- at $R \simeq 40.5R_c$, frequency locking of ω_2 on $\omega_1/2$ (period $2T_1$);
- onset of frequency $\omega_1/4$ at $R \simeq 42.7R_c$; the period is now $4T_1$, a period doubling has occurred;
- after a few similar period doublings, for $R > 43R_c$, the signal is chaotic.

4. Thermal Convection

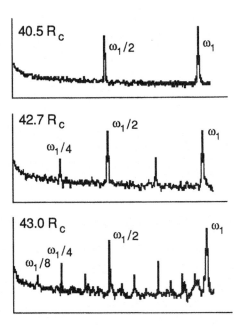

Fig. 15. Typical period-doubling scenario observed at low Prandtl number in confined geometry, after experimental results of Libchaber and Maurer (1980); Fourier spectra of temperature fluctuations close to the top plate for increasing Rayleigh numbers.

Fourier spectra corresponding to some of these regimes are given in Fig. 15. As long as the signal is periodic, fine spectral lines are obtained with a very large signal-to-noise ratio. In contrast, the chaotic signal is associated with broad band noise of intrinsic origin.

4.2.4. Weakly Confined Systems

We now consider what happens when the aspect ratio increases. The starting point is an early experiment performed by Ahlers and Behringer (1978) showing that, still working with liquid helium but in cylindrical geometry at large aspect ratio $\Gamma = D/2h = 57$ ($D =$ diameter), convection is already turbulent at $R = 1.27R_c$ whereas, according to available stability analyses, it should be steady. This puzzling result is confirmed for $\Gamma = 12$ by Libchaber and Maurer (1978) who observe (see Fig. 16):

- steady convection from R_c up to $3.3R_c$;
- at about $3.3R_c$, the onset of a weakly turbulent state character-

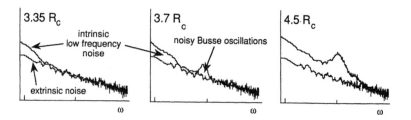

Fig. 16. Transition to turbulence at low Prandtl number and intermediate aspect ratio, after Libchaber and Maurer (1978); notice that the low-frequency broad-band noise appear below the threshold for Busse's oscillations.

ized by a low-frequency broad-band noise;
- at higher Rayleigh number, the onset of (noisy) Busse oscillations
- a regularly increasing amount of low-frequency noise as R is increased further.

It is crucial to observe that weak turbulence seems to appear before any trace of secondary instability (here Busse oscillations) and, furthermore, that disorder grows irreversibly. However the origin of the low-frequency broad-band noise is still unexplained. The clue can be found in a more recent experiment developed by Pocheau, Croquette and Le Gal (1985) using a similar geometry (cylindrical box, $\Gamma \simeq 15$) but with gaseous argon under pressure at ambient temperature replacing liquid helium at $4K$. The Prandtl number of argon is still 0.7 but using it under pressure makes its optical index variations larger, which makes visualization easier. The following sequence is observed:

- at threshold, convection sets in as a system of straight rolls well aligned along a diameter of the box;
- at $1.14 R_c$, a periodic process is observed, involving the birth of a pair of dislocations in the middle of the box, a breaking of the pair, a migration of the dislocations first toward the lateral walls and then toward diametrically opposed regions where they disappear, which brings the system back to its initial configuration

4. Thermal Convection   133

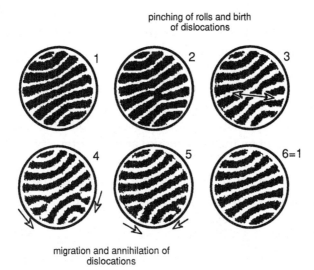

Fig. 17. Periodic motion of dislocations at the origin of low-frequency time-dependence in low Prandtl number fluids (after Pocheau, Croquette and Le Gal, 1985).

after some slow rearrangement (Fig. 17);

- when R is increased further, first the periodic nucleation/motion/annihilation of dislocations becomes less regular, and later Busse oscillations set in, which corroborates the findings of Libchaber and Maurer.

The origin of this process can be traced back to antagonistic conditions imposed on convection rolls by nonlinear interactions above the threshold. First, rolls have a tendency to arrive perpendicularly to the walls, which gives them the shape of divergent lenses, wide at the lateral walls, narrow in the middle. Second, when the local wavelength at the center of the box becomes short enough, the ideal pattern becomes unstable against the skewed varicose mode (see Fig. 10). A pinch then appears that further degenerates into a pair of dislocations. When the two dislocations separate from each other the structure in the center has lost a pair of rolls so that the average wavelength is larger and the structure is brought back inside the Busse balloon where rolls are stable. At the same time, the curvature induces a large scale secondary flow that helps the migration of the dislocations and takes part in the global rearrangement. After the dislocations have disappeared, the structure has returned to its initial configuration. The duration of a complete cycle can be very large and not very reproducible since fine details influence the instant of the dislocation nucleation (a local process linked to the amplification of an unstable mode) and the intensity of the secondary flow.

Such a weak turbulence can be opposed to the stronger turbulence appearing after a transition to "hard" time periodicity. In the present case, frustration effects linked to the interplay between geometry and dynamics seem to play a major role. This may be diffucult to handle theoretically owing to the somewhat contradictory features of the ingredients of the problem: coherence at the scale of the box (slow modulations) and localized singularities (defects).

4.3. Concluding Remarks

In summary, for *confined systems*, several scenarios can been observed. The precise nature of the physical mechanisms involved seems secondary, therefore validating the abstract approach in terms of *dynamical systems*, more concerned with the universal nature of nonlinear couplings. However, the structure of the phase space and its evolution when control parameters are varied may be complicated. This is attested to by the fact that, from one experiment to another apparently identical, one may not even be able to predict the kind of scenario that will be observed. Nevertheless, once the system is engaged in a given scenario results are remarkably reproducible. Moreover the existence of lockings and "re-entrant" regimes (i.e., the occurrence of a laminar regime beyond a chaotic one) proves that temporal chaos, which has no reason to increase monotonically and irreversibly, is not synonymous with disorder: determinism implies *order within chaos* as will be seen in the few next chapters.

The specific role of spatial dependence frozen by confinement effects is recovered, at least partially, in *extended geometry*. The

4. Thermal Convection

increase of the number of degrees of freedom can take place in different ways, but is often linked to the position of rolls and other structure elements (plumes/defects in large/small Prandtl number fluids), hence the importance of the concept of *phase turbulence*. In addition, the hydrodynamic origin of the problem cannot always be left out and supplementary difficulties linked to the presence of induced secondary flows can appear. In any case, the theoretical approach in terms of low-dimensional dynamical systems is no longer well suited to account for the progressive and rather irreversible loss of spatial coherence which characterizes weak turbulence at large aspect ratios. Chapters 8 to 10 will be devoted to a presentation of the new tools designed to deal with this situation.

5. Bibliographical Notes

A thorough presentation of the linear stability analysis of convection with a discussion of the Boussinesq approximation can be found in Chandrasekhar's book already quoted, see BN **3** [3]. See also:

[1] J. Wesfreid, Y. Pomeau, M. Dubois, C. Normand, and P. Bergé, "Critical effects in Rayleigh-Bénard convection," J. de Physique **39**, 725 (1978).

Two old general review articles on nonlinear aspects remain of interest:

[2] C. Normand, Y. Pomeau, M. G. Velarde, "Convective instability: a physicist approach," Rev. Mod. Phys. **49**, 581 (1977).

[3] F. H. Busse, "Nonlinear properties of thermal convection," Rep. Prog. Phys **41**, 1929 (1978).

or Busse's contribution in BN **1** [9].

The stability analysis of straight rolls, theory and experiments, has been developed by Busse and co-workers, see e.g.,

[4] F. H. Busse and R. M. Clever, "Instabilities of convection rolls in a fluid of moderate Prandtl number," J. Fluid Mech. **91**, 319 (1979).

The reference to "old experimental results" on the transition to turbulence in convection can be found in:

[5] R. Krishnamurti, "Some further studies on the transition to turbulent convection," J. Fluid Mech. **60**, 285 (1973).

[6] E.L. Kochmieder: "Bénard convection," Advances in Chem. Phys. vol.XXVI, 177 (1973).

Our presentation of more recent experiments on the routes to turbulence in convection is far from exhaustive. Many comprehensive reviews have appeared, it suffices to scan, e.g., Review of Modern Physics or conference proceedings in the Springer Synergetics series for a few years to get some of the most interesting. Here we just cite references relevant to the experiments described. The two first examples are adapted from:

[7] M. Dubois and P. Bergé, "Instabilités de couche limite dans un fluide en convection: évolution vers la turbulence," J. Physique **42**, 167 (1981).

[8] P. Bergé, M. Dubois, P. Manneville, and Y. Pomeau, "Intermittency in Rayleigh-Bénard convection," J. Physique Lettres **40**, L-505 (1979).

Experiments on the sub-harmonic cascade can be found in:

[9] A. Libchaber, J. Maurer, "Une expérience de Rayleigh-Bénard en géométrie réduite; multiplication, accrochage et démultiplication de fréquences," J. de Physique Colloques **41-C3**, 51 (1980).

In extended geometry the first experiment quoted is by:

[10] G. Ahlers and R.P. Behringer, "Evolution of turbulence from the Rayleigh-Bénard instability," Phys. Rev. Lett. **40**, 712 (1978).

whereas results from which Figures 16 and 17 are adapted have been published by:

[11] A. Libchaber and J. Maurer, "Local probe in a Rayleigh-Bénard experiment in liquid helium," J.de Physique Lettres **39**, 69 (1978).

[12] A. Pocheau, V. Croquette, and P. Le Gal, "Turbulence in a cylindrical container of argon near threshold of convection," Phys. Rev. Lett. **55**, 1094 (1985).

Large scale flows associated with curvature in convecting low Prandtl number fluids has been observed experimentally by:

[13] V. Croquette, P. Le Gal, A. Pocheau, and R. Guglielmetti, "Large-scale flow characterization in a Rayleigh-Bénard convective pattern," Europhys. Lett. **1**, 393 (1986).

Chapter 5

Low-Dimensional Dynamical Systems

Fluids are continuous macroscopic systems described by fields. The degrees of freedom are therefore of a functional nature and the phase space is infinite-dimensional. In contrast, an interpretation *à la Ruelle-Takens* of the transition to turbulence implies a small number of modes in effective interaction. Some sort of reduction of the number of degrees of freedom is thus expected to take place as a result of the macroscopic coherence inherent in the instability mechanism (normal modes have a collective character). When projecting the dynamical equations on the basis of normal modes, we will take advantage of this coherence and moreover we will be able to isolate *nearly marginal modes* which, evolving slowly, will drive the rapidly relaxing stable modes. A low-dimensional effective dynamics will result from the *adiabatic elimination* of these *slaved modes*. This will be shown explicitly on a partial differential equation modeling convection, Section 1.1, further used to introduce the basic concepts and techniques involved in the reduction to a *center manifold* and to the *normal form* of the resulting effective dynamical equations (Section 1.2, and at a more technical level, Section 2). Then we will examine the general features of the dynamics in one and two dimensions. From a general viewpoint, we will see that the asymptotic time evolution remains "simple", i.e., stationary or periodic (Section 3.1 and 5.3). But we will also spend some time to describe bifurcations experienced by these systems since it is important to master their modeling at an elementary level (Section 3.2 to 3.6 and 5.4) and to be able to recognize their occurrence in concrete experimental situations.

1. Dimension Reduction: A Case Study

1.1. The Model and Its Normal Modes

Let us consider the partial differential equation in one space dimension

$$\partial_t w + w\partial_x w = r\,w - (\partial_{x^2} + 1)^2 w\,, \tag{1}$$

supplemented by the boundary conditions

$$w = \partial_{x^2} w = 0 \qquad \text{at } x = 0, \ell\,. \tag{2}$$

As will be discussed later (Chapter 8), this model is a one-dimensional member of a wider family of semi-realistic two-dimensional models of convection adapted to the study of pattern formation in extended geometry. Here we use it in a different context where finite size effects play a dominant role.

The relevant basic state is of course the trivial solution $w \equiv 0$. Infinitesimal perturbations are governed by the linearized system obtained from (1) by dropping the $w\,\partial_x w$ term. Looking for normal modes in the form $w = W(x)\exp(st)$, we are led to solve the homogeneous linear differential boundary value problem

$$\left[\partial_{x^4} + 2\partial_{x^2} + (1 + s - r)\right]W = 0$$

with the boundary conditions

$$W(0) = W(\ell) = 0 = \partial_{x^2}W(0) = \partial_{x^2}W(\ell)\,.$$

The general solution reads

$$W = \sum_{\pm} A_{\pm}\sin(k_{\pm}x) + B_{\pm}\cos(k_{\pm}x)\,,$$

5. Low-Dimensional Dynamical Systems

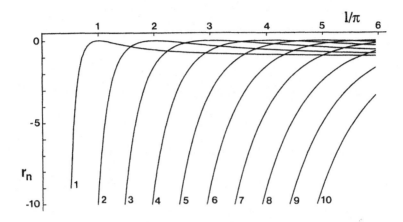

Fig. 1. Evolution of the marginal stability conditions for eigenmodes of the linearized model as a function of the length ℓ of the interval of definition.

where the $k_\pm$'s are the roots of

$$k^4 - 2k^2 + (1 + s - r) = 0 \quad \Rightarrow \quad k_\pm^2 = 1 \pm \sqrt{r - s}, \tag{3}$$

but it is easily checked that the boundary conditions imply:

$$W = A\sin(k_n x) \quad \text{with } k_n = n\pi/\ell.$$

Once inserted into (3), this yields

$$s_n = r - \left(k_n^2 - 1\right)^2.$$

Individual modes are marginal when $s_n = 0$, i.e., $r_n = (k_n^2 - 1)^2$, and, at a given ℓ, the basic state becomes unstable when r crosses the value of the smallest r_n. The variation of r_n with ℓ is displayed in Fig. 1. Resonances, i.e., $r_n = 0$, are observed every time ℓ is an integer multiple of π, whereas two neighboring modes are nearly degenerate when $\ell \simeq \pi(n + 1/2)$ for n large enough. At a given (r, ℓ) the spectrum $\{s_n\}$ can be read from Fig. 1 by subtracting r_n from r.

1.2. Elimination of Slaved Modes: Heuristic Approach

1.2.1. Projection

The saturation of unstable modes is insured by the nonlinear term that, here, looks like the advection term of Navier-Stokes equations. We seek the solution of the complete model in the form

$$w(x,t) = \sum_{n=1}^{+\infty} A_n(t)\sin(k_n x),$$

where A_n is called the *amplitude* of mode n.

Inserting this expansion in (1) and separating the different harmonics we get

$$\frac{dA_n}{dt} = s_n A_n - \frac{1}{2}\sum_{m=1}^{n-1} k_m A_m A_{n-m} + \frac{1}{2} k_n \sum_{m=1}^{\infty} A_m A_{n+m}. \quad (4)$$

The partial differential equation has been converted into an infinite system of ordinary differential equations.

1.2.2. Elimination

As seen in Fig. 1, when ℓ is small, the most unstable mode is well isolated from his immediate neighbors. In contrast, when ℓ becomes larger, the first eigenmodes (say $n = 1$ to 5 for $\ell = 4\pi$) stay well grouped, whereas for n large ($n \geq 6$) we have still isolated values varying asymptotically as $-k_n^4$. Here, we consider only the first case and, for more specificity, we choose $\ell = \pi$, whose spectrum is given in Table 1. Moreover we assume that r remains small when compared to $|s_2|$, the relaxation rate of the second mode. Therefore, all modes are strongly damped, except the first one that remains nearly marginal, either damped (r slightly negative) or amplified ($r > 0$).

The few first equations read

$$\frac{dA_1}{dt} = rA_1 + \frac{1}{2}(A_1 A_2 + A_2 A_3 + A_3 A_4 + \ldots), \quad (5a)$$

5. Low-Dimensional Dynamical Systems

n	s_n
1	0
2	-9
3	-64
4	-225
5	-576
6	-1225

Table 1. Spectrum for $\ell = \pi$ and $r = 0$.

$$\frac{dA_2}{dt} + (9-r)A_2 = -\frac{1}{2}A_1^2 + (A_1 A_3 + A_2 A_4 + \ldots), \quad (5b)$$

$$\frac{dA_3}{dt} + (64-r)A_3 = -\frac{3}{2}A_1 A_2 + \frac{3}{2}(A_1 A_4 + \ldots). \quad (5c)$$

Formally, the equations for $n > 1$ can be written as

$$\frac{dA_n}{dt} + |s_n|A_n = f_n(t), \quad (6)$$

which can be solved in two steps (variation of the constant):

(1) determine the solution of the homogeneous problem, here

$$A_n = \exp(-|s_n|t)A_{n,\text{h}}, \quad (7)$$

where $A_{n,\text{h}}$ is an integration constant;

(2) insert (7) into (6) assuming that $A_{n,\text{h}}$ is a function of time, which gives

$$\frac{dA_{n,\text{h}}}{dt} = \exp(|s_n|t)f_n(t)$$

$$\Rightarrow \quad A_{n,\text{h}} = A_{n,\text{nh}} + \int_{t_0}^{t} \exp(|s_n|t')f(t')dt',$$

where $A_{n,\text{nh}}$ is the integration constant for the inhomogeneous problem.

Finally, the complete solution reads:

$$A_n(t) = \exp(-|s_n|t)A_{n,\text{nh}} + \int_{t_0}^{t} \exp\left(-|s_n|(t-t')\right) f_n(t')\, dt'. \quad (8)$$

Now, adding the condition $s_n < 0$ with $|s_n| \gg 1$, we see that, except during a short transient, the first term can be neglected. Moreover, the exponential kernel in the integral term is short-ranged and, for $t - t' \gg 1/|s_n|$ the contribution of $f_n(t)$ is washed out. The integral can be estimated approximately as

$$\int_{t_0}^{t} \exp\left(-|s_n|(t-t')\right) f_n(t')\, dt'$$
$$\simeq \int_{t-1/|s_n|}^{t} \exp\left(-|s_n|(t-t')\right) f_n(t')\, dt'$$
$$\simeq \frac{1}{|s_n|} f_n(t) + \mathcal{O}\left(1/|s_n|^2\right).$$

Neglecting the transient, we see that we could have assumed $dA_n/dt \equiv 0$ directly (as suggested in Chapter 1, Section 2).

At this stage, it is interesting to evaluate the order of magnitude of the different modes; from Equation (5b), it is easily seen that $A_2 \sim A_1^2$; from (5c), that $A_3 \sim A_1 A_2 \sim A_1^3$, and so on, so that we obtain, at lowest order (r small),

$$9A_2 = -\frac{1}{2}A_1^2, \quad (5b')$$

$$64A_3 = -\frac{3}{2}A_1 A_2 \quad (5c')$$

$$\vdots$$

This defines a one-dimensional manifold in the space spanned by the whole series of amplitudes $\{A_n; n = 1, \infty\}$, whereas

$$\frac{dA_1}{dt} = rA_1 + \frac{1}{2}A_1 A_2 \quad (5a')$$

5. Low-Dimensional Dynamical Systems

determines the effective dynamics on this manifold. Indeed, eliminating A_2 from (5a') using (5b') we get

$$\frac{dA_1}{dt} = rA_1 - \frac{1}{36}A_1^3. \tag{9}$$

Notice that the orders of magnitude estimates derived above allow the truncation of System (5) just above A_2, easily seen to be consistent at order three in A_1 since neglected terms in (5a) and (5b) are of the order of A_1^5 and A_1^4, respectively.

1.2.3. Formal Generalization

The separation between A_1 and all the other amplitudes at the linear stage is an example of splitting of the *tangent space* into a central component and a stable component. The tangent space is the space spanned by all the possible infinitesimal perturbations, themselves governed by the *tangent operator*, i.e., the primitive system linearized around the basic state. More generally, the tangent space at a fixed point can be split into three components, the stable/central/unstable components associated with eigenvalues $s_n = \sigma_n + i\omega_n$ having negative/vanishing/positive real parts σ_n. The situation of interest is illustrated in Fig. 2 that displays a group of nearly marginal eigenvalues, well isolated from those corresponding to strongly stable modes. The dimension of the central subspace is given by the number ν_c of eigenvalues having simultaneously a (nearly) vanishing growth rate ($\sigma_n \simeq 0$). Indexing the central modes with a "c" and the stable ones with an "s", defining the restrictions L_c and L_s of L operating on the central and stable subspaces respectively, we can write formally

$$\frac{dA_c}{dt} = L_c A_c + N_c(A_c, A_s), \tag{10a}$$

$$\frac{dA_s}{dt} - L_s A_s = N_s(A_c, A_s) \tag{10b}$$

and reproduce the line of reasoning followed above. The result of the adiabatic elimination can then be written as a relation between

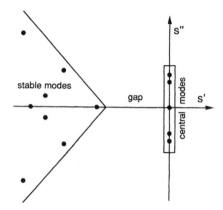

Fig. 2. Typical spectrum of a dynamical system with ν_c central modes well isolated from stable modes.

stable and central coordinates A_s and A_c,

$$A_s = G(A_c) \qquad (11)$$

generalizing $(5a', b', \ldots)$, so that the formal extension to (9) will read

$$\frac{dA_c}{dt} = L_c A_c + N_c(A_c, G(A_c)) = L_c A_c + \tilde{N}_c(A_c). \qquad (12)$$

Relations (11) and (12) define a manifold in phase space and an effective dynamics on that manifold. As can be anticipated from the form of Equations $(5b', c' \ldots)$, relation (11) is at least quadratic in A_c. At lowest order, this means simply $A_s = 0$, i.e., the center manifold is tangent to the center subspace.

2. Center Manifold and Normal Forms

2.1. Perturbative Approach to the Center Manifold

The basic idea is to extend at the nonlinear stage the decoupling of the dynamics between the stable and marginal variables valid at the linear stage by a nonlinear change of variables. Indeed, after the transients have decayed exponentially to zero, the evolution of infinitesimal perturbations is confined to the central subspace, i.e., no

5. Low-Dimensional Dynamical Systems

component in the stable subspace, $A_s \equiv 0$. At the nonlinear stage, there should exist a change of variables $A \to B$, such that, in the new variables, the center manifold should be characterized by $B_s \equiv 0$, asymptotically at the limit of long times. This nonlinear change of variable must obviously preserve the linear decoupling and automatically include the fact that the center manifold is tangent to the linear subspace spanned by marginal variables. Therefore, we must have $B = A + N_A(A, A)$, where $N_A(A, A)$ is a nonlinear function to be determined. In practice, it is often preferable to express "old" variables as functions of the "new" variables, say $A = B + N_B(B, B)$, $N_B(B, B)$ being another nonlinear function. The main idea is then to determine the nonlinear parts $N_A(A, A)$ or $N_B(B, B)$ iteratively by their Taylor expansions beginning with quadratic terms. In the following we will adopt the second starting point.

Example 1. We first consider the simplest possible application, using system (5) truncated above A_2 for $r = 0$:

$$\frac{dA_1}{dt} = \frac{1}{2} A_1 A_2, \tag{13a}$$

$$\frac{dA_2}{dt} + 9A_2 = -\frac{1}{2} A_1^2. \tag{13b}$$

We assume

$$A_n = B_n + \alpha_n B_1^2 + \beta_n B_1 B_2 + \gamma_n B_2^2 \qquad (n = 1, 2) \tag{14}$$

and concentrate our attention on the equation for B_2 which should reduce to an expression independent of the other variable. We have

$$\frac{dA_2}{dt} = \frac{dB_2}{dt} + 2\alpha_2 B_1 \frac{dB_1}{dt} + \beta_2 \left(B_1 \frac{dB_2}{dt} + B_2 \frac{dB_1}{dt} \right) + 2\gamma_2 B_2 \frac{dB_2}{dt}. \tag{15}$$

Inserting (14) and (15) into (13), replacing A_n by B_n when needed in quadratic terms, and neglecting contributions containing dB_1/dt which are cubic since dA_1/dt is quadratic, we obtain

$$\frac{dB_2}{dt} + 9B_2 + 9\gamma_2 B_2^2 = -\left(9\alpha_2 + \frac{1}{2} \right) B_1^2,$$

n	s_n
1	0
2	0
3	−6.4
4	−28.8
5	−80.6
6	−179.2

Table 2. Spectrum for $\ell = \pi\sqrt{5/2}$ and $r = 9/25$.

which is independent of B_1 provided that we choose $\alpha_2 = -1/18$. Moreover, we see that β_2 and γ_2, which are still free parameters, can be taken equal to zero. This leads to

$$A_2 = B_2 - \frac{1}{18}B_1^2$$

and

$$\frac{dB_2}{dt} + 9B_2 = 0.$$

Variables have been decoupled as expected, and after the extinction of the transient we have $B_2 = 0$, that is to say

$$A_2 = -\frac{1}{18}B_1^2 = -\frac{1}{18}A_1^2 \qquad \text{(at lowest order)},$$

which is of course identical to the result obtained previously.

Example 2. The number of marginal modes that can be handled in this way is not limited. For example, let us consider the first degenerate case displayed by model (1) when $s_1 = s_2 = 0$, that is to say $\ell = \pi\sqrt{5/2}$ and $r = 9/25$. The corresponding spectrum is given in Table 2.

Order of magnitude estimates show that a consistent truncation at order three in the two marginal modes A_1 and A_2 must include

5. Low-Dimensional Dynamical Systems

A_3 and A_4, which leads to

$$\frac{dA_1}{dt} = \frac{1}{\sqrt{10}}(A_1 A_2 + A_2 A_3),$$

$$\frac{dA_2}{dt} = -\frac{1}{\sqrt{10}}A_1^2 + \sqrt{\frac{2}{5}}(A_1 A_3 + A_2 A_4),$$

$$\frac{dA_3}{dt} + \frac{32}{5}A_3 = -\frac{3}{\sqrt{10}}A_1 A_2,$$

$$\frac{dA_4}{dt} + \frac{144}{5}A_4 = -\sqrt{\frac{2}{5}}A_2^2.$$

Proceeding as before, we obtain

$$A_3 = -\frac{3}{32}\sqrt{\frac{5}{2}}A_1 A_2$$

$$A_4 = -\frac{1}{72}\sqrt{\frac{5}{2}}A_2^2$$

and:

$$\frac{dA_1}{dt} = \frac{1}{\sqrt{10}}A_1 A_2 - \frac{3}{64}A_1 A_2^2, \qquad (16a)$$

$$\frac{dA_2}{dt} = -\frac{1}{\sqrt{10}}A_1^2 - \frac{3}{32}A_1^2 A_2 - \frac{1}{72}A_2^3. \qquad (16b)$$

Examples given up to now have concerned stationary modes. The extension to oscillatory modes can be handled similarly.

2.2. Normal Forms of Dynamical Equations

As seen above, the change of variable required to perform the elimination bears only on the stable modes and involve marginal modes only in the nonlinear part. The result is a set of dynamical equations governing the evolution of the marginal modes. Its expression may be complicated owing to the presence of so called *non-resonant terms*. A further nonlinear change of variables, this time within the set of marginal amplitudes, can help simplifying the evolution equations by eliminating all non-resonant terms that do not influence the qualitative dynamics of the system. The result is called a *normal form*. We illustrate the reduction to normal form using the van der Pol oscillator introduced at the end of Chapter 2. Of course, we consider the marginal case $\epsilon = 0$. Instead of a single second-order differential equation, we write it here as a system of two first-order equations:

$$\frac{dA_1}{dt} = A_2,$$
$$\frac{dA_2}{dt} = -A_1^2 A_2 - A_1.$$

We first turn to complex variables, setting $X = A_1 + iA_2$, which yields

$$\frac{dX}{dt} = -iX - \frac{1}{8}\left(X^3 + X^2\bar{X} - X\bar{X}^2 - \bar{X}^3\right). \qquad (17)$$

The linear evolution is simply given by $X \sim \exp(-it)$, and out of the four cubic terms on the r.h.s. it is easily seen that only the term $X^2\bar{X} \sim \exp(-it)$ is resonant since $X^3 \sim \exp(-3it)$, $X\bar{X}^2 \sim \exp(it)$, and $\bar{X}^3 \sim \exp(3it)$. The transformation we are looking for is aimed at eliminating these last three terms. Since the nonlinear terms on the r.h.s. are cubic we assume directly

$$X = Z + \alpha Z^3 + \beta Z^2 \bar{Z} + \gamma Z \bar{Z}^2 + \delta \bar{Z}^3.$$

Inserting this assumption into (17), and neglecting terms of order four and beyond, we obtain

$$(1 + 3\alpha Z^2 + 2\beta Z\bar{Z} + \gamma \bar{Z}^2)\frac{dZ}{dt} + (\beta Z^2 + 2\gamma Z\bar{Z} + 3\delta \bar{Z}^2)\frac{d\bar{Z}}{dt}$$
$$= -i(Z + \alpha Z^3 + \beta Z^2 \bar{Z} + \gamma Z \bar{Z}^2 + \delta \bar{Z}^3)$$
$$- \frac{1}{8}(Z^3 + Z^2\bar{Z} - Z\bar{Z}^2 - \bar{Z}^3).$$

At the considered order, we can simply replace $d\bar{Z}/dt$ by iZ and we can isolate dZ/dt by multiplying both sides of this equation by

5. Low-Dimensional Dynamical Systems

$1 - (3\alpha Z^2 + 2\beta Z\bar{Z} + \gamma \bar{Z}^2)$. This yields

$$\frac{dZ}{dt} = -iZ + \left(2\alpha - \frac{1}{8}\right) Z^3 + \left(-\frac{1}{8}\right) Z^2 \bar{Z}$$
$$+ \left(-2\gamma + \frac{1}{8}\right) Z\bar{Z}^2 + \left(-4\delta + \frac{1}{8}\right) \bar{Z}^3.$$

As expected, the resonant term can never be eliminated since β has disappeared from the equation (and can then be set equal to zero without damage). Taking $\alpha = \gamma = 2\delta = 1/16$ leads to the normal form

$$\frac{dZ}{dt} = -iZ - \frac{1}{8}|Z|^2 Z. \tag{18}$$

More formally, let us assume that, after the iterative elimination of the terms of order up to $k-1$ inclusive, we are left with the evolution equations for the marginal amplitudes in the form

$$\frac{dA}{dt} = LA + N_k(A, \ldots, A), \tag{19}$$

where N_k is a homogeneous polynomial of degree k with given coefficients. Writing the change of variables formally as

$$A = B + f_k(B),$$

where f_k is a homogeneous polynomial of order k in B with unknown coefficients to be determined, and further differentiating this relation, we get

$$\frac{dA}{dt} = \frac{dB}{dt} + \frac{\partial f_k}{\partial B}\frac{dB}{dt} = \left(I + \frac{\partial f_k}{\partial B}\right)\frac{dB}{dt}$$

(where I is the identity operator) which we insert in (19) to get

$$\frac{dB}{dt} = \left(I + \frac{\partial f_k}{\partial B}\right)^{-1} \left(L(B + f_k(B)) + N_k(B + f_k(B))\right)$$
$$= LB - \frac{\partial f_k}{\partial B} LB + L f_k(B) + N_k(B).$$

150   DISSIPATIVE STRUCTURES AND WEAK TURBULENCE

It is easily seen that the last three terms are all of order k. As in the example above, the free coefficients introduced in f_k are then chosen in order to suppress as many terms as possible.

2.3. Normal Forms and Symmetries

Returning to the primitive system (4), we can easily check that the amplitudes behave differently according to the parity of their indices and that the system is globally invariant under the simultaneous changes $A_{2n+1} \to -A_{2n+1}$ and $A_{2n} \to A_{2n}$. This property, which derives from the parity of the initial differential problem around the middle point $\ell/2$, must be shared by the normal forms. This is indeed the case in both examples treated, Equation (9) and System (16). In particular, this explains the absence of a term $A_1^2 A_2$ in (16a) since such a term would keep its sign under the replacement of A_1 by $-A_1$. Though detailed calculations are required to get the values of the coefficients in a normal form corresponding to a given problem, the general structure can be derived by symmetry considerations. So, in the total absence of symmetries, for a (diagonalizable) pair of real modes, the normal form is nontrivial already at second order:

$$\frac{dA_1}{dt} = a_{20} A_1^2 + a_{11} A_1 A_2 + a_{02} A_2^2,$$
$$\frac{dA_2}{dt} = b_{20} A_1^2 + b_{11} A_1 A_2 + b_{02} A_2^2$$

(coefficients a_{02} and b_{20} vanish if we impose the separate persistence of solutions $A_1 = 0$ and $A_2 = 0$, which leads to so-called Lotka-Volterra equations used for example in population dynamics). When the system is invariant under the changes $A_1 \to -A_1$ and $A_2 \to -A_2$, quadratic terms vanish identically. Moreover, if the symmetry property applies on the two amplitudes separately, it is easily seen that the normal form reads

$$\frac{dA_1}{dt} = A_1 \left(a_{30} A_1^2 + a_{12} A_2^2 \right),$$
$$\frac{dA_2}{dt} = A_2 \left(b_{21} A_1^2 + a_{03} A_2^2 \right).$$

2.4. Slightly Off the Critical Point

Up to now, we have assumed that the system was at a critical point with one (Example 1) or two (Example 2) marginal modes. The corresponding values of ℓ and r are special, say r_0 and ℓ_0, and we must worry about what happens in their vicinity since the complete problem depends on these two control parameters. Expanding the growth rate in Taylor series, for $r = r_0 + \delta r$ and $\ell = \ell_0 + \delta \ell$, we obtain at lowest order

$$s_n(r, l) = \left[s_n(r_0, \ell_0) \equiv 0 \right] + \frac{\partial s_n}{\partial r} \delta r + \frac{\partial s_n}{\partial \ell} \delta \ell,$$

where partial derivatives are evaluated at (r_0, ℓ_0). For model (1) we get

$$s_n(r, \ell) = \delta r + 4 \left(\frac{n\pi}{\ell_0} \right)^2 \left[\left(\frac{n\pi}{\ell_0} \right)^2 - 1 \right] \frac{\delta \ell}{\ell_0}.$$

Let us examine first the case of a single marginal mode at $\ell_0 = \pi$. Considering the enlargement of Fig. 1 presented in Fig. 3, we see that there is no qualitative difference between crossing the marginal stability curve by increasing r at $\ell = \pi$ and at neighboring values. At constant ℓ, the system can leave the critical point in a single way, i.e., upon variation of r; δr is called the *unfolding parameter*. In fact one can vary r and ℓ simultaneously, but it remains that the marginal stability curve can be crossed upon variation of a single combination of parameters. A problem that requires a single unfolding parameter is said to be of *codimension one*. Turning to the second example with two marginal modes at $\ell_0 = \pi\sqrt{5/2}$, we now see that the two parameters $\delta \ell$ and δr are necessary to unfold the singularity; we have

$$s_1 = \delta r - \frac{1}{\pi} \frac{24}{25} \sqrt{\frac{2}{5}} \delta \ell \quad \text{and} \quad s_2 = \delta r + \frac{1}{\pi} \frac{96}{25} \sqrt{\frac{2}{5}} \delta \ell,$$

and we see that crossing first the marginal curve for mode $n = 1$, i.e., $s_1 = 0$, is not equivalent to crossing first the curve for mode $n = 2$, i.e., $s_2 = 0$. The problem is then of *codimension two*. The codimension of a problem is thus the number of independent perturbations

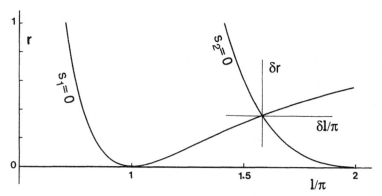

Fig. 3. Marginal stability curves of low-lying normal modes for model (1) with ℓ small; codimension one (two) problem for $\ell = \pi$ ($\ell = \pi\sqrt{5/2}$).

that have to be imposed to pass from the most general problem to the specific case considered (compatible with the same global constraints). This relates to the notion of *genericity* which we will not define mathematically in a proper way but take roughly synonymous of "greatest generality" (e.g., a polynomial equation of degree n has generically n distinct roots).

Now, let us return to the expressions for s_1 and s_2 and suppose that $\delta\ell$ is kept fixed at some nonvanishing but small and positive value. Mode 2 is then destabilized at δr_2 given by $s_2(\delta r_2, \delta\ell) = 0$, whereas the threshold of mode 1 given by $s_1(\delta r_1, \delta\ell) = 0$ is slightly larger

$$\delta r_1 = \delta r_2 + \frac{1}{\pi}\frac{24}{5}\sqrt{\frac{2}{5}}\,\delta\ell.$$

(Of course, the role of the two modes is interchanged when $\delta\ell$ is negative.) The important point to notice is that there is always a vicinity of δr_1 where mode 1 can be eliminated adiabatically to yield the normal form (9) of the codimension-one problem. Indeed, this is possible when the relaxation rate of mode 1 is sufficiently large, that is to say for $r - \delta r_2 \ll \delta r_1 - \delta r_2$. Our ability to tune all the control parameters therefore turns out to be an important matter; we can distinguish a coarse level, where the normal form of highest codimension is relevant, from a finer level (or even a hierarchy of

5. Low-Dimensional Dynamical Systems

finer levels), where the problem is effectively of lower codimension. In the rest of this chapter, we examine the general properties of the dynamics in the one- and two-dimensional cases.

3. Dynamics and Bifurcations in One Dimension

3.1. General Dynamics

When a single mode becomes marginal, the elimination of the slaved modes leaves us with an effective dynamical system for a single real variable that we may write as

$$\frac{dA}{dt} = F_r(A), \tag{20}$$

where F_r is some nonlinear function. Since the problem is of first order in time, the solution can be obtained explicitly:

$$\frac{dA}{F_r(A)} = dt \quad \Rightarrow \quad \int_{A_0}^{A} \frac{dA'}{F_r(A')} = t - t_0,$$

which needs only to be solved for A as a function of time t. The formal procedure above implies that the initial condition A_0 at time t_0 is not taken among the set $\{A_f\}$ of the fixed points of F_r given by $F_r(A_f) = 0$. In one dimension, the dynamics is therefore entirely determined by the nature and the position of the fixed points of F_r, which is better realized by writing (20) in the form

$$\frac{dA}{dt} = -\frac{\partial G_r(A)}{\partial A} \quad \text{with } G_r = -\int F_r \, dA, \tag{21}$$

where $G_r(A)$ is the potential from which (20) can be derived. G_r can only decrease during the evolution since

$$\frac{dG_r}{dt} = \frac{\partial G_r}{\partial A} \frac{dA}{dt} = -\left(\frac{dA}{dt}\right)^2 \leq 0.$$

In this new context, fixed points of F_r are stationary points of G_r. A local minimum of G_r is a stable fixed point (an attractor), a

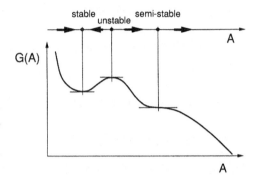

Fig. 4. The dynamics in one dimension amounts to a search for local minima of a potential from which the dynamical system can be derived.

local maximum is unstable (a repellor), and an inflection point with horizontal tangent is semi-stable (a vague attractor). When G_r is not bounded from below, some trajectories escape to infinity. These different situations are illustrated in Fig. 4.

The next interesting question relates to the dependence of the fixed points A_f upon control parameters. In fact, the fixed points must be understood as implicit functions of r and, to make this more apparent, we write $F_r(A)$ as $F(r, A)$ so that the fixed point condition now reads $F(r, A_\mathrm{f}) = 0$. Now, let $r = r_0$ and let A_0 be the corresponding fixed point under study, $A_0 = A_\mathrm{f}(r_0)$. To determine the fixed points at a neighboring value $r = r_0 + \delta r$, we expand this implicit relation around (r_0, A_0). At lowest order, with $A = A_0 + \delta A_\mathrm{f}$, we find

$$\left[F(r_0, A_0) \equiv 0\right] + \frac{\partial F}{\partial r}\delta r + \frac{\partial F}{\partial A}\delta A_\mathrm{f} + \ldots = 0,$$

where, here and later on, all partial derivatives are evaluated at (r_0, A_0). The *implicit function theorem* tells us that if at least one of the two first order partial derivatives of F is nonvanishing, the equation can be solved for one of the two variables δr or δA_f as a function of the other. The considered root (r_0, A_0) of the fixed point equation is then *regular* and the neighboring solution can be obtained with accuracy by a *regular perturbation expansion*.

If both partial derivatives are nonvanishing, the solution at lowest order reads

$$\delta A_\mathrm{f} = \frac{\partial F/\partial r}{\partial F/\partial A}\delta r,$$

5. Low-Dimensional Dynamical Systems

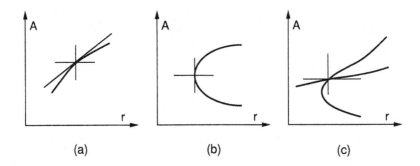

Fig. 5. At a regular point the implicit function theorem can be applied trivially when the first-order partial derivatives are different from zero, case (a), or less trivially when only one derivative is nonvanishing, case (b); usually two branches of solutions intersect at a singular point (c).

as illustrated in Fig. 5a. If $\partial F/\partial r = 0$ but $\partial F/\partial A \neq 0$, A_{f} simply does not depend on r at lowest order; in the opposite case, r remains constant with A_{f}. If we do not find these answers sufficiently accurate, we are obliged to expand the fixed point equation at a higher order. For example, in the second case, $\partial F/\partial r \neq 0$ and $\partial F/\partial A = 0$ but $\partial^2 F/\partial A^2 \neq 0$ and $\partial^2 F/\partial r \partial A = \partial^2 F/\partial r^2 = 0$, and we find

$$\frac{\partial F}{\partial r}\delta r + \frac{1}{2}\frac{\partial^2 F}{\partial A^2}\delta A_{\mathrm{f}}^2 = 0\,,$$

which yields $\delta A_{\mathrm{f}} \propto \pm\sqrt{\delta r}$, as illustrated in Fig. 5b, an example of a *turning point* to be analyzed in more detail later. If all second-order derivatives vanish we have to expand F up to a higher order but it is important to keep in mind that, in all cases, a single curve goes through the point (r_0, A_0) in the (r, A_{f})-plane.

When the two first-order partial derivatives vanish at (r_0, A_0), the point is *singular* and the implicit function theorem no longer applies. At lowest nontrivial order, we have

$$\frac{1}{2}\frac{\partial^2 F}{\partial A^2}\delta A_{\mathrm{f}}^2 + \frac{\partial^2 F}{\partial r \partial A}\delta r\,\delta A_{\mathrm{f}} + \frac{1}{2}\frac{\partial^2 F}{\partial r^2}\delta r^2 = 0\,. \qquad (22)$$

In general this quadratic form does not vanish identically. The intersting case is when the fixed point (r_0, A_0) actually belongs to at least one continuous branch, i.e., is not an isolated solution of the fixed point equation (which would correspond to (22) being nondegenerate with only $\delta r = \delta A_f = 0$ as a possible solution). Equation (22) can then be written as $\alpha_1^2 \delta r'^2 - \alpha_2^2 \delta A_f'^2 = 0$ through a change of variables $(\delta r, \delta A_f) \to (\delta r', \delta A_f')$, so that the set of solutions is indeed made of two intersecting branches $\alpha_1 \delta r' \pm \alpha_2 \delta A_f' = 0$ (Fig. 5c).

At this point, it is more instructive to leave the formal level and work out a specific model. Therefore we suppose that the fixed point of interest stands at the origin, i.e., $A_0 \equiv 0$ and that it becomes linearly unstable at $r = 0$, being stable below ($r < 0$) and unstable above ($r > 0$). In addition, for the growth rate s, the simplest analytic expression that changes its sign at $r = 0$, i.e., $s \propto r$, is thus a reasonable choice. Moreover, since when r is positive an infinitesimal perturbation to solution $A \equiv 0$ is exponentially amplified, we must add nonlinear terms to insure some kind of saturation. Finally we arrive at

$$\frac{dA}{dt} = rA + aA^2 + bA^3 + \ldots . \qquad (23)$$

Clearly

$$\frac{\partial F}{\partial r} = A, \qquad \frac{\partial F}{\partial A} = r + 2aA + 3bA^2 + \ldots$$

shows that $(r_0 = 0, A_0 = 0)$ is indeed critical. Furthermore,

$$\frac{\partial^2 A}{\partial r^2} = 0, \qquad \frac{\partial^2 A}{\partial r \partial A} = 1, \qquad \frac{\partial^2 A}{\partial A^2} = 2a + 6bA + \ldots ,$$

so that, at the critical point, (22) simply reads

$$2\delta A_f (\delta r + a \delta A_f) = 0 ,$$

which describes the intersection of branch $\delta A_f = 0$ with branch $\delta r + a \delta A_f = 0$, as expected. The discussion of the stability of these solutions require supplementary assumptions which will now be examined.

5. Low-Dimensional Dynamical Systems

3.2. Normal/Inverse Bifurcations

We suppose first that states with amplitudes A and $-A$ are physically equivalent. Then, if $A(t)$ is a solution, i.e., $dA/dt = F_r(A)$, $-A(t)$ is also a solution and $d(-A)/dt = F_r(-A) = -dA/dt = -F_r(A)$, so that F_r must be an odd function of A. This condition suppresses all even powers of A from the expansion (23). At lowest order we are left with

$$\frac{dA}{dt} = rA - bA^3. \tag{24}$$

Fixed points are given by

$$A_f = 0 \quad \text{and} \quad A_f = \pm\sqrt{r/b}. \tag{25}$$

The bifurcation is called *normal* or *supercritical* when $b > 0$ since the nontrivial solution exists for $r > 0$, i.e., above the linear threshold. When $b < 0$, the nontrivial solution exists for $r < 0$, i.e., below the threshold, and the bifurcation is called *inverse* or *subcritical*.

The stability of the bifurcated solutions is analyzed by inserting $A = A_f + A'$ in (24) with A_f given by (25), which yields

$$\begin{aligned}\frac{dA'}{dt} &= r(A_f + A') - b(A_f + A')^3 \\ &= rA_f - bA_f^3 + (r - 3bA_f^2)A' + \mathcal{O}(A'^2).\end{aligned}$$

Terms independent of A' cancel exactly by virtue of the fixed point condition that can be used also to simplify the expression of the coefficient of A', so that we simply get

$$\frac{dA'}{dt} = -2rA'.$$

Therefore, when the bifurcation is supercritical, fixed points exist for $r > 0$ and perturbations are damped ($-2r < 0$); the nontrivial fixed points A_f are linearly stable. In the opposite case (subcritical bifurcation) the A_fs are unstable. Remember that the trivial fixed point $A_f = 0$ is stable for $r < 0$

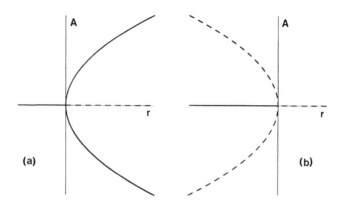

Fig. 6. a) supercritical bifurcation; b) subcritical bifurcation; thick (broken) lines indicate branches corresponding to stable (unstable) solutions.

and unstable for $r > 0$. The bifurcation diagram is given in Fig. 6.

Equation (24) depends on several parameters that can be removed by changing scales for t and A. Assuming $b > 0$ and $r > 0$, inserting $t = \tau \tilde{t}$ and $A = \alpha \tilde{A}$ in (24) with

$$\tau = 1/r \quad \text{and} \quad \alpha = \sqrt{r/b} \qquad (26)$$

helps us writing (24) in a form that contains no free parameters: $d\tilde{A}/d\tilde{t} = \tilde{A} - \tilde{A}^3$. Fixed points are now $\tilde{A} = 0$ and $\tilde{A} = \pm 1$ whereas the evolution from a (scaled) initial condition $\tilde{A}_0$ is easily obtained explicitly by a straightforward integration:

$$\frac{d\tilde{A}}{\tilde{A}(1-\tilde{A}^2)} = \frac{d\tilde{A}}{\tilde{A}} + \frac{d\tilde{A}}{2(1-\tilde{A})} - \frac{d\tilde{A}}{2(1+\tilde{A})} = d\tilde{t}$$

$$\Rightarrow \quad \frac{\tilde{A}}{\sqrt{|1-\tilde{A}^2|}} = \frac{\tilde{A}_0}{\sqrt{|1-\tilde{A}_0^2|}} \exp(\tilde{t}).$$

Notice the *slowing down* of the dynamics since, from (26), when $\tilde{t}$ varies by a quantity of order one, the physical time t varies by a factor of order $1/r \gg 1$ for $r \ll 1$.

5. Low-Dimensional Dynamical Systems

3.3. Conditional Stability, Hysteresis, and Turning Points

When the coefficient b is negative, the cubic term is insufficient to insure the saturation of the amplitude and we must add a higher order term, here of order five, to fulfill the parity requirement. The corresponding coefficient must be negative since we ask for an effective damping at large amplitudes, which leads to

$$\frac{dA}{dt} = rA - bA^3 - dA^5 \quad \text{with} \quad b < 0 \quad \text{and} \quad d > 0. \qquad (27)$$

Figure 7a displays the bifurcation diagram, the determination of which is straightforward. The remarkable fact is that stable branches appear at a finite distance from the A-axis and that a change of stability takes place at *turning points* (subscript "tp" in the following). When r is varied, the system can describe a hysteresis cycle with jumps from one stable branch to the other. In the interval $r_{\rm tp} < r < 0$ the trivial fixed point $A = 0$ remains stable against finite amplitude perturbations smaller than that corresponding to the nontrivial unstable fixed points but unstable against larger amplitude perturbations. It is said to be *conditionally stable*. The nontrivial unstable fixed points mark the boundary of the basin of attraction of the trivial and nontrivial stable fixed points.

The coordinates of the turning points are easily found to read

$$r_{\rm tp} = -\frac{b^2}{4d} \quad \text{and} \quad A_{\rm tp}^2 = -\frac{b}{2d}$$

(remember that $b < 0$ and $d > 0$). Choosing one of the two equivalent turning points, say $A_{\rm tp} = +\sqrt{-b/2d}$, we set $A = A_{\rm tp} + A'$ and $r = r_{\rm tp} + \delta r$ and expand (27) at lowest order, obtaining

$$\frac{dA'}{dt} = A_{\rm tp}(\delta r + 2bA'^2).$$

Changing the time scale and setting $a = -2b > 0$, we get

$$\frac{dA'}{dt} = \delta r - aA'^2, \qquad (28)$$

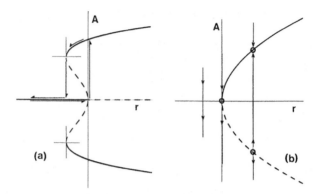

Fig. 7. a) Bifurcation diagram for model (27); the nontrivial stable branch appears at finite amplitude, the trivial fixed point is metastable so that hysteresis cycles can be described. b) Stability change at a turning point; the stable and unstable fixed points collapse and disappear upon decreasing the control parameter.

which yields the bifurcation diagram of Fig. 7b (notice that such a turning point is a regular point; indeed we have $\partial F/\partial \delta r = 1$, whereas $\partial F/\partial A' = -2aA'$ which goes through 0 at $A' = 0$).

When $\delta r > 0$, we have two fixed points: one stable, the other unstable. The two fixed points merge at $\delta r = 0$ and for $\delta r < 0$ the time derivative keeps a constant sign (here < 0) so that there is no longer any fixed point in the neighborhood of $A' = 0$ (remember that we have performed a translation to get Equation (28); returning to the original formulation, we see that in fact the system decays to the trivial fixed point of (27)).

This process is called a *saddle-node* bifurcation for a reason to be explained in Section 5.4. Notice that Equation (28) does not fulfill the assumption of fixed-point persistence underlying the derivation of model (23), which will be discussed further in Section 3.6.

3.4. Transcritical Bifurcation

In the absence of special symmetry, the quadratic term in (23) has no reason to be absent. Then, at lowest order we have

5. Low-Dimensional Dynamical Systems

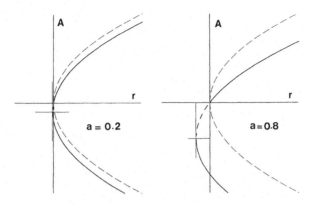

Fig. 8. Transcritical bifurcation in model (30): a) for $a = 0.2$ the symmetry breaking term is comparatively small; b) strong symmetry breaking for $a = 0.8$; dotted line → symmetrical case.

$$\frac{dA}{dt} = rA - aA^2, \qquad (29)$$

whose fixed points are $A_{f,t} = 0$ (trivial) and $A_{f,nt} = r/a$ (nontrivial). This bifurcation is called *transcritical* since the trivial and nontrivial roots are present on both sides of the bifurcation point. For the same reason it is also called sometimes "two-sided" as opposed to the "one-sided" case where nontrivial fixed points exist either above or below the threshold but not on both sides.

It is easily seen that the trivial solution is stable for $r < 0$ and unstable for $r > 0$ whereas the nontrivial fixed point is stable/unstable in the reverse cases. Clearly there is an "exchange of stability" at threshold but this expression should be avoided since it has been used also for supercritical stationary instabilities such as Rayleigh-Bénard convection.

Since its quadratic term is not sufficient to insure the saturation of the amplitude, Equation (29) has to be completed by a cubic term $-bA^3$ with some coefficient $b > 0$ that can be set equal to 1 by rescaling the amplitude. Figure 8 displays the bifurcation diagram of the resulting model

$$\frac{dA}{dt} = rA - aA^2 - A^3. \qquad (30)$$

The position of the turning point that appears is controlled by parameter a: $r_{\rm tp} = -a^2/4$ and $A_{\rm tp} = -a/2$, i.e., $r_{\rm tp} = -A_{\rm tp}^2$, which helps us distinguish weak from strong symmetry breaking.

3.5. Imperfect Bifurcation

The amplitude A may be coupled with external perturbations which force the response of the system. Calling H such a perturbation, in the context of thermodynamics we would define a *susceptibility* χ through $A = \chi H$ and, according to the theory of linear irreversible processes, we would understand this relation as resulting from an evolution equation $dA/dt = -M\big((A/\chi) - H\big)$ with M some relevant kinetic coefficient. Here, we simply admit a nonlinear generalization of this approach to the case of the supercritical bifurcation described by model (24) with $b > 0$ which, after proper rescaling of variables, reads

$$\frac{dA}{dt} = rA - A^3 - H \qquad (31)$$

(the other cases can be analyzed in the same way).

Figure 9 displays the bifurcation diagram modified with respect to that in Fig. 6. It is now disconnected with one branch joining the displaced trivial solution, $(r \to -\infty, A \to 0)$, to the positive nontrivial solution of the unperturbed problem, asymptotically for $(r \to +\infty, A \to +\infty)$. The second branch connects the remnant of the unstable trivial solution to the remnant of the second stable nontrivial solution through a turning point. Notice that the singularity at $r = 0$ has disappeared. The whole process is called an *imperfect bifurcation*.

The linear susceptibility defined above, here $\chi(r) = 1/|r|$, is seen to diverge as $r \to 0$ from below. On the other hand, the strength of the imperfection can be measured by the value of A at $r = 0$. From the steady state equation $dA/dt = 0$, we get $A \sim H^{1/3}$. These features are well known in Landau's theory of phase transitions in thermodynamic systems with an order parameter. On general grounds, the *critical behavior* at second-order phase transitions is characterized by sets of *critical exponents* $(\beta, \gamma, \delta, \ldots)$ controlling the power-

5. Low-Dimensional Dynamical Systems

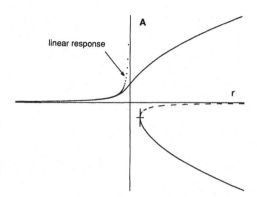

Fig. 9. Imperfect bifurcation displayed by (31) obtained from (24) by coupling amplitude A with its conjugate variable H.

law variation of thermodynamic quantities close to the transition point. Here A plays the role of the order parameter and H is the conjugate field, whereas the control parameter r corresponds to the temperature. Result (25) can therefore be read as $A \propto r^\beta$ with $\beta = 1/2$; further on we have $\chi \propto r^{-\gamma}$ with $\gamma = 1$, $A \propto H^{1/\delta}$ with $\delta = 3$, etc. These values are called *classical*. In the case of thermodynamic phases transitions, microscopic fluctuations cannot be neglected, which leads to nontrivial corrections to the classical values depending on the system and the dimension of physical space (e.g., $\beta \approx 1/3$ for magnetism in dimension 3). In contrast, the nature of the modes involved in bifurcations is genuinely macroscopic so that the system should not be very sensitive to Brownian motion and corrections to the classical values are unobservable in practice.

3.6. Mathematical Context: Unfolding of Singularities

"At" a singular point, the dynamical system is structurally unstable; a small change in the control parameters may lead to qualitatively different regimes. The global behavior is basically determined by the the degree of the singularity, which says how the system responds at "large" values of the amplitude. The most general set of small perturbations of lower degree then allows us to unfold the singular behavior, i.e., to distinguish between different possibilities "close to" the singular point.

As a first example, let us consider a degree-two singularity, i.e., $F_0 = -A^2$. The most general perturbed system reads:

$$F = F_0 + \delta F = -A^2 + (r_0 + r_1 A + r_2 A^2)$$

with r_0, r_1, r_2 "small." However, perturbing the coefficient of A^2 leaves the qualitative dynamics unchanged so that r_2 can be dropped. Moreover, the term linear in A can be removed through the change of variable $A = \tilde{A} + r_1/2$, so that the set of unfolding parameters is reduced to a single effective parameter $\tilde{r} = r_0 + r_1^2/4$, the most general unfolded vector field being written as

$$F = \tilde{r} - \tilde{A}^2.$$

This shows that the most general dynamics around a degree-two singularity is given by Equation (28) and not by Equation (29). The reason is that (29) implicitly assumes the persistence of the fixed point, which is an additional condition.

Considering now a degree-three singularity: $F_0 = -A^3$; we expect a perturbed system of the form

$$F = r_0 + r_1 A + r_2 A^2 - A^3$$

that can be reduced to

$$F = \tilde{r}_0 + \tilde{r}_1 \tilde{A} - \tilde{A}^3$$

by similar changes of variables.

From the correspondence $H \leftrightarrow \tilde{r}_0$, $r \leftrightarrow \tilde{r}_1$ we see that Equation (31) that accounts for the imperfect bifurcation is, in fact, the most general unfolding of a degree-three singularity. This does not mean that "perfect" bifurcations never occur but that special care must be taken for their observation. Examples can be found in Rayleigh-Bénard convection; imperfections are inherent to any experimental set-up and the question is whether or not convection induced by imperfections can be detected below threshold. In the same way, we must not deduce from the fact that (29) is not generic

5. Low-Dimensional Dynamical Systems

that transcritical bifurcations never occur since the persistence of the trivial fixed point may be imposed by the physics of the problem. Considering again the case of convection, we know that the trivial fixed point corresponds to the rest state which is a solution at all Rayleigh numbers. In the presence of top-bottom symmetry, the expected structure is made of rolls, the dynamics of which can be reduced to that of an amplitude governed by Equation (24); when the top-bottom symmetry is broken, hexagons are expected and the corresponding amplitude is now governed by (30) (the actual situation may be more complicated, see Chapter 9); we see however on this example that when a transcritical is to be found the explanation lies, not in the unfolding of the degree-two singularity as could be thought from (28), but in the unfolding of a higher degree singularity with respect to a special kind of perturbation, here a symmetry breaking term.

General considerations, involving possibly symmetry requirements, may help us to guess the degree of the singularity. The next step is then the derivation of the most general relevant perturbations and the determination of the corresponding behavior close enough to the singular point. Such a qualitative understanding of the dynamics is obviously fundamental. The quantitative approach is most often very tedious and may not be thought very rewarding; however it may be important to fix quantities such as the amount of symmetry breaking or the intensity of a (physical) perturbation.

4. Introduction to Higher Dimensional Problems

From a general viewpoint, the interpretation of "experimental" results in terms of an effective dynamics on a low-dimensional center manifold will, at one stage or another, meet practical difficulties linked to the presence of the neglected transverse coordinates. At lowest order, as long as the separation of time scales is sufficient, it is legitimate to account for them by a set of uncoupled complementary amplitudes, relaxing linearly to zero. For simplicity, let us first consider the case of a single stable mode A_2 added to a single bifurcating mode A_1:

$$\frac{dA_1}{dt} = F_r(A_1), \qquad \frac{dA_2}{dt} = -A_2. \tag{32}$$

Clearly, this can be written as

$$\frac{dA_n}{dt} = -\frac{\partial G}{\partial A_n} \qquad (n = 1, 2) \tag{33}$$

with

$$G(A_1, A_2) = G_r(A_1) + \frac{1}{2}A_2^2,$$

where $G_r = -\int F_r \, dA_1$ as formally defined previously. A system that can be written in the form (33) is called a *gradient flow*. The corresponding vector field is everywhere *perpendicular* to the level lines of the potential, $G(A_1, A_2) = \text{Cst}$. This property is illustrated in Fig. 10a that displays the simplest possible case, $G(A_1, A_2) = \frac{1}{2}(|s|A_1^2 + A_2^2)$ describing the vicinity of a stable fixed point at the origin.

For the same reason as before, the potential can only decrease during the evolution from all regular initial conditions. As illustrated in Fig. 10b, trajectories all end at one of the local minima of the potential (except those initiated at unstable points that remain there forever and those starting on the stable manifold of saddle points that are exceptional limit sets unattainable in practice owing to unavoidable external fluctuations).

As in the previous section, bifurcations of gradient flows can be analyzed using genericity considerations yielding what is known as *elementary catastrophe theory* (see Bibliographic Notes) but, in practice, physical systems far from equilibrium do not derive from a (thermodynamic) potential. The reduced dynamical system describing the effective dynamics close to a bifurcation point, say

$$\frac{dA_n}{dt} = F_n(r, \{A_{n'}\}) \qquad (n, n' = 1, \ldots),$$

has generically no reason to verify the cross-derivative property

$$\frac{\partial F_n}{\partial A_{n'}} = \frac{\partial F_{n'}}{\partial A_n},$$

5. Low-Dimensional Dynamical Systems

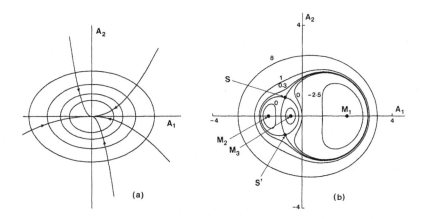

Fig. 10. a) Close to a stable fixed point of a gradient flow system, trajectories remain perpendicular to the level lines of the potential which assume an elliptical shape. b) Level lines of the two-dimensional gradient-flow system deriving from the potential $G(A_1, A_2) = -aA_1 - \frac{1}{2}(bA_1^2 + cA_2^2) + \frac{1}{4}(A_1^2 + A_2^2)^2$ for $a = 3/2$, $b = 13/4$, $c = 5/4$; $\mathbf{M}_1$ at $(2,0)$ corresponds to the absolute minimum $G = -5.5$) and $\mathbf{M}_2$ at $(-1.5, 0)$ to a relative minimum ($G = -0.140625$; $\mathbf{M}_3$ at $(-0.5, 0)$ is a relative maximum ($G = 0.359375$); the level line passing through the saddle points $\mathbf{S}$ and $\mathbf{S}$' at $(-0.75, \pm 0.8291562)$ corresponds to $G = 0.1718750$; this potential was introduced to account for the first bifurcations of the Taylor–Couette flow between concentric spheres (Manneville and Tuckerman, 1987).

which derives from the existence of a potential through

$$\frac{\partial^2 G}{\partial A_n \partial A_{n'}} = \frac{\partial^2 G}{\partial A_{n'} \partial A_n}.$$

The simplest nonpotential, two-dimensional system is obviously the harmonic oscillator governed by

$$p = \mu \frac{dx}{dt}, \qquad \frac{dp}{dt} = -\kappa x,$$

where x is the position and p the momentum, μ the mass and κ the elastic stiffness. The frequency is then given by $\omega = \sqrt{\kappa/\mu}$ and, as is well known, the gradient property is replaced by the *conservation* of total energy:

$$\mu\omega^2 x \frac{dx}{dt} + \frac{1}{\mu} p \frac{dp}{dt} = \frac{d}{dt}\left(\frac{\mu\omega^2}{2} x^2 + \frac{1}{2\mu} p^2\right) = \frac{dE}{dt} = 0.$$

In phase space, the vector field is therefore everywhere *parallel* to the level lines of the total energy. The harmonic oscillator is a trivial example of a *conservative dynamical system* governed by *Hamiltonian mechanics* which is time reversible. Adding a viscous damping force proportional to the velocity dx/dt, we obtain

$$\frac{dp}{dt} = -g \frac{dx}{dt} - \kappa x = -\frac{g}{\mu} p - \mu\omega^2 x,$$

where $g > 0$ is a friction constant. Energy conservation is lost since

$$\frac{dE}{dt} = -g \frac{p^2}{\mu^2} \leq 0.$$

Furthermore, writing the equation for the damped oscillator as

$$\mu \frac{d^2 x}{dt^2} + g \frac{dx}{dt} + \kappa x = 0$$

and letting μ tend to zero, we see that the one-dimensional case, whose dynamics is purely relaxational, can be understood as *fully dissipative*. Typical far-from-equilibrium dynamics in higher dimensional phase space is expected to be, in some sense, intermediate between relaxational and conservative. The most salient feature is of course the breakdown of the gradient flow property, though this is not a sufficient condition for nontrivial asymptotic time behavior.

5. Dynamics and Bifurcations in Two Dimensions

5.1. Linear Dynamics

The most general real linear dynamical system in two dimensions reads
$$\frac{dA_1}{dt} = aA_1 + bA_2, \qquad (34)$$
$$\frac{dA_2}{dt} = cA_1 + dA_2.$$

The eigenmodes fulfill
$$0 = (a-s)A_1 + bA_2, \qquad (35)$$
$$0 = cA_1 + (d-s)A_2,$$

which yields:

$$(s-a)(s-d) - bc = s^2 - (a+d)s + ad - bc = 0. \qquad (35')$$

Equation (35') has two roots in general, either real or conjugate complex.

5.1.1. Distinct Real Roots

The matrix can be put in diagonal form and, in the basis of its eigenvectors, system (34) reads

$$\frac{dA_1}{dt} = s_1 A_1, \qquad \frac{dA_2}{dt} = s_2 A_2. \qquad (36)$$

The phase portrait is the set of orbits obtained by eliminating t between $A_1(t) = A_1(0)\exp(s_1 t)$ and $A_2(t) = A_2(0)\exp(s_2 t)$, where $A_1(0)$ and $A_2(0)$ are the initial conditions. We have $(A_1/A_1(0))^{1/s_1} = (A_2/A_2(0))^{1/s_2}$ or, equivalently $A_2 \propto A_1^{s_2/s_1}$, with two sub-cases:
(1) $s_1 s_2 > 0$, e.g., $s_2 < s_1 < 0$; the fixed point is called a *stable node*; in its vicinity, trajectories have a parabolic aspect, Fig. 11(a1);

(2) $s_1 s_2 < 0$, e.g., $s_2 < 0 < s_1$; the fixed point is a *saddle*; in its vicinity, trajectories have a hyperbolic aspect, Fig. 11(a2).

5.1.2. Complex Roots

Let us denote $s = \sigma \pm i\omega$; there are no real eigenvectors but a change of variables helps us to write problem (34) in the form

$$\begin{aligned} \frac{dA_1}{dt} &= \sigma A_1 - \omega A_2 \,, \\ \frac{dA_2}{dt} &= \omega A_1 + \sigma A_2 \,, \end{aligned} \tag{37}$$

which gives

$$\begin{aligned} A_1(t) &= \exp(\sigma t)(A_1(0)\cos(\omega t) - A_2(0)\sin(\omega t)) \,, \\ A_2(t) &= \exp(\sigma t)(A_1(0)\sin(\omega t) + A_2(0)\cos(\omega t)) \,. \end{aligned}$$

The corresponding phase portrait is given in Fig. 11(b1) for $\sigma < 0$ (stable case); the fixed point is called a *spiral point* or a *focus* (the unstable case $\sigma > 0$ is obtained by reversing the arrows). The special case $\sigma = 0$ called an *elliptic point* or a *center*, corresponds to a structurally unstable system, Fig. 11(b2).

5.1.3. Double Roots

This case occurs when $(a-d)^2 + 4bc = 0$, which gives immediately $s = \frac{1}{2}(a+d)$. The components of the eigenvectors derived from (35) are given by $\frac{1}{2}(a-d)A_1 + bA_2 = 0$ and $cA_1 + \frac{1}{2}(d-a)A_2 = 0$. In general, these two equations define only one eigenvector, so that the matrix cannot be set in diagonal form by a linear change of variable. Let us suppose $b \neq 0$. Then, the two vectors $(-b, \frac{1}{2}(a-d))$ and $(0,1)$ are not parallel and thus form a basis. In this basis, system (34) assumes its *Jordan normal form*

$$\frac{dA_1}{dt} = sA_1 + A_2 \,, \tag{38a}$$

$$\frac{dA_2}{dt} = sA_2 \tag{38b}$$

5. Low-Dimensional Dynamical Systems

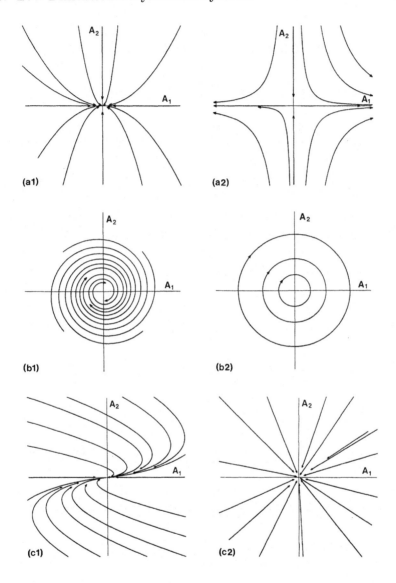

Fig. 11. Phase portrait in the vicinity of a fixed point: (a) two distinct real eigenvalues: a1) stable node, a2) saddle; (b) two complex conjugate eigenvalues: b1) stable spiral point, b2) center (marginal case); (c) double root: c1) nondiagonalizable case: improper node, c2) diagonalizable case. (The pictures corresponding to all unstable cases can be obtained by reversing arrows.)

(for more information on algebraic properties of linear systems, see Appendix 2 and Bibliographic Notes). System (38) is readily integrated by treating first Equation (38b), i.e., $A_2(t) = A_2(0)\exp(st)$, and then Equation (38a), which leads to $A_1(t) = A_1(0)\exp(st) + A_2(0)t\exp(st)$, where the second term is called a *secular term*. The fixed point, called an *improper node*, is stable when $s < 0$. The corresponding phase portrait, obtained by eliminating time between $A_1(t)$ and $A_2(t)$, is displayed in Fig. 11(c1).

When $bc = 0$ and b or c is different from 0, we have still an improper node but in a slightly less general context. Finally, the last possibility is with $b = c = 0$; the matrix is then diagonal and the eigenvector problem completely indeterminate —any vector is an eigenvector. This is a limiting case of node, stable if $s < 0$, unstable if $s > 0$ the corresponding phase portrait is trivial, Fig. 11(c2).

5.2. Nonlinear Dynamics, an Example: The Pendulum

Let us consider first the frictionless pendulum governed by

$$\frac{d^2 X}{dt^2} + \sin(X) = 0$$

or preferably

$$\frac{dX}{dt} = Y, \qquad \frac{dY}{dt} = -\sin(X).$$

The corresponding phase space is a cylinder $\mathsf{T}^1 \times \mathsf{R}$ periodic in the X-direction. Fixed points are easily found to lie at $Y = 0$ and $X = k\pi$, k integer. They are alternatively centers (k even, especially the origin for $k = 0$) and saddles (k odd). The phase portrait is depicted in Fig. 12a where we see that two successive saddle points are connected by *separatrices* which isolate *passing trajectories* from trajectories *trapped* around the centers. The periods of trapped trajectories increase with their amplitudes and tend to infinity when initial conditions get closer and closer to the separatrices that form a chain of *heteroclinic loops*.

Adding viscous damping, i.e.,

$$\frac{dX}{dt} = Y, \qquad \frac{dY}{dt} = -aY - \sin(X)$$

5. Low-Dimensional Dynamical Systems

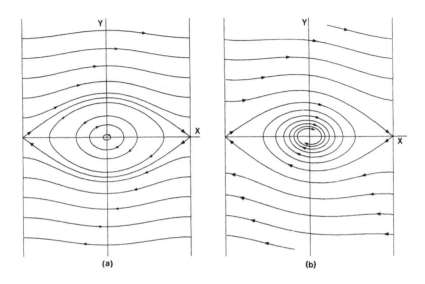

Fig. 12. Phase portraits of the classical pendulum: a) ideal, frictionless; b) realistic, weakly damped.

with $a > 0$, we obtain the modified phase portrait given in Fig. 12b. The position of the fixed points is unchanged, still $(k\pi, 0)$, but the eigenvalues are now given by

$$s^2 + as + (-1)^k = 0$$

and when dissipation is weak ($a \ll 1$) we have

points $(2k\pi, 0)$: $\qquad s = \pm i - a/2$,
points $((2k+1)\pi, 0)$: $\qquad s = \pm 1 - a/2$;

thus the saddles remain saddles (hyperbolic point, structurally stable) while the centers (nonhyperbolic) become stable spiral points so that the dynamics changes qualitatively in their neighborhood.

The nature of the separatrices also changes; for example, a trajectory starting along the unstable direction of point $(-\pi, 0)$ which reached point $(\pi, 0)$ in the absence of dissipation now misses its target owing to the energy loss and finally spirals around point $(0, 0)$; all heteroclinic loops are now open.

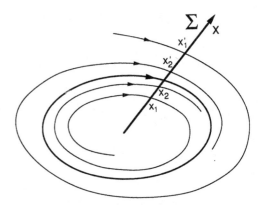

Fig. 13. Illustration of the Poincaré section technique in two dimensions.

As long as the dissipation remains small its precise value does not matter, the global aspect of the phase portrait remains unchanged, and the weakly damped pendulum is structurally stable, which was not the case of the frictionless pendulum. Two structurally unstable features have disappeared: the centers and the heteroclinic loops.

Another change takes place at $a = 2$, where the system has a double root $s = -1$ at $(0,0)$; the origin is then a stable improper node. Above this value, in the limit of large dissipation, the origin becomes a stable node and the pendulum reaches its equilibrium position without oscillating.

5.3. General Dynamics in Two Dimensions

An important result concerning general nonlinear systems defined on R^2 is the *Poincaré–Bendixon Theorem* which states that a limit set (which characterizes the asymptotic regime once transients have decayed) either positive or negative (i.e., forward or backward in time) which is non void, compact (i.e., closed and bounded), and contains no fixed point, is a a *limit cycle* (a closed orbit).

The rigorous derivation of this result (see Bibliographic Notes) involves the definition of: (1) a local *section* of the flow that, in two dimensions, is simply a piece of curve C transverse to the vector field, (2) a *first return map* or *Poincaré map* relating successive intersection of the trajectories with C.

5. Low-Dimensional Dynamical Systems

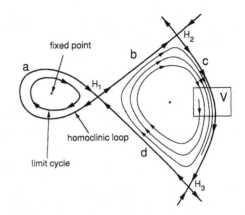

Fig. 14. In $\mathbb{R}^2$, non-wandering sets can be fixed points, limit cycles, heteroclinic or homoclinic loops joining saddle points; points belonging to homo/heteroclinic loops, though nonrecurrent, are nonwandering.

In two dimensions, trajectories form a set of lines that, owing to the existence-uniqueness property, are not allowed to intersect at ordinary (regular) points. Intersections with $\mathcal{C}$ can be ordered in a monotonous way along $\mathcal{C}$: crossing times $t_1 < t_2 < t_3$ implying either $x_1 < x_2 < x_3$ or $x_1 > x_2 > x_3$, where x is a curvilinear coordinate along $\mathcal{C}$, as illustrated in Fig. 13. Intersections of trajectories with $\mathcal{C}$ on one side of the limit set bound the series of intersections on the other side, and reciprocally. Therefore, each series admits a limit which is the intersection of the limit set itself with $\mathcal{C}$. This limit set cannot be more complicated than a limit cycle. If convergence occurs for $t \to +\infty$ (*positive limit set*) the limit cycle is stable. On the contrary, if convergence occurs for $t \to -\infty$ (*negative limit set*), this means that neighboring trajectories actually diverge from it when t increases, i.e., the limit cycle is unstable.

According to this theorem, in two dimensions, nonwandering sets (as opposed to transients) can be either *fixed points* or *limit cycles*, or else *cycles of saddle points* connected to each other (heteroclinic orbits) or to themselves (homoclinic orbits). This last case, also called a loop of *saddle connections* is depicted in Fig. 14. It can serve to illustrate the notion of nonwandering point when it is neither a fixed point nor a recurrent point belonging to a limit cycle. A point is *recurrent* if the trajectory that *goes through it* comes again in its vicinity. A nonwandering point is a point such that trajectories that *pass in its neighborhood* will pass indefinitely in that neighborhood. All points of such saddle connections are obviously nonwandering but they are not recurrent since homoclinic or heteroclinic trajectories take an infinite time to join fixed points standing at their extremities.

The existence of limit cycles is subjected to the *Bendixon criterion* which states that a simply connected region of the plane cannot contain a limit cycle if the divergence $\partial_X F_X + \partial_Y F_Y$ of the vector field $(F_X(X,Y), F_Y(X,Y))$ keeps a constant sign in that region. The reason is nearly obvious. Suppose that such a cycle exists. The area of the domain of phase space limited by this curve should be invariant under the evolution since the curve itself is invariant, but this is impossible if the divergence keeps a constant sign since the expansion rate of phase space elements in that region is not vanishing.

Generic properties of dynamical systems on a general two-dimensional manifold are given by *Peixoto's theorem*: nonwandering sets of a *structurally stable* vector field on a compact two-dimensional manifold are either *fixed points* or *limit cycles*, in finite number and all *hyperbolic*.

As already stated, nonhyperbolic limit sets, i.e., not strictly stable or unstable, are structurally unstable. The structural instability of homoclinic trajectories has the same origin as that of heteroclinic connections already evoked for the damped pendulum.

As an example of two-dimensional compact manifold, we may take a two-torus T^2, Fig. 15a, which can be parametrized by two periodic phase variables, (θ, ϕ) defined modulo 1 (or 2π). On such a torus, the simplest flow is induced by the constant field $(1, \alpha)$, i.e., $d\theta/dt = 1$ and $d\phi/dt = \alpha$ which generates sets of trajectories $\theta(t) = \theta(0) + t$ and $\phi = \phi(0) + \alpha t$ starting at given initial conditions $(\theta(0), \phi(0))$.

- When α is an irreducible rational number — $\alpha = p/q$, p and q relatively primes — the trajectories are *periodic* with period q, since $\theta(q) = \theta(0) + q = \theta(0)$ (mod 1) and $\phi(q) = \phi(0) + q(p/q) = \phi(0) + p = \phi(0)$ (mod 1), i.e., p complete turns for ϕ during q

5. Low-Dimensional Dynamical Systems

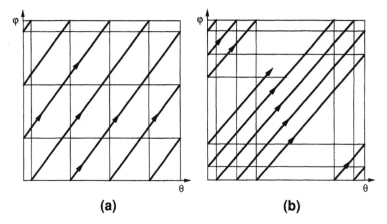

Fig. 15. a) Two phase variables θ and ϕ are required to define a point on a two-torus; b) when α is rational the trajectories are periodic; c) when α is irrational, trajectories induced by a constant vector field densely cover the torus.

complete turns for θ, see Fig. 15b.

- When α is irrational, the whole torus is covered densely by the orbit (a given trajectory approaches any point on the torus arbitrarily closely, Fig. 15c), the regime is said to be *quasi-periodic*, here simply doubly-periodic since we have only two incommensurate periods.

Peixoto's theorem tells us that two-periodicity is structurally unstable and that periodic regimes are structurally stable. This is consistent with the fact that any irrational number can be approached as close as wished by a rational number. The perturbations which bring the system from quasi-periodic (α irrational) to periodic (α rational) form a dense set (Q in R). We shall come back to this when we shall study the *locking* phenomenon.

5.4. Bifurcations in Two Dimensions

Returning to the results of the linear analysis in Section 5.1, we see that three different cases can occur at a bifurcation point:

(1) one critical mode and one stable mode both real, adiabatic elim-

ination of the stable mode yielding a codimension-one problem;
(2) two critical modes; at lowest order, we have a codimension-two problem;
(3) a pair of complex conjugate modes $s_\pm = \sigma \pm i\omega$, crossing the imaginary axis ($\sigma = 0$). The frequency ω remaining noncritical, a single control parameter is needed to monitor σ, so that we are left with a codimension-one problem.

Keeping in mind that the structure of normal form governing Case 2 depends on symmetry requirements (see Section 2.3), here we examine only Cases 1 and 3. Case 3 is genuinely two-dimensional and will bring in a nontrivial time dependence (i.e., a periodic behavior) in contrast with case 1, best understood as the direct product of two one-dimensional dynamics.

5.4.1. Saddle-Node Bifurcation

As a specific example of Case 1, we consider a system with a central mode A bifurcating through a turning point at $r = 0$, completed by a stable mode B uncoupled to A at lowest order:

$$\frac{dA}{dt} = r + A^2, \qquad \frac{dB}{dt} = -B.$$

Fixed points are given by $B_f = 0$ and $r + A_f^2 = 0$.

- When $r < 0$, we have two roots for A_f: $\pm\sqrt{-r}$. Point $(-\sqrt{-r}, 0)$ is a stable node; its eigenvalues are both negative: $-2\sqrt{-r}$ (small) and -1 (large); corresponding eigenvectors are along the A-axis and B-axis, respectively. Point $(+\sqrt{-r}, 0)$ is a saddle, stable along the B-axis (eigenvalue -1, negative and large) and unstable along the A-axis (eigenvalue $+2\sqrt{-r}$, positive and small). The basin of attraction of the stable fixed point is the whole open half-plane $A < +\sqrt{-r}$ (see Fig. 16a).
- When $r = 0$, we have a single semi-stable fixed point with mixed characteristics, resembling a node for $A < 0$ and a saddle for $A > 0$, hence the name of *saddle-node bifurcation* (Fig. 16b).
- When $r > 0$, all trajectories go to $+\infty$ (Fig. 16c) but they spend a long time in the vicinity of the origin.

5. Low-Dimensional Dynamical Systems

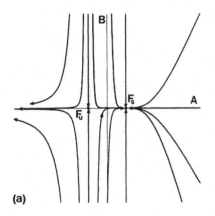

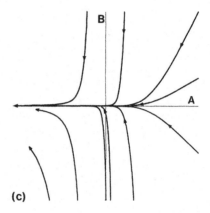

Fig. 16. Saddle-node bifurcation; (a) before the collapse of the two fixed points, one, $\mathbf{F_s}$, is a stable node and the other, $\mathbf{F_u}$ is a saddle; (b) at the bifurcation point, the fixed point $\mathbf{F_{sn}}$ has mixed characteristics; (c) after the collapse, trajectories all end at $-\infty$ but they spend a long time in the region where fixed points were lying.

5.4.2. Hopf Bifurcation

We now describe the bifurcation of a spiral fixed point that becomes unstable. The starting point is a generalization of normal form (18) to a slightly off-critical situation. We write it as

$$\frac{dZ}{dt} = (\sigma + i\omega)Z - g|Z|^2 Z. \qquad (39)$$

The bifurcation takes place when our control parameter σ goes through 0 from negative values. Let us specify $g = g' + ig''$ and insert $Z = |Z|\exp(i\phi)$ in (39). This gives

$$\exp(i\phi)\left\{\frac{d|Z|}{dt} + i|Z|\frac{d\phi}{dt} = (\sigma + i\omega)|Z| - (g' + ig'')|Z|^3\right\}.$$

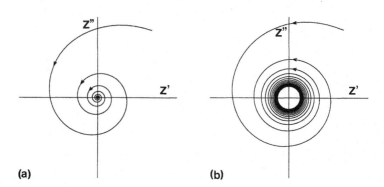

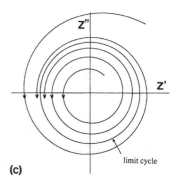

Fig. 17. Hopf bifurcation in the complex plane, $Z = Z' + iZ''$; (a) $\sigma < 0$, trajectories spiral exponentially fast toward the origin; (b) $\sigma = 0$, trajectories spiral algebraically slow toward the origin; (c) for $\sigma > 0$, when the bifurcation is supercritical, a limit cycle is obtained for initial conditions taken either close to the origin or far from it.

The evolution of modulus $|Z|$ and phase ϕ are therefore given by

$$\frac{d|Z|}{dt} = (\sigma - g'|Z|^2)|Z|, \tag{40a}$$

$$\frac{d\phi}{dt} = \omega - g''|Z|^2 \tag{40b}$$

(except at the singular point $|Z| = 0$).

We see from (40a) that the modulus is not coupled with the phase so that its evolution can be solved independently. This one-dimensional problem has already been considered in Section 3.2 from which we know that the nontrivial fixed point $|Z| = \sqrt{\sigma/g'}$ is stable when g' is positive, making the bifurcation *supercritical*, whereas it is unstable when g' is negative and the bifurcation *subcritical*.

5. Low-Dimensional Dynamical Systems

Once the variation of $|Z|$ is known, the evolution of the phase can be determined. Let us consider simply the asymptotic regime where $|Z|$ has reached the fixed point value. We have

$$\frac{d\phi}{dt} = \omega - \frac{g''\sigma}{g'} \quad \Rightarrow \quad \phi = \phi_0 + \tilde{\omega}t$$

with $\tilde{\omega} - \omega \propto |Z|^2 \sim \sigma$, i.e., a periodic motion with a frequency slightly corrected from the nonlinear contribution. The birth of the limit cycle is illustrated in Fig. 17.

6. Conclusion

Bifurcations in one and two dimensions studied in the second part of this chapter only lead to regular asymptotic motion. This is basically related either to the gradient-flow property or to the possibility of the ordering of the successive intersections of a given trajectory with a given surface of section. More complicated behaviors can be found only in higher dimensional systems resulting from the reduction process described in Section 1 and 2 when one or more new modes become marginal. In dimensions greater than two, the surface of section is at least two-dimensional, and chaotic behavior will be seen to derive from the fact that trajectories have sufficient space to turn around each other without crossing during time of flight between two successive intersections with the surface of section.

7. Bibliographical Notes

In addition to Part 1 of the book by Abraham and Shaw BN 2 [5], for a mathematically oriented approach to the contents of the present chapter, one should consult:

[1] J. Guckenheimer and P. Holmes, *Nonlinear oscillations, dynamical systems and bifurcation of vector fields* (Springer, 1983).

and more particularly Chapters 1 to 3 (center manifolds, normal forms, theorems in two dimensions).

Two more ancient mathematics books are also of particular interest:

[2] M.W. Hisch and S. Smale, *Differential equations, dynamical systems and linear algebra* (Academic Press, 1974).

for its progressive but thorough presentation of nonlinear differential systems, from linear algebra to Poincaré theorems, and:

[3] S. Lefshetz, *Differential equations: geometric theory* (Dover, 1977).

Another interesting reference is:

[4] G. Iooss, "Reduction of the dynamics of a bifurcation problem using normal forms and symmetries."

in:

[5] E. Tirapegui and D. Villaroel, eds., *Instabilities and Nonequilibrium structures* (Reidel, 1987).

that contains several other contributions to be mentioned later.

The model used in Section 1 and 2 was introduced in:

[6] Y. Pomeau, P. Manneville, "Wavelength selection in cellular flows," Physics Lett. **75A**, 296 (1980).

A sufficiently self-contained presentation of catastrophe theory first introduced by R. Thom in the sixties has been given by:

[7] I. Steward, "Application of catastrophe theory to the physical sciences," Physica **2D**, 245 (1981).

Chapter 6
Beyond Periodic Behavior

The introduction of the Poincaré surface of section technique has proven to be an essential stage in the understanding of dynamics in two dimensional phase spaces. In Section 1, we extend this technique to higher dimensional cases and give a first example of its use for analyzing chaotic orbits generated by the celebrated Lorenz model. We then develop a step by step approach of the birth of temporal chaos, i.e., weak turbulence in a strongly confined context, by discussing the stability of a limit cycle in terms of Poincaré maps (Section 2). Next, we examine the main cases of codimension-one bifurcations of limit cycles, presenting in some detail the derivation of the relevant normal forms which relies on the notion of resonance (Section 3). Introducing the concept of *scenario* underlying the Ruelle–Takens approach (Section 4) we turn to a review of the essential properties of the three "classical" routes to turbulence, each associated with one of the three kinds of elementary bifurcations of limit cycles: the sub-harmonic, intermittent, and quasi-periodic scenarios (Section 5, 6, 7). We conclude this chapter by opening the perspective to other processes leading to time-chaotic behavior.

1. Poincaré Maps

1.1. Surface of Section and First Return Map

When more than two modes are implied in the dynamics, the reduction to a two-dimensional manifold is no longer legitimate but we may still make use of the idea of considering successive intersec-

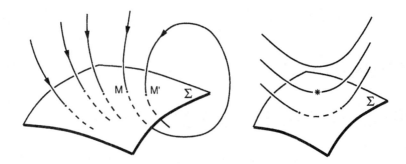

Fig. 1. The intersections of a trajectory with the surface of section define a first return map but the chosen surface should be transverse to the vector field to avoid difficulties with near-tangent trajectories.

tions of trajectories with a given hypersurface in phase space. In this way, properties of the d-dimensional, continuous-time, primitive system are converted into those of a $(d-1)$-dimensional discrete-time system.

Let Σ be a $(d-1)$–dimensional surface in the d–dimensional phase space. It can define a useful *surface of section* only if it intersects the nonwandering set (otherwise arbitrary trajectories only meet the surface a finite number of times during the transient). Successive intersections with a given trajectory, $\mathbf{M}_n$, $n = 0, 1, \ldots$, can then be viewed as iterates of some initial condition $\mathbf{M}_0$ under a map of Σ onto itself, $\mathbf{M} \to \mathbf{M}' = \Phi(\mathbf{M})$ (Fig. 1). This map, called the *Poincaré* or *first return map*, can be determined analytically in special cases only. Most often it is obtained by integrating the equations of motion numerically.

The choice of Σ is a matter of convenience but care should be taken in keeping it transverse to the vector field so that trajectories never arrive tangentially close to the surface but always at some finite angle. This condition is important in preventing difficulties in interpreting the properties of the Poincaré map.

When $d = 2$, Σ is one-dimensional and the first return map is an iteration of a single variable. When $d = 3$, Σ is two-dimensional and graphical representations are still easy to interpret. Here we

6. Beyond Periodic Behavior

illustrate the technique with the Lorenz model introduced in Chapter 1 as a concrete example of a three-dimensional system displaying chaotic behavior.

1.2. Application to the Lorenz Model

The Lorenz model is a three-dimensional autonomous differential system deriving from a clever truncation of an expansion of the equations of convection (stress-free top/bottom plates and periodic lateral boundary conditions). It reads

$$\begin{align}
\frac{dA_1}{dt} &= \sigma(A_2 - A_1), \\
\frac{dA_2}{dt} &= rA_1 - A_2 - A_1 A_3, \\
\frac{dA_3}{dt} &= -bA_3 + A_1 A_2.
\end{align} \quad (1)$$

Variable A_1 is the amplitude of the first horizontal harmonic of the vertical velocity, A_2 the amplitude of the corresponding temperature fluctuation, and A_3 a uniform correction to the temperature field. σ is a notation for the Prandtl number, r is a reduced Rayleigh number ($r = R/R_c$) and b is a parameter related to the horizontal wavevector.

Fixed points of system (1) are easily obtained. In addition to the trivial fixed point at the origin, $A_{1,f} = A_{2,f} = A_{3,f} = 0$, corresponding to the conduction regime, for $r > 1$ we have a pair of nontrivial fixed points: $A_{3,f} = r - 1$, $A_{1,f} = A_{2,f} = \pm\sqrt{b(r-1)}$, corresponding to steady convection.

Here, our purpose is not to describe the successive bifurcations when r is varied but to present the surface-of-section technique as a global tool for characterizing the temporal regime for $\sigma = 10$, $b = 8/3$, $r = 28$, the specific values chosen by Lorenz to demonstrate the existence of "aperiodic" (= chaotic) trajectories (Fig. 2).

The chosen surface of section is the $A_3 = r - 1$, parallel to the $A_1 A_2$-plane and containing the nontrivial fixed points. The main

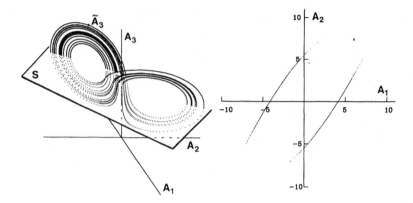

Fig. 2. Left: Typical trajectory for the Lorenz model with $r = 28$, $\sigma = 10$ and $b = 8/3$ (the maximum reached by A_3 during each revolution around the unstable nontrivial fixed points is recorded to build the Lorenz map in Fig. 3a below). Right: trace of the successive intersections of a trajectory with the surface of section defined by $A_3 = r - 1$ and the additional condition $dA_3/dt < 0$.

reason is that, except at these points, the vector field has a non-vanishing A_3-component, which means that trajectories which hit the plane never arrive tangentially. Finally, to get unambiguous results we must specify the direction of crossing, say from below, i.e., $dA_3/dt > 0$. The return map associated with the surface defined by $A_3 = r - 1$ for the canonical parameters of the Lorenz model is displayed in Fig. 2b.

The reduction to a two-dimensional map may still be insufficient to gain a real understanding of what is going on. To go further, Lorenz considered the reduced map $\tilde{A}_{3,n+1} = f(\tilde{A}_{3,n})$, where $\tilde{A}_{3,n}$ denotes the nth value of the maximum reached by variable A_3 along a trajectory (see Fig. 3a). This corresponds to a different choice for the surface of section, namely $dA_3/dt = 0$, i.e., $A_3 = A_1 A_2 / b$ (with the additional condition $d^2 A_3/dt^2 < 0$). Though derived from a genuine surface of section, the map obtained in this way is not a Poincaré map but only the projection of a Poincaré map. In particular, it is not invertible since specifying A_3 is not sufficient to define a point on a two-dimensional surface. How-

6. Beyond Periodic Behavior

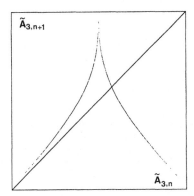
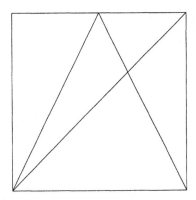

Fig. 3. The Lorenz map (left) obtained by plotting successive maxima of A_3, $\tilde{A}_{3,n+1} = F(\tilde{A}_{3,n})$, is reminiscent of the tent map (right). Since it has no stable fixed points and is everywhere expanding (absolute value of the slope larger than 1), the resulting attractor is chaotic.

ever, it turns out that this map, known as a *Lorenz map*, is useful for understanding the origin of chaos since it closely resembles a tent map (see Fig. 3b), a simple example of chaotic map similar to the diadic map used in Chapter 1 to visualize the divergence of trajectories. Long term unpredictability is then seen to result from the fact that the Lorenz map is everywhere expanding (absolute value of the slope larger than 1). Owing to the large dissipation in the Lorenz model, the correspondence between discrete trajectories and their projections is nearly one-to-one, which gives a justification for the success of this supplementary reduction.

2. Stability of a Limit Cycle

Let us first consider the simplest three-dimensional model displaying the Hopf bifurcation analyzed at the end of Chapter 5. We take a pair of nearly marginal oscillating modes $Z = A_1 + iA_2$ and $\bar{Z} = A_1 - iA_2$, with eigenvalues $s_\pm = \sigma \pm i\omega$ ($\omega \neq 0$ and σ positive or negative but close to 0) interacting with a real stable mode A_3 with eigenvalue

s_2 real and strongly negative ($|s_2| \gg |\sigma|$):

$$\frac{dZ}{dt} = (\sigma + i\omega)Z + g_1 Z A_3 ,$$
$$\frac{d\bar{Z}}{dt} = (\sigma - i\omega)\bar{Z} + \bar{g}_1 \bar{Z} A_3 , \qquad (2)$$
$$\frac{dA_3}{dt} + |s_2|A_3 = -g_2|Z|^2 .$$

In the first two equations, combinations introduced are resonant at frequency ω since A_3 is real, whereas in the third equation oscillations at frequency ω are removed by taking the modulus, which also implies g_2 real. Since the evolution rate of $|Z|^2$ is $2\sigma \ll s_2$ at lowest order, the adiabatic elimination of A_3 is straightforward; we get

$$A_3 = -\frac{g_2}{|s_2|}|Z|^2 . \qquad (3)$$

This is the equation of a two-dimensional surface in a three-dimensional space spanned by $(Z, \bar{Z}, A_3)$ or, equivalently, (A_1, A_2, A_3). Inserting this condition in the two first equations, we get

$$\frac{dZ}{dt} = (\sigma + i\omega)Z - g_{\text{eff}}|Z|^2 Z \qquad (4)$$

with $g_{\text{eff}} = g_1 g_2/|s_2|$ (+ the complex conjugate equation). According to the picture given in Fig. 4, it is legitimate to consider the newborn limit cycle Γ as drawn on the two-dimensional center manifold **S** defined by (3). More generally, as long as the supplementary modes are strongly stable, trajectories converge rapidly to the two-dimensional center manifold and perturbations transverse to it can be neglected. In fact, the stability analysis developed in the previous chapter was restricted to perturbations along this manifold, i.e., in the radial direction, so that the one-dimensional surface of section $\mathcal{C}$ used at that time is nothing but the trace on **S** of a $(d-1)$-dimensional surface Σ transverse to Γ.

A general stability analysis will therefore have to test for perturbations in $d-1$ directions. Before developing this geometrical approach, let us see how the problem is posed from an analytical point

6. Beyond Periodic Behavior

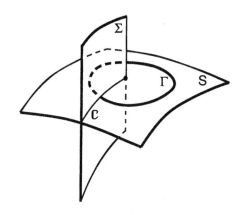

Fig. 4. The limit cycle Γ emerging from a Hopf bifurcation must be understood as being drawn on a two-dimensional center manifold **S** embedded in a d-dimensional phase space, ($d = 3$ for model (2)); stability must be checked in the whole $(d-1)$ dimensions of the surface of section Σ and no longer only along its trace $\mathcal{C}$ on **S**.

of view. We take for granted that the limit cycle is well installed, i.e., not to close to the Hopf bifurcation point so that the "radial" perturbations no longer play a special role. The basic state is then a time-periodic solution $V_0(t)$ with period T such that $V_0(t+T) \equiv V_0(t)$. The system governing an arbitrary perturbation $V' = V - V_0(t)$ is therefore periodically forced with period T. After linearization, this yields a linear differential system with periodic coefficients.

Clearly, perturbations must "feel" the oscillatory background and their amplification or decay must be estimated only after everything at the period of the limit cycle has been subtracted. The natural extension of $V'(t) = \exp(st)AX$ where X was time-independent, is obviously $V'(t) = \exp(st)AX(t)$, where $X(t)$ is a periodic function of t with period T, $X(t+T) = X(t)$ containing the somewhat trivial contribution of the forcing (*Floquet stability theory*). The exponential term accounts for the divergence from a strictly periodic behavior at period T. Measuring the perturbation every T interval, we get

$$V'(t+T) = \exp\bigl(s(t+T)\bigr)AX(t+T)$$
$$= \exp(sT)\bigl(\exp(st)AX(t)\bigr) = \exp(sT)V'(t).$$

The evolution of the perturbation therefore depends on the value of the real part σ of the *Floquet exponent* $s = \sigma + i\omega$ or, equivalently,

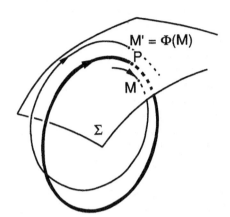

Fig. 5. Poincaré surface of section and first return map in the general case.

on the modulus of the quantity $\lambda = \exp(sT)$, called the *Floquet multiplier*. The basic state (limit cycle) is unstable when $\sigma > 0$ as usual, which yields $|\lambda| > 1$; the frequency ω of the eigenmode has no reason to be related to the basic frequency in general.

Let us return to the geometrical approach in phase space. The general case is depicted in Fig. 5. Clearly, a trajectory starting on the limit cycle Γ itself hits the surface of section Σ at point **P** periodically. A trajectory starting sufficiently close to Γ intersects Σ at a series of points $\mathbf{M}_n$, $n = 0, 1, \ldots$, which may remain in the neighborhood of **P** for some time. Writing $\mathbf{M}_{n+1} = \Phi(\mathbf{M}_n)$, we derive immediately that **P** is a fixed point of the map, and that the stability of the limit cycle Γ can be decided from the stability of **P**, i.e., according to whether points $\mathbf{M}_n$ converge to or diverge from **P**. In turn, this convergence/divergence can be inferred from the spectrum $\{\lambda\}$ of the map Φ linearized around **P**, which is a $(d-1) \times (d-1)$ matrix.

The linearized first return map has $d - 1$ eigenvalues which may be real or complex, degenerate or not, but which can always be ordered by decreasing values of their modulus. The limit cycle is stable when all the eigenvalues are inside the open unit disc $\{|\lambda| < 1\}$. A perturbation is marginal if $|\lambda| = 1$, and the limit cycle is unstable when at least one eigenvalue lies outside the unit disc, $|\lambda| > 1$; see Fig. 6.

Notice that a limit cycle generated by an autonomous d-

6. Beyond Periodic Behavior

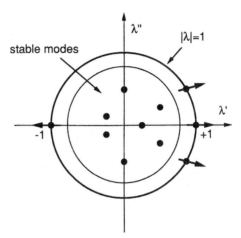

Fig. 6. Eigenmodes of the tangent map are stable when $|\lambda| < 1$, marginal when $|\lambda| = 1$, unstable when $|\lambda| > 1$. Bifurcations occurs when they leave the unit disc either through ± 1 or a pair of complex conjugate numbers.

dimensional system has d Floquet exponents among which one is vanishing. In tangent space, the direction associated with this vanishing eigenvalue is the direction of the vector field itself, which corresponds to the fact that the distance between trajectories with initial conditions taken exactly on the limit cycle neither decreases nor increases *on average over a period* of the motion. Obviously, this special eigenvalue has no counterpart in the discrete-time picture, which leaves $d - 1$ nontrivial eigenvalues as expected. The stability condition above, $|\lambda| < 1$, is then merely the translation of the condition $\sigma < 0$ to the nontrivial part of the Floquet spectrum.

3. Bifurcations of a Limit Cycle

A bifurcation takes place when, as a function of control parameters, at least one eigenvalue crosses the unit circle $\{|\lambda| = 1\}$. The simplest cases correspond to the crossing of a *nondegenerate* pair of complex eigenvalues or a *single* real eigenvalue, either $+1$ or -1. In order to describe these bifurcations, as in the continuous-time case, we have to perform the elimination of "stable modes" with eigenvalues strictly inside the unit disc. Corresponding to *strong resonances*, the two real cases yield one-dimensional iterations that will be studied separately later. In the complex case, this elimination leads to a two-dimensional effective map that has first to be reduced to its normal form.

3.1. Normal Form in the Complex Case

When the marginal eigenvalue is complex, it is preferable to represent the resulting two-dimensional map using complex variables $Z, \bar{Z}$. On general grounds we expect

$$Z_{n+1} = \lambda Z_n + N(Z_n, \bar{Z}_n) \tag{5a}$$

with

$$N(Z_n, \bar{Z}_n) = \sum_{2 \leq m; 0 \leq j \leq m} a_{mj} Z^{m-j} \bar{Z}^j \tag{5b}$$

and, by assumption, $|\lambda| = 1$ but $\lambda \neq \pm 1$. As in the continuous-time case, the reduction to normal form is obtained by successive nonlinear changes of variables. Here we shall consider only the reduction of the quadratic terms, $r = 2$. Expressing the old variables in terms of the news

$$Z = Z' + \sum_{0 \leq j \leq 2} c_{2j} Z'^{2-j} \bar{Z}'^j .$$

Upon insertion in (5), we obtain

$$Z'_{n+1} + \sum_j c_{2j} Z'^{2-j}_{n+1} \bar{Z}'^j_{n+1} = Z'_{n+1} + \sum_j c_{2j} \lambda^{2-j} Z'^{2-j}_n \bar{\lambda}^j \bar{Z}'^j_n$$

$$= \lambda \left(Z'_n + \sum_j c_{2j} Z'^{2-j}_n \bar{Z}'^j_n \right) + \sum_j a_{2j} Z'^{2-j}_n \bar{Z}'^j_n .$$

The quadratic terms can be suppressed if

$$a'_{2j} = a_{2j} + \left(\lambda - \lambda^{2-j} \bar{\lambda}^q \right) c_{2j}$$

can be set equal to zero, which will be the case if the coefficient of c_{2j} does not vanish. This gives the *nonresonance condition*

$$\lambda - \lambda^{2-j} \bar{\lambda}^j \neq 0 . \tag{6}$$

6. Beyond Periodic Behavior

Terms corresponding to $j = 0$ or $j = 1$ can always be eliminated since this condition leads to $\lambda \neq 1$ and $\bar{\lambda} \neq 1$, respectively, which is excluded by assumption. The third term corresponding to $j = 2$ can be eliminated only when $\lambda - \bar{\lambda}^2 \neq 0$, or using the fact that, at criticality, $|\lambda|^2 = 1$, i.e., $\bar{\lambda} = \lambda^{-1}$ when $\lambda^3 \neq 1$.

When $\lambda = \exp(2\pi i p/3)$, this term cannot be eliminated so that we obtain

$$Z_{n+1} = \lambda Z_n + a\bar{Z}_n^2 + \mathcal{O}(|Z_n|^3) \tag{7a}$$

(primes indicating the new variables have been dropped). This is a new case of *strong resonance*, to be added to the cases $\lambda = \pm 1$. Indeed, we can readily check that if $\lambda = \exp(\pm 2\pi i/3)$, the two contributions on the r.h.s. of (7a) rotate synchronously.

When λ is not a cubic root of unity, all quadratic terms can be eliminated and the procedure can be repeated to eliminate the largest possible number of cubic terms. When $\lambda^4 = 1$, i.e., $\lambda = \exp(2\pi i p/4)$, two cubic terms remain and one gets

$$Z_{n+1} = \lambda Z_n + a|Z_n|^2 Z_n + b\bar{Z}_n^3 + \mathcal{O}(|Z|^4), \tag{7b}$$

but if this is not the case, b can also be set equal to zero. If $\lambda^5 = 1$, i.e., $\lambda = \exp(2\pi i p/5)$, the normal form reads

$$Z_{n+1} = \lambda Z_n + a|Z_n|^2 Z_n + c\bar{Z}_n^4 + \mathcal{O}(|Z|^5).$$

In all other cases, the fourth order term can be dropped, so that the normal form reads

$$Z_{n+1} = \lambda Z_n + a|Z_n|^2 Z_n + \mathcal{O}(|Z|^5). \tag{8}$$

When $\lambda^q = 1$, i.e., $\lambda = \exp(2\pi i p/q)$, with $q \geq 5$ we have what is called a *weak resonance*. Using $|\lambda|^2 = \lambda\bar{\lambda} = 1$, the nonresonance condition that generalizes condition (6) for a term of order $|Z|^m$ can be written as $\lambda = \lambda^{m-j}\bar{\lambda}^j$ or $\lambda^{m-2j-1} = 1$, which implies $m - 2j - 1 = lq$, where l is an integer, $0 \leq j \leq m$, and $m \geq 2$ for a truly nonlinear contribution. This condition is discussed graphically in Fig. 7. Firstly, for all q with $l = 0$ we get $m = 2j + 1$,

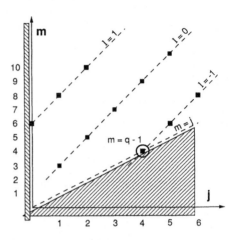

Fig. 7. Example for $q = 5$ of the resonance condition $m = lq + 2j + 1$ with $m \geq 2$, $j \geq 0$, and $j \leq m$, l integer; $l = 0$ gives the term $|Z|^{2q}Z$ which is always present; $l = -1$, $m = j = q - 1$ is the lowest degree nontrivial contribution.

which corresponds to the "trivial" term $|Z_n|^{2j}Z_n$ (at least cubic since $j \geq 1$, as already seen above). Secondly, we see that the lowest order "nontrivial" contribution (smallest m) at given q is obtained for $l = -1$ and, subsequently, that $m = j = q - 1$. Strongly resonant cases are readily recovered, e.g., for $q = 3$ we obtain $j = 2$, i.e., the $\bar{Z}_n^2$ term. In case of weak resonance of order q, i.e., $q \geq 5$, the normal form then reads

$$Z_{n+1} = \lambda Z_n + \sum_{j, 2j+1 < q} a_j |Z_n|^{2j} Z_n + b \bar{Z}_n^{q-1} + \mathcal{O}(|Z_n|^q). \qquad (9)$$

The *nonresonant* case corresponds to $\lambda = \exp(2\pi i \alpha)$ with α irrational.

3.2. Hopf Bifurcation for Maps

Here, we consider the bifurcation of a limit cycle through a pair of complex eigenvalues $(\lambda_0, \bar{\lambda}_0)$, $\lambda_0 = \exp(2\pi i \alpha_0)$, either nonresonant, α_0 irrational, or weakly resonant, $\alpha_0 = p_0/q_0$, $q_0 \geq 5$. To describe the bifurcation, the normal form (9) will be truncated just above the cubic term so that the difference between weak resonance and no resonance will disappear. Higher order corrections responsible for the *locking phenomenon* will be examined later.

To account for off-marginal conditions at lowest order, we have then simply to correct the linear term by specifying how the unit

6. Beyond Periodic Behavior

cycle is crossed as a function of the control parameter. The result of linear stability analysis is a pair of eigenvalues whose path in the complex plane requires two small parameters to be described close to the threshold: $\lambda = \exp(2\pi i\alpha)$ with $\alpha = \alpha_0 + r'' - ir'$. The angular variation is accounted for by r'' and the growth of the modulus by r'. They are both functions of the control parameter r and, by proper rescaling, we can write: $\lambda = \lambda_0 + \delta\lambda(r) = \lambda_0(1 + r(1+iu))$, (where u is some constant real scalar constant related to the angular variation). The starting iteration then reads

$$Z_{n+1} = \big(1 + r(1+iu)\big)\lambda_0 Z_n - g|Z_n|^2 Z_n \qquad (10)$$

where g is the coupling constant calculated at $r = 0$ (a in the preceding section).

The analysis closely follows that for of continuous-time Hopf bifurcation. Setting $Z = \rho\exp(2\pi i\theta)$, we derive the evolution of ρ from

$$\rho_{n+1}^2 = |\lambda|^2 \rho_n^2 - (\lambda\bar{g} + \bar{\lambda}g)\rho_n^4 + \mathcal{O}(\rho^5).$$

Using $|\lambda|^2 = |1 + r(1+iu)|^2 = 1 + 2r + r^2(1+u^2) \simeq 1 + 2r$ and $(\lambda\bar{g} + \bar{\lambda}g) = 2g'$, we obtain the fixed-point equation

$$\rho_*^2 = (1 + 2r)\rho_*^2 - 2g'\rho_*^4.$$

As expected, in addition to the trivial fixed point $\rho = 0$ which is stable for $r < 0$ and unstable for $r > 0$, we find a nontrivial solution $\rho_* = \sqrt{r/g'}$. When $g' > 0$, the bifurcation is *supercritical*: ρ_* exists for $r > 0$ and is easily checked to be *stable*. When $g' < 0$ the bifurcation is *subcritical* and the corresponding solution *unstable*.

The evolution of the modulus being independent of the value of the angle at lowest order, we can consider simply the asymptotic evolution of the phase variable θ after the decay of the transient, i.e., replace ρ_n by ρ_*. In this limit, the evolution of the phase is simply given by

$$\exp(2\pi i\theta_{n+1}) = C\exp(2\pi i\theta_n),$$

where, from (10), $C = (1 + r(1+iu))\lambda_0 - g\rho_*^2$ is a complex constant. Inserting the expression obtained for ρ_*^2 in this constant shows that

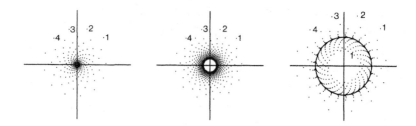

Fig. 8. Birth of the torus resulting from the Hopf bifurcation of a limit cycle: a) below the bifurcation point, iterates spiral exponentially fast toward the origin; b) at the threshold, convergence is algebraic; c) above the threshold, if the bifurcation is super-critical, iterates converge to a circle both from inside and from outside; this circle is the section of a two-torus round which trajectories are wound.

$|C| = 1$ and can be written as $\exp(2\pi i \alpha_*)$ with $\alpha_* = \alpha_0 + r\delta\alpha$. The asymptotic evolution of the angle then amounts to a rotation

$$\theta_{n+1} = \theta_n + \alpha_*.$$

When α_* is irrational, θ sweeps the unit circle uniformly and, in turn, the successive intersections of the trajectories with the surface of section sweep a circle of radius ρ_*, so that trajectories are drawn on a two-dimensional torus T^2 with radius ρ_* surrounding the limit cycle that bifurcates. When $g' < 0$, ρ_* exists for $r < 0$ (subcritical bifurcation) but is unstable so that the torus itself is unstable. On the other hand, it is stable when the bifurcation is supercritical. The birth of a stable torus is illustrated in Fig. 8.

The motion is *quasi-periodic* with two frequencies; observables evolve as doubly periodic functions of time, i.e., as $f(\omega t, \omega' t)$, f periodic with period 2π in each of its arguments, ω the frequency of the limit cycle that becomes unstable and $\omega' = \alpha_* \omega$ the second frequency.

We remark that when r is varied slowly, by continuity α_* assumes both irrational and rational values. According to the assumptions made, if around the bifurcation point, α_* is rational, i.e., $\alpha_* = p/q$, it cannot be strongly resonant and thus must remain sufficiently far

6. Beyond Periodic Behavior

from a rational with a denominator smaller than 5. The corresponding attractor is then a closed trajectory drawn on the torus, making p turns along the cycle which has lost stability and q turns in the complementary direction. At first sight, one could think that rational values are exceptional since they form a countable set immersed in the uncountable set of irrational values. In fact this is not the case owing to the *locking phenomenon* to be analyzed later.

3.3. Bifurcations at Strong Resonances 1/1 and 1/2

Let us now turn to the important and frequent special cases of strong resonance leading to one-dimensional problems that can be formulated in terms of real variables. We consider first the case of a bifurcation through $\lambda = 1$. At the bifurcation point, the normal form obviously reads

$$X_{n+1} = X_n - aX_n^2 + \ldots \qquad (11)$$

(the minus sign for convenience only) and, close to the bifurcation point, in the absence of symmetry we expect naively

$$X_{n+1} = (1+r)X_n - aX_n^2 + \ldots, \qquad (12)$$

which is easily seen to account for the exchange of stability between two limit cycles. One, corresponding to the root $X_* = 0$ of the fixed point equation $X_*(r - aX_*) = 0$, is stable when r is negative and unstable when r is positive; the other, corresponding to the nontrivial root $X_* = r/a$ has opposite stability properties. This bifurcation is pictured in Fig. 9 which makes the concept of 1/1-resonance clear since limit cycles which are approaching each other must have the same frequency, owing to the smoothness of the underlying vector field.

When writing down Equation (12) it is implicitly assumed that the original cycle survives. In fact, this may not be the case since there is another way to perturb Equation (11) and to unfold the degeneracy at the bifurcation point for which the map is tangent to the first diagonal. Instead of rotating the curve as in Fig. 9, we may translate it a little bit thus passing from a situation with two

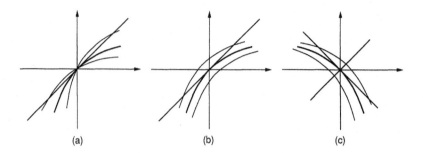

Fig. 9. Unfoldings of critical iterations at real eigenvalues ±1. When $\lambda = +1$, two cases can occur according to whether the persistence of the fixed point is imposed (a) or not (b), yielding either an exchange of stability or a saddle-node bifurcation, respectively. The case $\lambda = -1$ does not present such an alternative; the fixed that becomes unstable can do nothing but persist.

intersections to a situation where contact is completely lost, which is obviously the generic case (see Fig. 10). The corresponding normal form reads

$$X_{n+1} = r + X_n - aX_n^2 + \ldots, \qquad (13)$$

which describes the *saddle-node* bifurcation of a pair of limit cycles that collapse and disappear. Indeed, the fixed-point equation $X_* = r + X_* - aX_*^2$ gives us $X_* = \pm\sqrt{r/a}$, existing for $r/a \geq 0$, one of the fixed points being stable and the other unstable. For r/a negative, though the fixed points (and the corresponding limit cycles) have disappeared, the system keeps a memory of their presence for neighboring values of the control parameter and trajectories spend a long time in the region of phase space where the limit cycles were located, if they happen to visit that neighborhood. In this respect, saddle-node bifurcations of fixed points for maps are thus strictly similar to their counterpart for fixed points of flows (see Chapter 5, Section 5.4).

To conclude this section, we consider the case of an eigenvalue $\lambda = -1$. Close to the bifurcation point, the normal form reads

$$X_{n+1} = f_r(X) = -(1+r)X_n + aX_n^2 + bX_n^3 + \ldots, \qquad (14)$$

6. Beyond Periodic Behavior

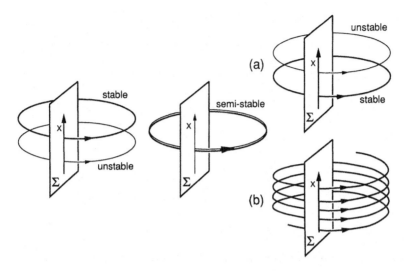

Fig. 10. Sketch of the exchange of stability (a) and saddle-node bifurcation (b) of a pair of limit cycles.

and one can see (Fig. 11) that the fixed point always persists while becoming unstable when r goes through zero. Here, the difficulty does not come from the coexistence of several possible unfoldings but from the fact that, under such a map, iterates alternate between positive and negative values as time goes on. This is clearly impossible if the limit cycle belongs to a piece of orientable two-dimensional manifold since this would imply intersection of trajectories (Fig. 11a). However, if we remember that the cycle is embedded in a larger space and that in the weakly or nonresonant case bifurcated orbits are wound around the limit cycle, we can imagine what happens in the strongly resonant case—one turn is performed around the cycle during the time necessary for two turns along it (1/2 resonance). In fact, trajectories belong to a twisted surface, i.e., a Mœbius band, Fig. 11b.

The fixed-point equation corresponding to (14),

$$X_* = -(1+r)X_* + aX_*^2 + bX_*^3$$

yields

$$(2 + r - aX_* - bX_*^2)X_* = 0.$$

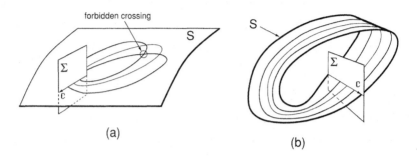

Fig. 11. a) On a nontwisted surface, the alternation of iterates of the first return map with eigenvalue -1 would imply a forbidden self-crossing of trajectories; b) nothing such happens on a twisted surface.

In addition to $X_* = 0$, this equation can have other roots that are irrelevant to the present problem since they remain at finite distance from the origin for r small (e.g., $X_* \approx 2/a$ for $b = 0$). When r is negative, the origin is stable and iterates converge to it with the characteristic alternating behavior allowed by the twisting of the center manifold. When r is positive, the iterates diverge from the origin but we see immediately that the divergence/convergence of a specific series of iterates has to be appreciated after grouping them two by two. We are therefore interested in the properties of the compound map

$$f^2(X) = f \circ f(X) = f(f(X)) = (1+r)^2 X + ar(1+r)X^2 \\ - (1+r)(2a^2 + b(2 + 2r + r^2))X^3 + \mathcal{O}(X^4),$$

the fixed points of which are given by

$$X_*\Big(r(2+r) + ar(1+r)X_* \\ - (1+r)(2a^2 + b(2+2r+r^2))X_*^2\Big) = 0, \tag{15}$$

i.e., $X_* = 0$ (a fixed point of f is of course also a fixed point of $f \circ f$) and two other solutions that are easily checked to be of order $\sqrt{r}$. Keeping only terms of order $r^{3/2}$ for consistency, we obtain the simplified evolution equation

$$X_{n+2} = (1+2r)X_n - 2(a^2 + b)X_n^3, \tag{16}$$

6. Beyond Periodic Behavior

where we can notice the absence of quadratic term (rX^2 of order r^2 in (15)) and the presence of a contribution from the cubic term in (14), which, being of order $r^{3/2}$, is indeed relevant.

The simplified fixed point equation reads

$$r - (a^2 + b)X_*^2 = 0 \qquad (17)$$

so that nontrivial fixed points are given by $X_\pm = \pm\sqrt{r/(a^2+b)}$. It may be remarked at this stage that if we want a higher order approximation, e.g., exact at order r, we must take into account not only the term of order rX^2 neglected above but also a term of order X^4 including the contribution of a supplementary term cX^4 in (14).

Fixed points of $f \circ f$ define a period-two (discrete time) orbit for f since $f(X_+) = X_-$ and $f(X_-) = X_+$ as trivially checked at lowest order since, in (14), quadratic and cubic terms are already of order higher than $\sqrt{r}$ and thus have to be dropped. The stability of this two-cycle is straightforward. Assuming $X_0 = X_+ + \delta X_0$ and using (16–17), we readily get the perturbation after two iterations of the original map f: $\delta X_2 = (1 - 4r)\delta X_0$ so that the two-cycle is stable if $|1 - 4r| \leq 1$, i.e., $-1 \leq 1 - 4r \leq 1$, i.e., $0 \leq r \leq 1/2$. For r small, only the first inequality is relevant and we see that the two-cycle is stable if it is *supercritical*. From (17), this requires $(a^2 + b) > 0$. Note that if the bifurcation is well described with cubic and higher order terms negligible in (14), the bifurcation is always supercritical and the bifurcated two-cycle is stable.

Returning to the continuous-time system of which (14) is the first return map, we note that the attractor corresponding to the two-cycle studied above is a limit cycle with a period twice as long as that of the limit cycle that becomes unstable.

We will not study the two remaining cases of strong resonance, $p/3$ and $p/4$, described by normal forms $(7a,b)$, leaving them as exercises to the reader.

4. Nature of Turbulence and Transition Scenarios

As seen above, the temporal behavior resulting from the first bifurcation of a limit cycle is still not complicated and, to summarize, up to now we have *time-independent* → *time-periodic* → *doubly periodic* or *periodic* (sub-harmonic). But when stresses are increased, more modes are expected to play a role and we must examine what can happen then.

According to Landau (1944), each new unstable mode brings a new frequency in the system, incommensurate with the set of frequencies of the previously installed modes (i.e., there is no relation of the form $\sum n_i \omega_i = 0$ with n_i nonvanishing integer numbers, positive or negative). The motion is quasiperiodic and observables vary as $g(\omega_1 t + \phi_2, \omega_2 t + \phi_2, \ldots)$, g being periodic with period 2π in each of its arguments. At a given stage of the *cascade of bifurcations* the set of frequencies $\{\omega_i\}$ can be determined, at least in principle, but the set of phases $\{\phi_i\}$ remains a function of initial conditions and, as such, completely out of control. Turbulence is then the regime reached after an infinite number of such oscillatory bifurcations.

As formulated by Ruelle and Takens (1971), objections can be raised to this *scenario*. First, from a physical point of view, though multi-periodic motion can look very complicated, correlations do not decay at large times. The Fourier spectrum of any temporal signal is made of sharp lines corresponding to the basic frequencies and their arithmetic combinations. This is contrary to the belief that turbulence is characterized by a decay of correlations (mixing) and by Fourier spectra of physical observables displaying continuous parts. From a mathematical point of view, Landau's scenario is in some sense *too linear* since a simple superposition of motions with different time-scales is assumed whereas an true *nonlinear* interaction is more likely. For two-periodicity, the result of such an interaction is the locking phenomenon to which we shall come back. In higher dimensions, other similarly nontrivial consequences can be expected.

As a concrete alternative, Ruelle and Takens proposed a modified scenario involving only a finite and small number of oscilla-

6. Beyond Periodic Behavior

tory bifurcations before the onset of *temporal chaos* understood as the very *nature of turbulence*. Their scenario rests on theorems far beyond our purpose. These theorems (Ruelle–Takens, 1971; Ruelle–Takens–Newhouse, 1978) state that dynamical systems with a three-frequency (four-frequency) quasiperiodic attractor on a three-torus (four-torus) are structurally unstable against perturbations of class $\mathcal{C}^2$ ($\mathcal{C}^\infty$). The perturbed systems have generically *strange attractors*, i.e., attractors with time-dependent orbits that are neither periodic nor quasi-periodic. The resulting aperiodic motion is chaotic in the sense that it is unpredictable in the long term owning to the *sensitive dependence of trajectories on initial conditions and infinitesimal perturbations*. Strange attractors, also called *chaotic* or *stochastic*, are characterized by correlation functions that decay at large times and by Fourier spectra of physical observables displaying continuous parts. In addition, these strange attractors are structurally stable (*robust*), i.e., the system can be perturbed while remaining chaotic.

Concretely, this means that after two bifurcations from a time-independent regime (fixed-point attractor) to time-periodic behavior (limit cycle) and then to a doubly periodic behavior (two-torus), chaotic behavior can be expected if allowed perturbations have continuous second order derivatives at most, whereas a bifurcation to three-frequency quasi-periodicity may be required if perturbations are smooth. That perturbations lacking smoothness favor chaotic behavior can be easily understood from a physical point of view, but we should insist on the fact that these theorems say nothing about the intensity of the required perturbations, and the probability of finding strange attractors.

The main merit of Ruelle–Takens theory is first to shed some light on the problem of the "nature of turbulence" in reconciling determinism and stochasticity *via* long term unpredictability associated with the instability of trajectories on a strange attractor. But beyond that, in proposing a specific scenario of transition to turbulence, it opened the way to the discovery of different routes which will now be examined.

At this stage, it should be noted that a scenario is nothing more than an attempt to extrapolate the behavior of a given system, making a plausibility assumption on a *simplified* description of the neighborhood of some observed state in phase space, and pushing this assumption at the level of a conjecture of more universal range, based on a *genericity* argument. In the following, we will point out these (often left implicit) assumptions systematically since this may be important to understand the behavior of the specific (experimental) system under study. Here, we now examine scenarios associated with the destabilization of a limit cycle and begin with the most celebrated, the *sub-harmonic cascade* issued from a 1/2-resonance, then continue with *intermittency*, which happen at a saddle-node bifurcation, and with the transition *via quasi-periodicity*. We will conclude this chapter by sketching a few other scenarios of similar importance just to give an idea of the complexity of behavior implied by the presence of nonlinearity even in systems with a small number of degrees of freedom.

5. The Sub-Harmonic Route to Turbulence

5.1. The Modeling Issue

In the spirit of what we have just said about scenarios, we return to the expression (14) of the map governing a sub-harmonic bifurcation. We suppose further that cubic and higher order terms can be neglected and that a is negative (this is just for specificity since the sign of a is irrelevant). After rescaling the X-variable, we start with

$$X_{n+1} = f_r(X_n) = -(1+r)X_n - X_n^2. \qquad (18)$$

A bifurcation toward a two-cycle takes place at $r = 0$ and the bifurcated periodic points are stable. Now, we consider this *local* model valid for r small as a *global* model acceptable in a sufficiently large neighborhood of $(r = 0, X = 0)$ and we examine the fate of this two-cycle when r is increased. A difficulty appears immediately if (18) is still to be considered as the first return map of some physical system since iteration $X_{n+1} = f_r(X_n)$ can no longer be inverted,

6. Beyond Periodic Behavior

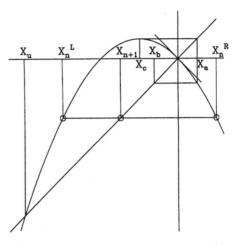

Fig. 12. Iteration (18) is non-invertible since any point X_{n+1} has two preimages, X_n^R and X_n^L, on both sides of the critical point, X_c. $f_r(X_c) = X_a$ and $f_r(X_a) = f_r^2(X_c) = X_b$ define the two ends of an interval that remains invariant as long as $X_a > X_u$.

a given X_{n+1} having two preimages in general, see Fig. 12. The difficulty arises from the reduction to a single variable X and the neglect of all transverse coordinates which play the role of internal degrees of freedom able to distinguish between states with the same value of X. As long as r remains sufficiently small, interesting phenomena occur close to the X origin; the fact that f_r is not invertible is irrelevant and additional variables are useless. This is obviously the case as long as the periodic points emerging from the sub-harmonic bifurcation remain on the right hand side of the maximum, called the *critical point* of the map (X_c in the following). Indeed, by means of simple geometrical representations of the iteration process, an *invariant interval*, i.e., an interval $I = [X_a, X_b]$ such that $f_r(I) \subset I$, can be determined with the first iterate of the maximum $X_b = f_r(X_c)$ as an upper end and its second iterate $X_a = f_r(X_b) = f_r^2(X_c)$ as a lower end (notation: $f_r(f_r(X)) = (f_r \circ f_r)(X) = f_r^2(X)$). Clearly, the lack of invertibility becomes relevant when $X_c \geq X_a$. Therefore, when the maximum belongs to the invariant interval, it is sufficient to find a way to specify which of X_n^R and X_n^L ("R" for "right" and "L" for "left" of the maximum) is the relevant preimage of a given X_{n+1}. From Fig. 12, it can be seen that the knowledge of the state at time $n - 1$ answers the question.

Therefore, we are led to add onto the right hand side of (18) a contribution proportional to X_{n-1}, i.e., we pass from a *first order* to a *second-order* difference equation

$$X_{n+1} = f_r(X_n) + bX_{n-1}, \tag{19}$$

which can be transformed to an equivalent *first-order* two-dimensional system through the introduction of an auxiliary variable $Y_n = X_{n-1}$. Equation (19) then reads

$$Y_{n+1} = X_n,$$
$$X_{n+1} = f_r(X_n) + bX_{n-1} = f_r(X_n) + bY_n,$$

which is invertible as soon as $b \neq 0$ since

$$X_n = Y_{n+1},$$
$$Y_n = \frac{1}{b}(X_{n+1} - f_r(Y_{n+1})).$$

Returning to the initial map and performing the change of variable $X + (1+r)/2 \to X$, we obtain

$$X_{n+1} = a - X_n^2 \tag{20}$$

with $a = (1+r)(3+r)/4$. This form makes explicit the fact that f_r is the prototype of functions with a *quadratic* maximum (the generic case for a maximum). Further rescaling the variable by a factor $1/a$ leads to an equivalent form to be used later,

$$X_{n+1} = 1 - aX_n^2. \tag{21}$$

Finally, including the contribution of X_{n-1}, we arrive at the *Hénon map* written here as

$$Y_{n+1} = X_n,$$
$$X_{n+1} = 1 - aX_n^2 + bY_n. \tag{22}$$

6. Beyond Periodic Behavior

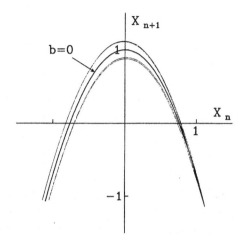

Fig. 13. The Hénon attractor for $a = 1.707, b = 0.15$, and the corresponding one-dimensional map at the limit of infinite dissipation, $b = 0$.

The Jacobian J of a map governs the expansion rate of a volume element in phase space. Here, from

$$ J = \begin{vmatrix} 0 & 1 \\ b & -2aX \end{vmatrix} = -b $$

we see that J does not depend on the point, and that the map is expanding when $|b| > 1$, which should be discarded, area preserving when $|b| = 1$ and contracting or *dissipative* when $|b| < 1$. Figure 13 displays the Hénon attractor obtained for $b = 0.15$ and $a = 1.707$ superimposed with the parabola that would be obtained for $b \to 0$ corresponding to the original noninvertible one-dimensional iteration. Provided that we keep in mind the nature of the process that leads to this reduced system, we can work without worrying further about noninvertibility.

5.2. The Sub-Harmonic Cascade

Let us notice first that for the parabolic map the second intersection of $f_r(X)$ with the diagonal also plays an important role since the very existence of an invariant interval depends on its position. Indeed, as can be checked easily, this fixed point at $X_u = -(2+r)$ is unstable for r in the range of interest (hence the subscript "u") and trajectories with initial condition $X_0 < X_u$ all diverge to $-\infty$. The invariant

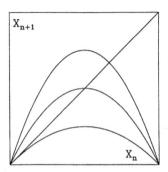 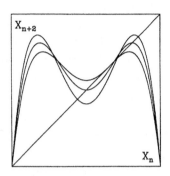

Fig. 14. The two first sub-harmonic bifurcations. Left: as deduced from the graph of $f_r(X)$, the one-cycle appears at $r = 1/4$, is superstable at $r = 1/2$ and becomes unstable at $r = 3/4$. Right: the two-cycle (the set of fixed points $f_r^2(X)$) becomes stable at $r = 3/4$, superstable at $r = 0.809\ldots$, and unstable at $r = 0.862\ldots$ with respect to a four-cycle.

interval exists as long as $X_{\rm a}$ remains larger than $X_{\rm u}$. Performing the change of variable that places this point at the origin of coordinates and rescaling the interval $[X_{\rm u}, X_{\rm v}]$ to unity (see Fig. 12), we arrive at

$$X_{n+1} = 4rX_n(1 - X_n), \qquad (23)$$

often called the *logistic map*. Parameter r in (23) above is related to the r in (18) by $r_{\rm new} = (3 + r_{\rm old})/4$ so that the sub-harmonic bifurcation that occurred at $r = 0$ for (18) now takes place at $r = 3/4$. The main advantage of this formulation is to keep the invariant set included in $[0, 1]$ and the critical point of the map (we assume $r > 0$) at a fixed location, $X_{\rm c} = 1/2$, with $f(X_{\rm c}) = r$.

Here we follow first the nontrivial fixed point of (22) given by $X_{(0)} = 1 - 1/4r$. Exchanging its stability with the origin, it becomes stable at $r = 1/4$. From $f'(X_{(0)}) = 4r(1 - 2X_{(0)}) = 2 - 4r$ we see that $f'(X_{(0)})$ decreases from 1 at $r = 1/4$ to 0 at $r_{(0)}^{\rm ss} = 1/2$, a value at which the cycle is said to be *superstable* since the convergence toward the fixed point is easily seen to be faster than geometric. At $r_{(1)} = 3/4$ the slope reaches -1 and the one-cycle becomes unstable with respect to the *two-cycle* $\{X_{(1)}^{\pm}\}$ as discussed in the previous section. (Notice that the subscript (n) indexing r and X is related

6. Beyond Periodic Behavior

n	$T_{(n)}$	$r_{(n)}$	$r_{(n)}^{ss}$
0	1	1/4	1/2
1	2	3/4	0.80901...
2	4	0.86237...	0.87464...
3	8	0.88602...	0.88866...
4	16	0.89218...	...
...	...	...	...
∞	∞	0.89248...	0.89248...

Table 1. Sub-harmonic cascade; bifurcation thresholds and superstable points.

to the period by $T_{(n)} = 2^n$.)

When r is increased, $(f^2)'$ decreases from 1 at the bifurcation point and $X_{(1)}^{(\pm)}$ migrate away from $X_{(0)}$. One of them reaches X_c for $r_{(1)}^{ss} = (1 + \sqrt{5})/4$, the value at which the two-cycle is superstable. Then, $(f^2)'$ becomes negative and reaches -1 for $r_{(2)} = (1 + \sqrt{6})/4$. Beyond this value, each fixed point is split in two and we get a *four-cycle* (see Fig. 14).

The process repeats itself again and again, the period increasing by a factor of two at each bifurcation. Therefore, we have an infinite cascade of sub-harmonic—or *period-doubling*—bifurcations. Thresholds accumulate geometrically at $r_{(\infty)} = 0.89248...$; results are summarized in Table 1 and illustrated in Fig. 15.

5.3. Universality and Renormalization

The geometric convergence described above can be characterized by its decrement

$$\delta_{(n)} = \frac{r_{(n)} - r_{(n-1)}}{r_{(n+1)} - r_{(n)}}.$$

(It could have been defined also from the superstable points.) From the table above, for the two first estimates of this ratio, using the bifurcation points (third column), we obtain $\delta_{(1)} = 4.449$ and $\delta_{(2)} = 4.751$, while the asymptotic value obtained by a careful numerical

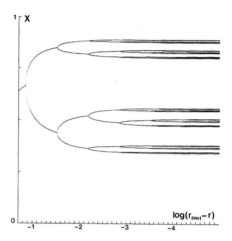

Fig. 15. Sub-harmonic cascade; periodic points are plotted against the logarithm of the distance to the accumulation point $r_{(\infty)}$.

study (Feigenbaum, 1979) reads

$$\delta_{(n)} \to \delta = 4.6692\ldots.$$

This constant is universal and depends on the map only through the order of its critical point, i.e., on the value of the exponent m in $f_r(X) \sim \left(|X - X_c|\right)^m$. The generic case corresponds to $m = 2$ which yields the value quoted above but higher order critical points can be considered, e.g., $m = 4$ or even noninteger positive ms.

In the vicinity of $r_{(\infty)}$, the period of the cycles diverges as a power of $(r - r_{(\infty)})$. Let $T(r)$ be the period of the cycle stable at a given r, and ν the exponent defined by

$$T(r) \sim (r_{(\infty)} - r)^{-\nu};\qquad(24)$$

then, numerically, one finds

$$\nu = 0.4498\ldots,$$

which can be related to δ by the following argument: T being given by (24), we have

$$\frac{dT}{dr} = \nu(r_{(\infty)} - r)^{-(\nu+1)} \quad\text{or}\quad \frac{dr}{dT} = \frac{1}{\nu}T^{-(1+1/\nu)}.$$

6. Beyond Periodic Behavior

Now, let us consider two successive period doublings. For the first one, $T/2 \to T$, we have $dT \sim T/2$ so that

$$dr(T/2 \to T) \sim \frac{1}{\nu}(T/2)^{-1/\nu};$$

for the second period doubling, $T \to 2T$, we have $dT = T$ and

$$dr(T \to 2T) \sim \frac{1}{\nu}(T)^{-1/\nu}.$$

This leads to the estimate

$$\delta = \frac{dr(T/2 \to T)}{dr(T \to 2T)} = \frac{(T/2)^{-1/\nu}}{(T)^{-1/\nu}} = 2^{1/\nu},$$

so that

$$\nu = \frac{\log(2)}{\log(\delta)}.$$

This scaling argument suggests a more sophisticated *renormalization* procedure that basically consists of making the $((n) \to (n+1))$ bifurcation look like the $((n-1) \to (n))$ bifurcation. Using this procedure recursively from infinity back to the first bifurcation should help solving the problem since the latter is easily tractable. In addition this would prove the universality of the terminal phase of the cascade. The detailed theory is somewhat beyond the scope of the present course; here we shall just give an introductory, mostly graphical, presentation.

Of course, the whole graph of $f_r^{2^{n+1}}$ cannot look like the graph of $f_r^{2^n}$ but, all their fixed points playing equivalent roles, we can restrict the study to the *local properties* of the iterated map around a single series of homologous fixed points. From Fig. 16, which displays the graphs of f_r, f_r^2, and f_r^4 at values of r which correspond to the respective superstable cycles (period 1, 2, and 4), we guess that a good choice will be given by the series of fixed points which are closest to $X_c = 1/2$. Indeed, it is easily checked that, apart from a simple flip, the deformation of f_r^2 in the boxed region of Fig. 16b for $r_{(1)} = 3/4 < r < r_{(2)}$ is similar to the deformation of f_r for $r_{(0)} =$

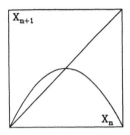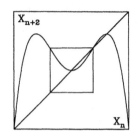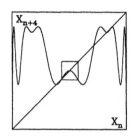

Fig. 16. Graphs of f_r at $r = 1/2$ (a), f_r^2 at $r = 0.809\ldots$ (b), and f_r^4 at $r = 0.874\ldots$ (c). The renormalization procedure maps boxed regions one on top of the other, back to (a).

$1/4 < r < r_{(1)}$ (Fig. 16a), and that the same is true for f_r^4 in the boxed region of Fig. 16c in the corresponding range $r_{(2)} < r < r_{(3)}$.

The hope is then to find a change of variable and a change of parameters that ensure a *quantitative* correspondence between maps at two successive steps of the cascade. The constant δ introduced above obviously relates to the asymptotic scaling of r; a second universal Feigenbaum constant will correspond to the scaling of the X variable, i.e., the asymptotic value of the expansion factor that allows us to map the extremum at step $(n+1)$ onto the extremum at step (n). Since the scaling affects the X variable and thus bears directly on the phase space, this procedure is called *direct-space* or *real-space* renormalization (as opposed to *reciprocal-space* renormalization introduced first in the theory of phase transitions). The first estimate obtained by rescaling the box in Fig. 16b onto Fig. 16a can be computed explicitly to yield $2/(3-\sqrt{5}) \simeq 2.618$. Asymptotically, one gets

$$\alpha_{(n)} \to \alpha = 2.5029\ldots.$$

Up to now, only a purely *phenomenological renormalization* has been performed. The aim of a real theory is then to predict the asymptotic shape of the function and the values of the two universal constants that govern the terminal phase of the period-doubling cascade. Let us sketch briefly the simplest argument and first of all perform the translation $X = \widetilde{X} + 1/2$ which places the critical point at $\widetilde{X}_c = 0$.

6. Beyond Periodic Behavior

This is desirable since the graphs of $f_r^{2^n}$ all remain symmetrical with respect to the line $X = 1/2$ so that the series of homologous fixed points considered will all stay nearest to $\widetilde{X} = 0$. The transformed iteration then reads

$$\widetilde{X}_{n+1} = r - \frac{1}{2} - 4r\widetilde{X}_n^2$$

which can be written further under the form (21) by performing the change $\widetilde{X} \to (r - 1/2)\widetilde{X}$ (except for $r = r_{(0)}^{ss} = 1/2$). Dropping the tildes, we obtain

$$X_{n+1} = 1 - aX_n^2 = f_a(X_n) \quad \text{with } a = 4r(r - 1/2).$$

The second iterate is given by

$$X_{n+2} = f_a^2(X_n) \simeq 1 - a + 2a^2 X_n^2 \tag{25}$$

(the fourth-order term $-a^3 X_n^4$ can be dropped since we try to map the neighborhood of the extremum of f_a^2, at $X = 0$, onto that of f_a). Now, we search a change of scale for X, $\widehat{X} = kX$ (more generally, an invertible change of variables $\widehat{X} = h(X)$) such that, apart from a minus sign accounting for the flip, $f_a^2(X)$ can be written as $f_{\hat{a}}(\widehat{X})$, for some new value $\hat{a}$ to be determined, i.e.,

$$h^{-1}\big(f_{\hat{a}}(\widehat{X})\big) \approx -f_a^2\big(h^{-1}(\widehat{X})\big). \tag{26}$$

Formally, we can write this *functional relation* as

$$f_{\hat{a}} = h \circ f_a^2 \circ h^{-1}.$$

Here, we obtain

$$\widehat{X}_{n+2} = k(a - 1) - \frac{2a^2}{k}\widehat{X}_n^2,$$

hence the identification

$$k = 1/(a - 1) \quad \text{and} \quad \hat{a} = 2a^2(a - 1). \tag{27}$$

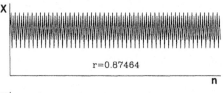
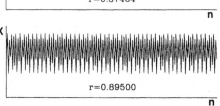

Fig. 17. Superstable four-cycle in the direct cascade compared to a noisy period-four trajectory in the inverse cascade.

This last relation defines a correspondence in parameter space, so that the phase space (the variable X) and the parameter space (the curvature of the parabola controlled by a) are both *renormalized* in the process. The phenomenological approach suggests that the shape of the graphs converges to a limit when more steps are performed down the cascade, so that when $n \to \infty$, a must be a fixed point of (27). Therefore, we obtain $a_* = 2a_*(a_* - 1)$ whose positive root reads $a_* = (1 + \sqrt{3})/2 = 1.366\ldots$. From $k = 1/(a-1)$, we have the estimate

$$\alpha = 1/(a_* - 1) = 2/(\sqrt{3} - 1) = 2.732\ldots$$

in reasonable agreement with the exact value.

The estimate for the decrement δ follows from the remark that the fixed point a_* of (27) is unstable. This relates directly to the fact that the renormalization procedure is basically a step backward in the cascade so that trajectories of (27) (in parameter space) that start in the vicinity of a_* diverge from a_*. An approximation to δ can be obtained from two successive renormalization steps $((n + 1) \to (n))$ and $((n - 1) \to (n))$ sufficiently deep in the cascade (n large, $a_{(n-1)}$, $a_{(n)}$, and $a_{(n+1)}$ close to a_*). Differences $\Delta_{(n+1)} = a_{(n+1)} - a_{(n)}$ and $\Delta_{(n)} = a_{(n)} - a_{(n-1)}$ are small and the relation between them can be determined from (27) linearized around a_*: $\Delta_{(n)} = 2a_*(3a_* - 2)\Delta_{(n+1)}$, which leads directly to

$$\Delta_{(n)}/\Delta_{(n+1)} = (1 + \sqrt{3})(3\sqrt{3} - 1)/2 = 5.732\ldots \approx \delta.$$

Approximate values obtained by this simple analysis are rather

6. Beyond Periodic Behavior

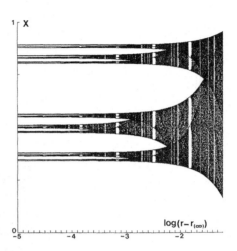

Fig. 18. In the inverse cascade, chaotic attractors (plotted against the logarithm of the distance to the accumulation point $r_{(\infty)}$ as in Fig. 15) are made of bands that merge progressively as r is increased; note the presence of periodic windows.

crude. Though the whole procedure remains basically the same, more accurate estimates can be obtained by choosing higher order fits of the extrema, coordinate changes more complicated than the simple scaling used here and specific rules to adjust the set of free parameters introduced in that way; for references to more sophisticated presentations, see Bibliographic Notes.

5.4. Beyond the Accumulation Point

The sub-harmonic cascade is a spectacular but, in some sense, a rather simple phenomenon. Let us return to map (22). Beyond $r_{(\infty)}$, the situation is rather complicated and the nature of the attractor depends sensitively on the value of r. First of all, "at" $r_{(\infty)}$, the attractor though of infinite period is not chaotic (no divergence of trajectories). It is still ordered, owing to its construction process by successive splittings. For r just below $r_{(\infty)}$, the attractor is a 2^n-cycle (Fig. 17a), whereas just above trajectories appear as *noisy 2^n-cycles* (Fig. 17b). The corresponding attractor forms a set of 2^n bands that are visited successively in the same order as the points of the corresponding periodic 2^n-cycle. The trajectories diverge from each other but chaos is very weak since their wandering is confined within the bands. When r is increased, an *inverse cascade* is observed, characterized by the merging of bands two by two (Fig. 18). The

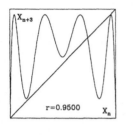
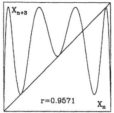
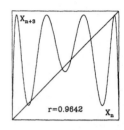

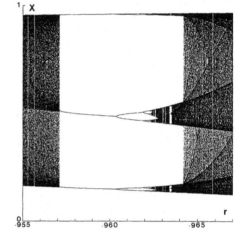

Fig. 19. Top: birth of the stable three-cycle around $r = 0.9571\ldots$. Right: evolution of the attractor as a function of r through the period-three window.

thresholds of band merging also occur in a geometrical progression accumulating at $r_{(\infty)}$ and governed by the same decrement δ.

In fact, above $r_{(\infty)}$, one can find windows in r where the attractor is periodic as exemplified in Fig. 19a for the period 3. Simple geometrical constructions show that stable cycles with odd periods cannot occur for $r < 0.9196\ldots$, and that stable cycles with even periods, no longer necessarily of the form 2^n, can be observed in the range $0.8924\ldots < r < 0.9196\ldots$. Odd periods then appear by decreasing values, the three-cycle being the last for $r = (1 + 2\sqrt{2})/4$. The order of appearance of the cycles is not random but deeply rooted in number theory; it depends simply on the fact that there is a single extremum, not even on the order of the extremum (*structural universality*).

A cycle appears *via* a saddle-node bifurcation, is stable on a limited r-interval and becomes unstable *via* a sub-harmonic bifurcation

6. Beyond Periodic Behavior

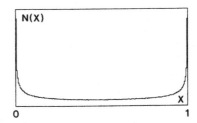

Fig. 20. Left: time series of the beginning of a very long trajectory at $r = 1$ displaying a strongly chaotic behavior. Right: histogram of the repartition of the iterates on $[0, 1]$ recorded during the simulation; notice the nearly uniform distribution in the middle of the interval and the peaks at both ends of the interval (the analysis of such distributions will be developed in the next chapter).

at the origin of a cascade strictly similar to that studied above, except that the basic period is some integer T and that the subsequent periods are of the form $T \times 2^n$. This is exemplified in Fig. 19b which displays the evolution of the attractor in the window corresponding to the three-cycle (which is the widest and the easiest to detect).

Periodic windows can be difficult to identify among the set of chaotic regimes which has positive measure in the parameter space. This complicated behavior persists up to $r = 1$, at which the whole interval $[0, 1]$ is visited as illustrated in Fig. 20.

When $r > 1$, the interval $[0, 1]$ is no longer invariant and iterates always escape to $-\infty$ through a narrow gate of width $\mathcal{O}(\sqrt{r-1})$ opened around the critical point; this is the *crisis* phenomenon to be examined in Section 8. A very long chaotic transient with properties analogous to that at $r = 1$ can be observed before the iterates leave the interval.

5.5. Concluding Remark

The universal properties of the "ideal" cascade presented in previous sections has much more than an academic interest. Indeed, many systems can become chaotic *via* a sub-harmonic cascade. When the system is conservative different exponents are found, e.g.,

$\delta = 8.721\ldots$ instead of $4.669\ldots$, but as soon as some dissipation is introduced, results obtained in the infinite dissipation limit become relevant since the effective dissipation increases dramaticly with the order of the cycle as the cascade proceeds (in the Hénon model, with $|b| < 1$, for a cycle of period 2^n, the effective Jacobian b^{2^n} tends to zero when n increases). A crossover from a nearly conservative situation to a fully dissipative situation is expected so that the terminal phase of the cascade is always governed by the one-dimensional results.

This scenario has been observed many times in laboratory experiments (convection, chemistry of oscillatory reactions), in analog computations (more or less *ad hoc* electronic circuits simulating Josephson junctions), and in digital computations (numerical simulations of differential systems, e.g., the Lorenz model for r around 145), and its universal characteristics *both below and beyond* the first accumulation point are now well documented.

6. Temporal Intermittency

6.1. Modeling of Intermittent Behavior

Up to now, we have assumed that the sub-harmonic bifurcation was supercritical. This is the case as long as the third-order term in (14) is either negligible or such that $(b + a^2) > 0$. When this is not true, the bifurcation is subcritical and no stable two-cycle exists above the threshold. For the sub-harmonic cascade only an extension of the validity range of map (18) was required but everything could be described *locally* in phase space. The scenario associated with the subcritical case is, in some sense, of a different nature since it calls for assumptions on the *global structure* of the phase space. Since nothing stable exists *in the neighborhood* of the fixed point that loses its stability two cases can occur: either some (simple) attractor exists elsewhere and we observe a discontinuous transition to the corresponding new regime, or no such thing exists and trajectories wander in phase space, visiting from time to time the neighborhood

6. Beyond Periodic Behavior

of the fixed point. This last condition is easily accounted for by using some periodic manifold, e.g., a circle since we deal here with a single real variable. On the other hand, the wandering in phase space, supposed to uncorrelate trajectories in the long term, can easily be modeled by a permanent stretching, e.g., $X \to 2X$ (mod 1). What is basically needed is only the possibility of a reinjection close to the fixed point, so that different maps having this property can be invoked to explain the concrete behavior observed in specific experiments. They are often only conjectured but sometimes they can be empirically determined, e.g., for the Lorenz model.

When the loss of stability of a limit cycle is considered and the bifurcation is not supercritical three main situations can occur, each being related to one of the three generic ways the eigenvalues of the linearized Poincaré map have to leave the unit disc. Type I intermittency will correspond to the saddle-node bifurcation that generically occurs when the eigenvalue is real and gets out at $+1$, type II to a subcritical Hopf bifurcation at a non-resonant eigenvalue with unit modulus, and type III to a subcritical sub-harmonic bifurcation (real eigenvalue -1). Note that type I intermittent behavior is not bound to chaos but can also appear in the context of quasi-periodicity (see Section 7). Intermittency can also appear in case of subcritical strong resonance, especially at a resonance $1/3$ since no stable periodic cycle exists generically above the threshold but we shall only consider the three "classical" situations, beginning with the two real cases (I and III), which are easier to identify experimentally, and examining the complex case afterwards.

6.2. Type I Intermittency

The *local* expression of the first return map governing the saddle-node bifurcation of a pair of $1/1$ resonant limit cycles generically reads

$$X_{n+1} = f_r(X) = r + X_n + aX_n^2 \tag{28}$$

with, e.g., $a > 0$. Rescaling both X and r we can set $a = 1$ without loss of generality. This will be assumed tacitly in the following.

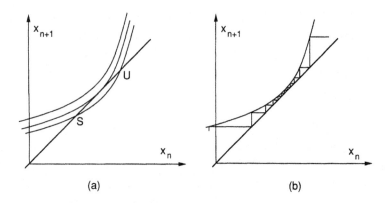

Fig. 21. The pair of stable and unstable fixed points present below the bifurcation threshold, merge and disappear (a). After the saddle-node bifurcation, trajectories visiting the vicinity of $X = 0$ evolve very slowly in the narrow channel between the graph of f_r and the first diagonal (b).

When r is negative, map (28) has two fixed points given by $X_\pm = \pm\sqrt{-r}$. Their stability is controlled by $f'_r(X_\pm) = 1 + 2X_\pm$ from which it follows that X_- is stable and X_+ unstable. The stable fixed point attracts trajectories starting with $X_0 < X_+$ (Fig. 21a, curve 1). When $r = 0$ the two fixed points merge into a semistable fixed point. Trajectories converge to it for $X_0 \leq 0$ and diverge from it for $X_0 > 0$ (curve 2). This asymptotic state is then particularly sensitive to fluctuations and can be considered as physically unstable since infinitesimal fluctuations are unavoidable. Finally, when r is positive, the fixed points have disappeared and a narrow "tunnel" has opened between the first diagonal and the graph of f_r (Fig. 21a, curve 3). In this latter case, iterates now travel slowly through the tunnel, first approaching the origin ("ghost" of the fixed point) and then escaping (Fig. 21b).

As stated above, what happens next depends on the global structure of the phase space. If reinjection at the entrance of the tunnel is allowed and the dynamics far from the origin is mixing, then the global behavior is chaotic with a succession of long regular *intermissions*, also called *laminar phases*, interrupted by short, seemingly random, *turbulent bursts*.

6. Beyond Periodic Behavior

Close to the intermittency threshold, the system spends most of its time in the laminar phases and the turbulent bursts appear comparatively short. The distribution of the lengths of laminar phases drops to zero above an upper bound easily obtained by the following argument:

In the tunnel, successive iterates are very close to each other so that it is legitimate to replace the original difference equation by a differential equation

$$X_{n+1} - X_n = \frac{\delta X}{(\delta n = 1)} = r + X_n^2 \quad \Rightarrow \quad \frac{dX}{dn} = (r + X^2),$$

n being understood as a continuous variable. This equation can be integrated to yield:

$$n\sqrt{r} = \arctan(X/\sqrt{r}).$$

The beginning of a laminar phase corresponds to X "far" from the origin, i.e., $X \to -\infty$, and the end to $X \to +\infty$. Since $\arctan(x \to \pm\infty) = \pm\pi/2$, the maximum length is given by

$$N_{\max} \sim \pi/\sqrt{r}.$$

Of course the numerical factor π must not be taken seriously but the dependence on r is correct. More generally, the duration of a laminar phase beginning at X_0 is given by:

$$N(X_0) = \frac{1}{\sqrt{r}}\left(A - \arctan(X_0/\sqrt{r})\right),$$

where A is a constant depending only on the criterion used to decide when a laminar phase has finished, e.g., simply taking the cut off at $X = 1/2$ yields $A = \arctan(1/2\sqrt{r}) \sim \pi/2$ for r small enough. The detailed statistics of the lengths of laminar phases depends on the reinjection process. As a first guess, one can a assume a uniform reinjection probability close to the origin. The average length of laminar phases is then given by $\langle N \rangle = \int N(X_0)\, dX_0$, where the integral extends on a neighborhood of the origin. Owing to the parity of the arctangent function, the contribution of the X_0-dependent term drops out for a neighborhood symmetrical with respect to the origin, e.g., $[-1/2, 1/2]$. A similar $r^{-1/2}$ behavior is then expected for the average length of the laminar phases. The frequency of the bursts increases as the inverse of this mean length, i.e., as $r^{1/2}$, and with it, the global amount of chaos since the turbulent bursts are supposed to uncorrelate two successive laminar phases (this will be made more precise in the next chapter, after the introduction of Lyapunov exponents).

This scenario has been observed in numerical simulations (e.g., the Lorenz model for $r \sim 166$) as well as in laboratory experiments (analog electronic devices, convection, ...). Predictions of this simple theory have been checked with success.

6.3. Type III Intermittency

The bifurcation through an eigenvalue -1 (resonance 1/2) is generically governed by an expression of the form

$$X_{n+1} = f_r(X_n) = -(1+r)X_n + aX_n^2 + bX_n^3 + \ldots$$

and the nature of the bifurcation depends on the sign of parameter $a^2 + b$. This parameter is related to the *Schwartzian derivative* defined as

$$Sf = \frac{f'''}{f'} - \frac{3}{2}\left(\frac{f''}{f'}\right)^2.$$

Here we have $(a^2 + b) = -Sf(X = 0)/6$. The interesting property of this somewhat strange quantity is the preservation of its sign under map composition when this sign is constant over an invariant interval, i.e., $\text{sgn}(Sf) \equiv \text{sgn}(S(f \circ f)) \equiv \text{sgn}(S((f \circ f) \circ (f \circ f)))\ldots$. When Sf is negative, the bifurcation is supercritical and this property explains why the sub-harmonic cascade of the logistic map proceeds up to infinity. When Sf is positive, the bifurcation is subcritical and type III intermittency occurs if reinjection close to the fixed point at $X = 0$ is allowed.

6. Beyond Periodic Behavior

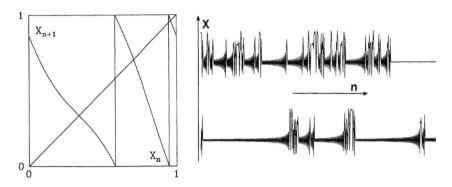

Fig. 22. Model (29) displays type III intermittency for $r > 0$: a) graph for $r = 0$; b1) chaotic transient below threshold, $r = -0.001$ b2) intermittent signal above the threshold, $r = +0.001$.

As a concrete example of a model including the reinjection property, we take

$$X_{n+1} = 1 - 2X_n - \frac{1}{2\pi}(1-r)\cos\left(2\pi(X_n - 1/12)\right) \pmod 1 \quad (29)$$

illustrated in Fig. 22.

This map has a fixed point at $X_* = 1/3$. Close to this fixed point, it expands as

$$\delta X_{n+1} = -(1+r)\delta X_n + b\delta X_n^3 \quad \text{with} \quad b = -\frac{4\pi^2}{6}$$

and, since b is negative the bifurcation is subcritical. X_* is linearly stable for $r < 0$ and unstable for $r > 0$. Once iterates have left the vicinity of X_*, they enter a chaotic burst governed by the expanding part of the map $1 - 2X \pmod 1$ which redistributes iterates evenly on $[0, 1]$ so that we can assume a uniform probability of reinjection in the neighborhood of X_*.

The signature of type III intermittency is readily obtained from considerations parallel to those used in the type I case, with one supplementary property—the existence of long-lived turbulent transients below the threshold associated with the metastable character of the fixed point for $r < 0$ and small. Indeed, in that case, trajectories converge to X_* only after one iterate has fallen between the two unstable fixed points of f_r^2. The probability of such an event is roughly proportional to the width of this interval which varies as $\sqrt{r}$ so that we expect the duration of transient to diverge as $1/\sqrt{r}$.

To determine statistical properties above the intermittency threshold we turn to f_r^2 which generically reads

$$X_{n+1} = (1+r)X_n + X_n^3 \qquad (30)$$

after appropriate rescalings of X, r, and n (taking one point every two to get rid of the alternating behavior typical of the 1/2-resonance). Notice that Equation (30) would account for a bifurcation at an eigenvalue $+1$ with the persistence of the fixed point at $X = 0$ and the absence of quadratic term imposed by a parity invariance.

The length of a laminar phase starting at $X = X_0$ can be found by integrating the differential equation equivalent to (30) as long as X_n remains close to the origin and thus evolves slowly

$$\frac{dX}{dn} = rX + X^3,$$

which gives

$$\frac{X}{\sqrt{r+X^2}} = \frac{X_0}{\sqrt{r+X^2}} \exp(rn).$$

Assuming a cut-off at $X = 1$ and neglecting r when compared with 1, we obtain

$$N(X_0) = \frac{1}{2r} \log\left(\frac{r+X_0^2}{X_0^2}\right) \qquad (31)$$

from which we infer the existence of two regimes according to whether $X_0 \ll r$ or X_0 "large". In the first case, we get

$$X_0 \ll r \quad \Rightarrow \quad N \sim \frac{1}{2r} \log\left(\frac{r}{X_0^2}\right)$$

6. Beyond Periodic Behavior

which means that, apart from a logarithmic correction, the length of the laminar phase is controlled by the linear term in (30). In the second case, expanding the logarithm we have simply

$$X_0 \gg r \quad \Rightarrow \quad N \sim 1/2X_0^2.$$

It should be noted that, in contradistinction with the case of type I intermittency, the maximum length is not bounded from above since the "initial condition" for any new laminar phase can fall very close to the unstable fixed point.

Provided that the reinjection is uniform in the neighborhood of the unstable fixed point, the distribution of lengths can be determined simply by writing that the number $\mathcal{D}(N)$ of laminar phases of length comprised between N and $N+dN$ is proportional to the interval $[X_0, X_0 + dX_0]$ of the corresponding initial conditions:

$$\mathcal{D}(N) = \frac{dN}{dX_0}, \tag{32}$$

N being given by (31). It may be more convenient to consider the number of laminar phases longer than a given value N defined formally as

$$\mathcal{N}(M > N) = \int_N^\infty \mathcal{D}(M)\,dM,$$

which correspond to initial conditions taken in the interval $[0, X_0(N)]$, where $X_0(N)$ is the inverse function of $N(X_0)$. The quantity $\mathcal{N}(N)$ is usually much less noisy than $\mathcal{D}(N)$, which may be interesting in an experimental perspective.

As for type I intermittency, the growth of chaos can be estimated from the frequency of the turbulent bursts, which can be inferred from the mean length of laminar phases. With the assumption of uniform reinjection, from (31) we obtain

$$\langle N \rangle = \frac{1}{2r} \int_{X_{\min}}^1 \log\left(\frac{X^2+r}{X^2}\right) dX$$

$$= \frac{1}{2r} \left[X \log\left(\frac{X^2+r}{X^2}\right) + 2\sqrt{r} \arctan\left(\frac{X}{\sqrt{r}}\right) \right]_{X_{\min}}^1,$$

where $[F(X)]_A^B$ means $F(B) - F(A)$ and $X_{\min}$ is the minimum value possible for X_0 according to the fact that the experiment lasts a finite amount of time. From the expression above, it can be seen that the value of $X_{\min}$ plays no role as long as r is not vanishingly small and that $\langle N \rangle \sim 1/\sqrt{r}$, so that the amount of chaos grows as $\sqrt{r}$.

All these predictions can be checked using model (29). Type III intermittency has been observed in convection experiments and in simulations of differential systems.

6.4. Type II Intermittency

The last remaining case corresponds to a subcritical Hopf bifurcation at a complex eigenvalue. As shown previously, the normal form reads

$$Z_{n+1} = \lambda Z_n - g|Z_n|Z_n \quad \text{with } |\lambda| = 1 + r \quad (33)$$

and the nature of the bifurcation depends on the sign of $\lambda g^* + \lambda^* g$. When this quantity is positive, the bifurcation is supercritical, which gives rise to a doubly-periodic behavior on an invariant torus. When it is negative, the bifurcation is subcritical and, if reinjection is allowed close to the origin, we have type II intermittency. This reinjection can occur either isotropically or anisotropically close to the origin. Isotropy was assumed in the first studies but requires more stringent conditions on the dimension of the phase space since it implies that the stable manifold of the system at the origin is at least two-dimensional, so that the minimum dimension for a continuous time system is $2 + 2 + 1 = 5$ (center manifold + stable manifold + 1 for the direction normal to the section). In the anisotropic case, $d = 4$ is enough since reinjection takes place along a one-dimensional manifold (see Bibliographic Notes).

Below the threshold, the origin is locally stable, and, in some sense, isolated from the rest of the phase space by the (unstable) invariant circle. Trajectories which end at the origin are turbulent transients, the duration of which is expected to diverge since the basin of attraction of the origin shrinks as $\sqrt{r}$ when r tends to 0.

6. Beyond Periodic Behavior

The statistical signature of type II intermittency can be obtained by following the same lines as before if one can neglect the rotation around the origin of the complex plane (remember that $\lambda \sim \exp(2\pi i \alpha)$) and consider only the evolution of the modulus, which after rescaling, is governed by an expression of the form

$$\rho_{n+1} = (1+r)\rho_n + \rho_n^3.$$

The length of a laminar phase beginning at distance ρ_0 from the origin is then given by strictly the same formula as for type III intermittency. Moreover, for a one-dimensional reinjection, nothing changes. If reinjection is isotropic, we can assume that the probability of having a laminar phase of length greater than a given value N is proportional to the surface of radius $\rho_0(N)$, i.e., as $\rho_0(N)^2$, which leads to some modifications regarding the distribution of the lengths of laminar phases. With the surface element being $2\pi\rho\,d\rho$ in polar coordinates, we now have

$$\langle N \rangle \propto \frac{1}{2r} \int_{\rho_{\min}}^{1} \log\left(\frac{\rho^2 + r}{\rho^2}\right) \rho\,d\rho$$

which leads to

$$\langle N \rangle \propto \log(1/r).$$

Accordingly, the growth of chaos is expected to be very slow, of order $1/\log(1/r)$.

Models can be built to check this prediction. By analogy with those defined to test other types of intermittency, it can be convenient to use a periodic manifold on which the dynamics is governed by (33) close to the origin smoothly connected to an expanding map far from it, e.g., $Z \to 2Z$ in the example studied. Numerical simulations suggests that the naive theory is slightly incorrect, in that chaos grows as a power law with a small exponent rather than logarithmically. For this specific model the origin of the discrepancy may be attributed to the neglect of rotation since, by considering $X = \rho^2$ with weight $dX = \rho\,d\rho$ we are led to the strictly one-dimensional model:

$$X_{n+1} = (1+r)X_n + X^2 \quad \text{with} \quad X \geq 0, \tag{34}$$

for which the argument predicting the logarithmic behavior works perfectly (note that (34) can be viewed as a member of a whole family of nongeneric models with a nonlinear term of the form $|X|^q$). There are still no experimental examples of type II intermittency with isotropic reinjection. Statistical properties in the anisotropic case have been checked on a Poincaré map issued from a periodically forced three-dimensional differential system.

7. Quasi-Periodicity

7.1. Introduction to the Locking Phenomenon

The quasi-periodic case is more complicated than the strongly resonant cases studied up to now mainly because complex variables are needed. Here, we first describe weak resonances in more detail, analyzing the role of the correction which makes this case different from the non-resonant case.

Assuming that $\lambda = \lambda_0(1 + r(u+iv))$, where $\lambda_0 = \exp(2\pi\alpha_0)$ with $\alpha_0 = p/q$ and $q \geq 5$, we already know that, as long as we neglect the correcting term $b\bar{Z}_n^{q-1}$ in the normal form, we obtain the same description of the bifurcation as in the irrational case at lowest order. In this latter case, we have

$$Z_{n+1} = Z_n \left(\lambda - \sum g_m |Z_n|^{2m}\right).$$

Inserting $Z_n = \rho_n \exp(2\pi i \theta_n)$ into this equation, at lowest nontrivial order we obtain

$$\rho_{n+1}^2 = \rho_n^2 \left(|\lambda|^2 - \sum_m (\lambda \bar{g}_n + \bar{\lambda} g_n)\rho_n^{2m}\right),$$

which yields the fixed-point equation for ρ

$$|\lambda|^2 - 1 = \sum_m (\lambda \bar{g}_m + \bar{\lambda} g_m)\rho_*^{2m} + \ldots,$$

6. Beyond Periodic Behavior

so that, after the decay of transients, the phase $2\pi\theta$ of Z is seen to rotate at constant speed

$$\theta_{n+1} = \theta_n + \alpha$$

where $2\pi\alpha$ is the argument of the complex number $\lambda - \sum_m g_m \rho_*^{2m}$.

But since α_0 is rational, the neglected term becomes relevant and the equation for ρ_{n+1}^2 now depends explicitly on θ_n. We obtain:

$$\rho_{n+1}^2 = |\lambda|^2 \rho_n^2 - \sum_{m; 2m \leq q-2} (\lambda \bar{g}_m + \bar{\lambda} g_m) \rho_n^{2m+2}$$
$$+ \rho^q (\lambda \bar{b} \exp(-2\pi i q \theta_n) + \bar{\lambda} b \exp(2\pi i q \theta_n) + \mathcal{O}(\rho^{q+1}) \quad (35)$$

or, taking the square root,

$$\rho_{n+1} = |\lambda| \rho_n - \sum \tilde{g}_m \rho_n^{2m+1} + \tilde{b}(\theta_n) \rho_n^{q-1},$$

where coefficients $\tilde{g}_m$ can be calculated from combinations of λ and the g_ms, and where $\tilde{b}(\theta)$ is a periodic function of θ with period q issued from the exponential terms in (35). The evolution of the modulus thus remains dominated by the constant contribution of order $\sqrt{r}$ arising from the phase-independent terms but a slight modulation of absolute magnitude $r^{(q-1)/2}$ coming from the periodic term is superimposed to it. As can be easily understood, after the decay of transients, the evolution of the phase must also involve two corrections to the rotation at speed α_0 derived from the linear stability analysis. The first one, coming from the dominant contribution to ρ^2, is a constant of order r. The other derives from the modulation and, for a basically dimensional reason, only its relative magnitude, i.e., $r^{(q-1)/2}/r^{1/2} = r^{(q-2)/2}$, can appear in the effective evolution equation, so that

$$\theta_{n+1} = \theta_n + \alpha_0 + \alpha_1 r + r^{(q-2)/2} \tilde{c}(\theta_n). \quad (36)$$

The phase-independent correction $\alpha_1 r$ contains two contributions: one linear since the argument of the eigenvalue λ is a function of r

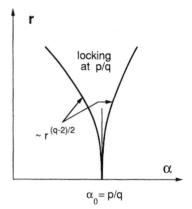

Fig. 23. Slightly above the threshold, the presence of a resonance of order q is felt in a region of width $\delta\alpha$ growing as $r^{(q-2)/2}$ around the resonant value $\alpha_0 = p/q$.

generically, and the other nonlinear coming from the ρ^2 term. Like the function $\tilde{b}$, $\tilde{c}(\theta)$ is periodic with period q.

At this stage, it is important to note that a supplementary reduction has been performed since variable the ρ no longer appears explicitly. The nature of the asymptotic regime can then be decided from the nature of the orbits of the dynamical system defined on T^1 by (36). This remains true as long as modulations are sufficiently small but, when nonlinearities become strong enough, the full bidimensional character of the problem has to be restored, especially to account for the transition to chaos, see Section 7.3. Considering a small neighborhood of the linear threshold, but assuming that the frequency can be controlled independently, from (36) we can estimate the range of α influenced by the presence of the resonance at $\alpha_0 = p/q$ to be $\mathcal{O}(r^{(q-2)/2})$. This defines *tongues* in the (α, r) parameter space as pictured in Fig. 23. A tongue is expected to start at any rational value, but from the power count above, it is easily understood that only the tongues associated with the simplest rationals, i.e., those with sufficiently small denominators, will be detectable.

Before discussing the behavior associated with this tongue structure in the next section, let us consider the following concrete example of two-dimensional map:

$$X_{n+1} = (C + r)X_n - SY_n$$
$$Y_{n+1} = SX_n + (C + r - X_n^2)Y_n \qquad (37)$$

6. Beyond Periodic Behavior

where $C = \cos(2\pi\alpha)$ and $S = \sin(2\pi\alpha)$. Its expression is analogous to that of a Van der Pol oscillator (see Chapter 2, Section 4.5). Trajectories starting in the neighborhood of the origin remain bounded for r sufficiently small and a Hopf bifurcation is easily seen to occur at $r = 0$. Results are presented in Fig. 24 for $r = 10^{-4} > 0$ and $2\pi\alpha = 0.5$ (i.e., $\alpha = 1/4\pi = 0.079577\ldots$, irrational) after the elimination of the transient. As expected, the attractor in Fig. 24a forms a closed continuous curve that is hardly distinguishable from a true circle. Modulations of the "radius" of this curve are presented as function of $2\pi\theta_n$ in Fig. 24b and deviations from a uniform rotation measured by $2\pi(\theta_{n+1} - \theta_n)$ in Fig. 24c. In fact, this value of α is roughly at equal distance from $1/12 = 0.083333\ldots$ and $1/13 = 0.076923\ldots$. From Fig. 24d in which the three attractors are displayed, we see that the resonant cases correspond to sets of 12 and 13 points, visited in a periodic way and it seems clear that when the attractor is periodic, its points stand on an invariant attracting closed curve depending smoothly on the parameters of the system. Moreover α can be varied slightly around these rational values while keeping a periodic attractor. This is precisely the *locking phenomenon* to be examined in the next section. Notice that we should not be surprised to see no trace of the "promised" periodicity of the corrections to the radius and the rotation rate in Fig. 24 that rather displays the order-two symmetry linked to the presence of the X^2 term in the primitive model. Indeed, this is simply due to the fact that the reduction to the normal form corresponds to a filtering process which cannot be performed directly on the raw results of a numerical simulation.

7.2. The Winding Number and the Structure of Lockings

In order to make the characterization of the attractor more precise we have to define a measure of the asymptotic rotation rate around the invariant attracting closed curve to which the dynamics can be restricted. The difference $\theta_{n+1} - \theta_n$ used above in Fig. 24c can be viewed as an instantaneous rotation rate, the temporal average of which will give us the required information. Therefore, the dynamics

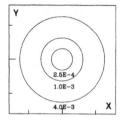 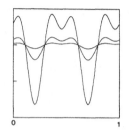 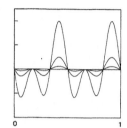

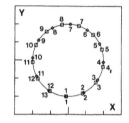

Fig. 24. a) For r small and $\alpha = 1/4\pi$, the attractor of model (37) is topologically a closed curve (a); moreover deviations from a true circle toured at constant speed are very small: b) radius variations; c) the rotation-rate variations. Resonances manifest themselves by converting the attractor into sets of points periodically visited; here, resonances $\alpha = 1/12$ ($\diamond$) and $\alpha = 1/13$ ($square$) with $\alpha = 1/4\pi$ (line) for $r = 0.001$ (d).

will be quasi-periodic or periodic according to whether the quantity

$$\hat{\alpha} = \lim_{N\to\infty} \frac{1}{N} \sum_{n=0}^{N-1} (\theta_{n+1} - \theta_n),$$

called the *winding number* or the *rotation number*, is irrational or rational.

The generic behavior of the winding number as a function of both the nominal or 'bare' rotation rate α and the intensity of nonlinear couplings is best analyzed using the map

$$\theta_{n+1} = f_{\alpha k}(\theta_n) = \theta_n + \alpha - \frac{k}{2\pi} \cos(2\pi\theta_n) \qquad (38)$$

proposed by Arnold. Here k measures the intensity of the nonlinearities. More general periodic perturbations could be added to the uniform rotation at rate α, e.g., that which has been extracted empirically for model (37), but, owing to the universality of the phenomenon, the simplest will be the best.

The mechanism of the *locking* can be attributed to the persistence of fixed points of the map upon perturbation of the system in

6. Beyond Periodic Behavior

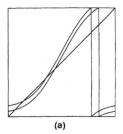 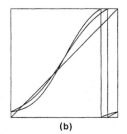 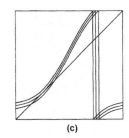

(a)   (b)   (c)

Fig. 25. The locking phenomenon results from the structural stability of the fixed points of map (38) describing the evolution of the phase; their existence is robust to sufficiently small variations of α or k: a) $k = 0.8$, $\alpha = 0.05$ and 0.1; b) $\alpha = 0.05$, $k = 0.5$ and 0.8; c) the unlocking threshold is determined from the tangency condition illustrated here for resonance $\alpha_0 = 0$ at $k = 0.8$ yielding $\alpha_{\lim} \simeq 0.1273$.

the (α, k) plane, as illustrated in Fig. 25a,b for $\alpha \sim 0$. From that figure, it becomes clear that fixed points arrive by pairs and that the *unlocking* results from the collapse of pairs of stable and unstable fixed points. Beyond the unlocking threshold, the quasi-periodic regime is characterized by a nonchaotic type I intermittent behavior, as expected at a saddle-node bifurcation with reinjection (Fig. 25c).

The threshold of this saddle-node bifurcation defines the boundary of the tongue. For resonance $\alpha_0 = 0$ it is obtained from the condition that the point with $f'_{\alpha k} = 1$, i.e., $\theta = 1/2$ lies right on the first diagonal $\theta = f_{\alpha k}(\theta)$, that is to say $\alpha_{\lim} = k/2\pi$ (see Fig. 25d). The case of a q-resonance can be solved numerically by a similar argument bearing on $f^q_{\alpha k}$.

The nature of the attractor as a function of the nominal winding number α can be inferred from the bifurcation diagram in Fig. 26 (top) which displays the set of points visited during a long numerical simulation (2000 iterations) after the elimination of transients. A quasi-periodic regime is represented by a continuous line (as far as it can be really featured in any numerical experiment) and a periodic attractor by a finite set of points. The occurrence of lockings is apparent. Quantitatively speaking, the "dressed" winding number $\hat{\alpha}$ which includes the time-averaged effect of nonlinearities is given

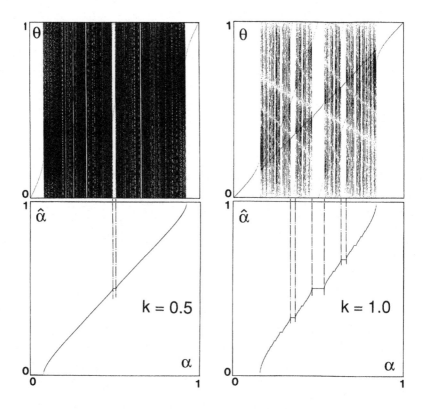

Fig. 26. Bifurcation diagrams (top) and winding numbers (bottom) for $k = 0.5$ (left) and $k = 1$ (right) as a function of the nominal winding number α. In the bifurcation diagrams, quasi-periodic regimes present themselves as continuous lines and periodic regimes as finite sets of points. Lockings appear as steps in the function $\hat{\alpha}(\alpha)$, called a *devil staircase*.

as a function α in Fig. 26 (bottom). This function is called a *devil staircase* since it can be shown to be continuous but with infinitely many discontinuities, one at every rational number. Plateaus are only clearly visible at the simplest rationals but the global structure is self-similar. For $k = 0$, the total length (the Lebesgue measure) of the plateaus is vanishing since there is no locking due to nonlinearities (rationals associated with periodic behavior form a set of measure zero among real numbers). For $k > 0$ and small, individual

6. Beyond Periodic Behavior

lockings acquire finite widths but a finite part of the unit interval remains left to true quasi-periodic regimes; the devil staircase is said to be *incomplete* (Fig. 26, left). When k increases, the fraction of unlocked regimes decreases and tends to zero for $k \to 1$. The devil staircase is then *complete* (Fig. 26, right). Quasi-periodic regimes then appear only for α belonging to a Cantor set of vanishing measure, which is consistent with the fact that continuous lines are no longer visible in the corresponding bifurcation diagram, Fig. 26 (top-right) The critical limit $k = 1$ displays universal features reminiscent of—and studied in the same renormalization framework as—those present at the accumulation point of the sub-harmonic cascade.

7.3. The Breakdown of a Two-Torus

As long as $k < 1$, map (38) is invertible and can account for the dynamics of the phase reliably since the *global* topological structure of the θ-space is preserved. At $k = 1$, the map develops an inflection point with horizontal tangent and for $k > 1$ it becomes noninvertible. Deep inside a resonance tongue, the overall behavior is only sensitive to the local shape of the map in the vicinity of the fixed points and not to its global aspect. The dynamics remain confined to invariant sub-intervals and we recover a situation best understood in terms of maps of a single variable as already studied in previous sections. Period doubling and subsequently chaos can then take place as before. Furthermore, beyond $k = 1$, resonance tongues of various orders overlap everywhere in the (α, k) plane, which results in a complicated behavior with the coexistence of attractors and chaos. However, as already discussed in the case of functions of a single variable displaying a maximum, though the study of the one-dimensional map (38) may still be interesting in itself, it cannot serve as a reasonable model of Poincaré map. Therefore, the transition to chaos *via* two-frequency *quasi-periodicity* cannot take place without some sort of *breakdown of the two-torus*. We have then to go back to a two-dimensional model restoring the invertibility by adding some finite dissipation. This can be done by passing to the

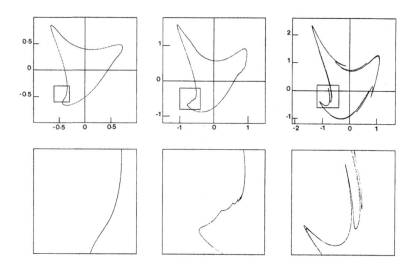

Fig. 27. Breakdown of a two-torus in the Curry-Yorke model for $\alpha = 2$ and $r = 0.27$ (left), 0.40 (middle), and 0.48 (right), with corresponding enlargements of the boxed areas (bottom).

so-called *dissipative standard map*

$$\theta_{n+1} = \theta_n + \alpha - \frac{k}{2\pi}\sin(2\pi\theta_n) + b\phi_n,$$
$$\phi_{n+1} = b\phi_n - \frac{k}{2\pi}\sin(2\pi\theta_n),$$

of which map (38) represents the infinite dissipation limit ($b = 0$).

Here, we present the transition observed in the *Curry-Yorke model* (1977) best defined as the composition of two maps: first, in polar coordinates, a rotation $\theta \to \theta + 2\pi\alpha$ and a variable stretching of the modulus $\rho \to (1+r)\log(1+\rho_n)$, and second, a coupling between the modulus and the phase written in Cartesian coordinates as $(X \to X, Y \to Y + X^2)$. The origin, which remains a stable fixed point for $r < 0$, bifurcates toward a limit cycle for $r > 0$, though not through a standard Hopf bifurcation since $\rho_* \sim r$ as easily seen from the form of $\rho_{n+1} = f_r(\rho_n)$.

In the strongly nonlinear domain, depending on the values of r and α different kinds of attractors can be observed, periodic, quasi-

6. Beyond Periodic Behavior   237

periodic or chaotic. For $\alpha = \ldots$, the transition to chaos taking place
around $r = \ldots$ is pictured in Fig. 27. Below threshold ($r = 0.27$) the
attractor is still topologically equivalent to a circle. At threshold,
infinitely many tiny wrinkles appear ($r = 0.40$). Beyond threshold,
these wrinkles develop into folds ($r = 0.48$) and the attractor that
has now gained a Cantorian transverse structure can be viewed as
a closed "curve" with finite thickness in much the same way as the
Hénon attractor is a thickened parabola. (in fact, several lockings
can be observed in the interval $0.27 < r < 0.40$.) We will stay at this
pictorial level since a detailed analysis of the process would require
a precise study of stable and unstable manifolds of fixed points of
the map.

7.4. Ruelle-Takens Scenario and n-Periodicity

One should not conclude from the previous discussion that the
Ruelle-Takens route to turbulence is irrelevant because it involves
more than two incommensurate frequencies. As a matter of fact,
it has been seen that, when nonlinearities are sufficiently weak, the
domain of true quasi-periodic regimes has finite measure, which is
equivalent to saying that quasi-periodicity has a real chance to exist
and serve as an intermediate step in the transition to chaos.

A simple way to get n-periodicity with $n > 2$ is to couple electronic oscillators, thereby testing the Ruelle-Takens route by analog
computations. Experimental examples from elsewhere in physics are
not numerous. In convection, quasi-periodic regimes with three or
more frequencies seem to be possible when nonlinear interactions
are weakened by a localization of the physical oscillators at different
places in physical space, which hinders synchronization.

The persistence of quasi-periodic behavior can easily be tested
numerically on simplified models of Poincaré maps generalizing (38)
in more than one dimension. For example, the Poincaré section of
a four-dimensional torus is a three-dimensional torus and the corresponding first return map can be expressed as a set of three coupled
iterations involving three independent phases (θ, ϕ, ψ). The unper-

turbed quasi-periodic dynamics is simply governed by:

$$\begin{aligned} \theta_{n+1} &= \theta_n + \omega_\theta & (\bmod\ 2\pi), \\ \phi_{n+1} &= \phi_n + \omega_\phi & (\bmod\ 2\pi), \\ \psi_{n+1} &= \psi_n + \omega_\psi & (\bmod\ 2\pi), \end{aligned}$$

i.e., a constant vector field on the three-torus with components: $(\omega_\theta, \omega_\phi, \omega_\psi)$ all incommensurate among themselves and with unity (which corresponds to the clock frequency of the Poincaré section), i.e., without relations of the form $n_\theta \omega_\theta + n_\phi \omega_\phi + n_\psi \omega_\psi = n$ with n_θ, n_ϕ, n_ψ, and n all integers. Perturbations to this system are then introduced in the form of additional periodic conveniently normalized functions of the three variables, $(g_\theta, g_\phi, g_\psi)$ with controllable intensity ϵ, i.e.,

$$\begin{aligned} \theta_{n+1} &= \theta_n + \omega_\theta + \epsilon g_\theta(\theta, \phi, \psi) & (\bmod\ 2\pi), \\ \phi_{n+1} &= \phi_n + \omega_\phi + \epsilon g_\phi(\theta, \phi, \psi) & (\bmod\ 2\pi), \\ \psi_{n+1} &= \psi_n + \omega_\psi + \epsilon g_\psi(\theta, \phi, \psi) & (\bmod\ 2\pi). \end{aligned}$$

The structure of the gs can be chosen at random (e.g., Fourier series with random coefficients) and, at given ϵ, a statistics can be made on the nature of the resulting attractor.

The increase of the dimension of the effective phase space opens more possibilities and in addition to total synchronization or chaotic behavior, partial synchronization can be observed, i.e., locking to quasi-periodicity with a smaller number of independent frequencies. As above, the fraction of unlocked regimes starts from unity for $\epsilon = 0$ and decays rapidly as the intensity of the perturbation is increased. The breakdown of a n-torus can then be understood by analogy with that of the two-torus.

8. Beyond "Classical" Scenarios

Scenarios discussed above are expected to be frequent since they rely on general assumptions, no degeneracy of the eigenvalues, no special symmetry, and accordingly, only one relevant control parameter (with possibly extra sign conditions bearing on the few first

6. Beyond Periodic Behavior

coefficients of some Taylor expansion, cf. period-doubling *versus* type III intermittency). These "codimension-one scenarios" can then be viewed as the elementary bricks appearing in any general theory of the behavior of systems, and especially their transitions to temporal chaos. However, they can be observed usually only at the most "microscopic" level in phase *and* parameter space, that is to say when all generic perturbations have been added to the leading singularity of the vector field in a limited region of interest in phase space, and only one unfolding parameter is varied. Very little is known at a more "macroscopic" level in parameter space, when higher codimension processes are involved, or when the global structure of the phase space is involved.

From the three elementary scenarios considered up to now, two have been seen to develop rather *continuously* in phase space: the sub-harmonic cascade and the bi-periodic route. By contrast, the third one involves a saddle-node bifurcation which, though a local process, the merging of a pair of fixed points, has necessarily global consequences since a whole basin of attraction suddenly disappears. In general, it is difficult to predict the nature of the regime beyond the threshold since this requires a knowledge of the global structure of the phase space. If reinjection is possible the overall behavior is intermittent (chaotic or not) but, more trivially, if an attracting fixed point exists in the neighborhood, it can capture trajectories; we have then the direct analogue of the saddle-node bifurcation for a one-dimensional differential system as studied in the previous chapter. The transition can well be termed *discontinuous* even in the intermittent case where the amount of chaos is seen to grow continuously. In the following, we shall present other examples of similar processes involving sudden morphological changes of chaotic attractors and formalize them under the concept of *crisis* first introduced by Grebogi, Ott, and Yorke (1982).

A simple example of *elementary crisis* is given by the tent map:

$$X_{n+1} = 2rX_n \quad \text{for} \quad X_n < 1/2$$
$$X_{n+1} = 2r(1 - X_n) \quad \text{for} \quad X_n > 1/2$$

For $r < 1$, the dynamics remain confined in an interval strictly in-

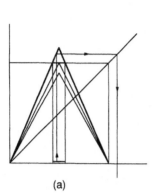 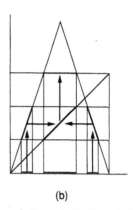

(a) (b)

Fig. 28. Elementary crisis in the tent map; a) for $r > 1$, iterates that fall in the interval of width $\mathcal{O}(r-1)$ around $X = 1/2$ diverge to $-\infty$; b) points that escape after exactly n iterations belong to the $(n-1)$th backward iterate of this interval, here for $r = 3/2$, $n = 1$ and 2 (at the limit $n \to \infty$ the classical triadic Cantor set would be obtained).

cluded in $[0, 1]$, but at $r > 1$, the $[0, 1]$ interval ceases to be invariant. The attractor has collided with the boundary of its basin of attraction. Indeed, for $r \to 1$ from below, the lower end of the attractor tends toward the lower end of its basin of attraction $X = 0$ (the unstable fixed point), hence the name of *exterior boundary crisis*.

Beyond the crisis point, all iterates diverge toward $-\infty$ (Fig. 28a) except those belonging to a Cantor set of vanishing measure complementary to the set obtained by iterating backward indefinitely the escape interval of width $\mathcal{O}(r-1)$ which opens around $X = 1/2$ (Fig. 28b).

The distribution of the duration of transients for initial conditions taken uniformly at random on the interval $[0.1]$ can be found easily. The probability of having a n-step transient is proportional to the cumulated length of sub-intervals, the nth iterates of which maps on the escape interval. Therefore, for one-step transients, we have $p_1 = (r-1)/r$; for two-step transients, $p_2 = (r-1)/r^2$, and for n-step transients, $p_n = (r-1)/r^n$ ($\sum_n p_n = 1$ since the Cantor set corresponding to infinite transients has vanishing measure). At given $r > 1$, the number of transients of length n then decays as $(1/r)^n$.

6. Beyond Periodic Behavior

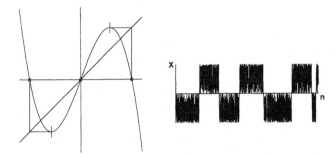

Fig. 29. Left: a crisis takes place when the two separate symmetrical attractors of the cubic map merge at $r = 3\sqrt{3}/2$. Right: the time series beyond the crisis point can be analyzed in terms of an intermittent switching between the two components of the attractor.

In a somewhat pictorial way, we can say that the dynamics during a *chaotic transient* is controlled by the existence of the limiting Cantor set defined above (a *strange repellor*), but that the system has a low probability to "learn" that it lives in the neighborhood of a set that is unstable, and that every iteration is an independent trial to find the escape window. The exponentially decaying probability distribution directly follows from this interpretation and we will not be surprised to see that the mean time-life that can be extracted from it, $\langle n \rangle = \sum_n n p_n = r/(r-1)$, diverges when the crisis point is approached from above.

The mechanism discussed above can also explain the sudden widening of a chaotic attractor or the intermittent switching between two chaotic components of an attractor. The logistic map offers examples of sudden widening after every periodic window (see Fig. 19b for the period-three window). Here we consider the case of the intermittent switching modeled by the cubic map $X_{n+1} = rX_n(1 - X_n^2)$ (Fig. 29a). For $r < r_c = 3\sqrt{3}/2$, two attractors coexist, one for $X > 0$, the other for $X < 0$, with separate basins of attraction symmetrical with respect to $X = 0$. At $r = r_c$, the unstable fixed point at $X = 0$ collides both basins which now merge into a single one. However, for r sufficiently close to r_c, an intermittent switching between the two parts of the attractor can be observed (Fig. 29b)

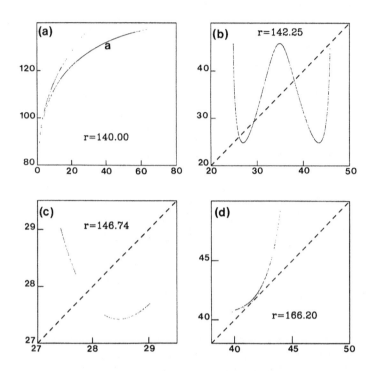

Fig. 30. Bifurcations of the Lorenz model for $\sigma = 10$, $b = 8/3$ and $140 < r < 170$: a) Poincaré map obtained by taking the $(X = 0)$ plane as a surface of section for $r = 140$, slightly below the crisis point where the attractor suddenly contracts; one-dimensional reduced maps $Y_{n+1} = f_r(Y_n)$ accounting for the bifurcation diagram: b) boundary crisis at $r = 142.24$; c) accumulation point of the sub-harmonic cascade ($r = 146.74$) for one of the possible attractors allowed by the multi-humped shape of the reduced map; d) type I intermittency at a tangent bifurcation for $r \simeq 166.07$.

with a statistical signature easily derivable from the previous analysis. Another illustration of this process can be found in the vicinity of the band merging points of the inverse cascade displayed by the logistic map.

The knowledge of properties of simple models can greatly help us to understand the nonlinear behavior of concrete systems. As an example, let us consider the Lorenz equations for $\sigma = 10$, $b = 8/3$ but for r in the range $[140, 170]$. Taking the $(X = 0)$ plane as a

6. Beyond Periodic Behavior

surface of section, we obtain a Poincaré map that looks locally one-dimensional (Fig. 30a). Globally, it displays two "sheets" (in fact two transversally fractal sets of closely packed sheets). A detailed understanding of the bifurcation diagram of the model can then be obtained from the consideration of the reduced map $Y_{n+1} = f_r(Y_n)$. As deduced from Fig. 30b, this map can reasonably be modeled by a fourth-order polynomial expression and the variations of its coefficients can "explain" the observed bifurcations of the attractor, e.g.,

- the sudden contraction of the attractor *via* a boundary crisis at $r \simeq 142.24$ (Fig. 30b);
- the sub-harmonic cascade for $r \sim 146$ (Fig. 30c);
- the explosion of the limit cycle *via* a tangent bifurcation yielding type I intermittency at $r \simeq 166.07$ (Fig. 30d).

To conclude, let us notice that for the moment we are still not able to derive from the first principles the main steps of the transition to chaos followed by a given nonlinear system. However, as soon as such a system is engaged in a specific route, it is bound to behave in a precise way, which can often be analyzed in terms of simple and generic reduced models. Therefore, it seems important to know the existence and main properties of lowest codimension scenarios and the potential sources of complications (crises) to be faced in concrete situations. In particular, the possibility of "discontinuous" transitions associated with attractor coexistence and *generalized metastability* should not be neglected besides the widely appreciated standard scenarios featuring a more "continuous" process of transition to turbulence.

9. Bibliographical Notes

General, mathematically oriented references for this chapter are:

[1] G. Iooss and D. D. Joseph, *Elementary stability and bifurcation theory* (Springer, Berlin, 1980),

[2] G. Iooss, *Bifurcation of maps and applications* (North-Holland, 1979),

[3] V. I. Arnold, *Geometrical methods in the theory of ordinary differential equations* (Springer-Verlag, 1983).

and again the book by Guckenheimer and Holmes, BN 5 [1]. The stability of limit cycles is treated in detail in the books by Arnold and Iooss (derivation of normal forms and subsequent codimension one bifurcations in all resonant cases). It may be found interesting to consult articles "Ordinary differential equations (qualitative theory)" and "Ergodic theory" in *Encyclopedic Dictionary of Mathematics* (MIT Press, 1980). At a more specialized level, one finds:

[4] G. Iooss, R. H. G. Helleman, R. Stora, eds., *Chaotic behaviour of deterministic systems*, Les Houches session XXXVI (North-Holland, 1983).

[5] G. I. Barenblatt, G. Iooss, D. D. Joseph, eds., *Nonlinear dynamics and turbulence* (Pitman, 1983).

The Lorenz model used to illustrate the recourse to Poincaré map has been defined in:

[6] E. N. Lorenz, "Deterministic nonperiodic flow," J. Atmospheric Sc. **20**, 130 (1963).

whereas its mathematical properties for a broad range of parameters are analyzed in:

[7] C. Sparrow, *The Lorenz equations: bifurcations, chaos, and strange attractors* (Springer-Verlag, Berlin, 1982).

Concerning the routes to temporal chaos, Schuster's monograph with its reference list and the collections of reprints edited by Citanović (BN 1 [5]) and Hao Bai-lin (BN 1 [6]) offer a comprehensive survey of the situation, both theoretical and experimental. For a shorter introduction, consult:

[8] J. P. Eckmann, "Roads to turbulence in dissipative dynamical systems," Rev. Mod. Phys. **53**, 643 (1981).

or

[9] H. L. Swinney, "Observation of order and chaos in nonlinear systems."

in [10] below, p.3.

[10] D. Campbell, H. Rose, eds., *Order in chaos*, Physica **7D**, Nos. 1–3 (1983).

that contains numerous interesting contributions.

6. Beyond Periodic Behavior

Seminal papers on the transition to turbulence are those of:

[11] L. D. Landau, "On the problem of turbulence," Akad. Nauk. Doklady **44**, 339 (1944). English translation in *Collected Papers of L.D. Landau*, D. ter Haar, ed. (Pergamon Press, Oxford, 1965), reprinted in Hao Bai-lin's book, BN 1 [6].

[12] D. Ruelle and F. Takens, "On the nature of turbulence," Comm. Math. Phys. **20**, 167 (1971) and Comm. Math. Phys. **23**, 344 (1971), also reprinted in BN 1 [6].

For definitions related to strange attractors, see, e.g., Guckenheimer and Holmes, Chapter 5. In fact, the Ruelle–Takens viewpoint was questioned by A. S. Monin in an article with the same title (Usp. Fiz. Nauk **125**, 97 (1978), English translation: Sov. Phys. Usp. **21**, 429 (1978)) who argued that strange attractors could not explain the presence of a continuous spatial spectrum for fluctuations in turbulent flows; hence the necessary distinction between temporal chaos and turbulence.

One should begin the study of one-dimensional maps by reading:

[13] R. May, "Simple mathematical models with very complicated dynamics," Nature **261**, 459 (1976).

A review of universal properties of the sub-harmonic cascade is given by:

[14] M. Feigenbaum: "Universal behavior in nonlinear systems," in [10] above, p.16.

Related mathematical technicalities are developed in:

[15] P. Collet and J. P. Eckmann: *Iterated maps on the interval as dynamical systems* Progress in Physics vol.1 (Birkhauser, 1980).

Hénon's model introduced here to insure the invertibility of the logistic map was originally derived and studied by:

[16] M. Hénon, "A two-dimensional mapping with a strange attractor," Comm. Math. Phys. **50**, 69 (1976).

The three generic types of intermittency are defined in:

[17] Y. Pomeau and P. Manneville, "Intermittent transition to turbulence in dissipative dynamical systems," Comm. Math. Phys. **74**, 189 (1980).

A review of properties of circle maps can be found in:

[18] P. Bak, T. Bohr, and M. H. Jensen, "Circle maps, mode-locking, and chaos," in [19] below, p.16.

[19] Hao Bai-lin, ed., *Directions in chaos*, vol.2 (World Scientific, Singapore, 1988).

see also Arnold's book. The break-down of a two-torus has been illustrated using the Curry–Yorke model defined and studied in:

[20] J. Curry, J. A. Yorke, "A transition from Hopf bifurcation to chaos: computer experiment with maps on R^2" Springer Notes in Mathematics **668** (Springer-Verlag, Berlin, 1977), p.48.

Higher dimensional problems related to the Ruelle–Takens scenario are analyzed by:

[21] C. Grebogi, E. Ott, and J. A. Yorke, "Attractors on a n-torus: quasiperiodicity versus chaos," Physica **15D**, 354 (1985).

who concentrate on the statistics of the chaotic, quasiperiodic, and periodic regimes as a function of the intensity of the perturbations.

For scenarios involving global bifurcations, see Guckenheimer–Holmes Chapter 6 and 7. The role of *homoclinic trajectories* in (at least) three-dimensional differential systems, close to a saddle-focus fixed point has been first stressed by:

[22] A. Arnéodo, P. Coullet, and C. Tresser, "Oscillators with chaotic behavior: an illustration of a theorem by Shil'nikov," J. Stat. Phys. **27**, 171 (1982).

A systematic approach to "discontinuous" scenarios has been given by:

[23] C. Grebogi, E. Ott, and J. A. Yorke, "Crises, sudden changes in chaotic attractors, and transient chaos," in [10] above, p.181.

The reduced description of the bifurcations of the Lorenz model adopted here has been initiated in:

[24] P. Manneville and Y. Pomeau: "Different ways to turbulence in dissipative dynamical systems," Physica **1D**, 219 (1980).

Chapter 7

Characterization of Temporal Chaos

In the preceding chapter, we have studied in detail the essential steps leading a given system to temporal chaos. Here we will attempt to characterize the resulting dynamics from two viewpoints. We will first look for a quantitative measure of the divergence rate of trajectories in terms of *Lyapunov exponents* (Section 1), beginning with the simplest case of a discrete-time one-dimensional system for which only stretching can take place, and then turning to higher dimensional maps and to time-continuous systems. Next, we will analyze the geometrical and probabilistic structure of the attractor in phase space, first defining a notion of *entropy* which can serve to measure the "complexity" of trajectories (Section 2.1). Then we will take a full statistical viewpoint. Such an approach aims at replacing *temporal averages* determined after having observed the evolution of the system by *ensemble averages* computed using a probability distribution in phase space; instead of following the actual system all along a single, very long trajectory, one considers the statistical evolution an *ensemble* of identically prepared systems according to a probability that gives to a given region of phase space a weight proportional to the fraction of time spent by a "typical" trajectory in that region. The values of the observables are then computed as averages according to this probability measure. The determination of such *invariant probability measures* (Section 2.2 to 2.4) is the subject of the *ergodic theory of strange attractors*.

Several illustrations of strange attractors have already been given. Their most striking feature is of course their transverse fractal structure. In Section 3 we will examine the different generalizations of the usual notion of *dimension* able to capture this aspect of their geometry. Finally, to conclude this chapter, we will consider experimental problems associated with the description of trajectories in phase space and the characterization of chaotic permanent regimes obtained in the laboratory by Lyapunov exponents and fractal dimensions (Section 4).

1. Divergence of Trajectories and Lyapunov Exponents

1.1. One-Dimensional Iteration

Let us consider the map

$$X_{n+1} = f(X_n) \tag{1}$$

and X_0 the initial condition generating the reference trajectory $\{X_i\}$. The stability of this trajectory is determined from the evolution of a neighboring trajectory starting at $\tilde{X}_0 = X_0 + \delta X_0$ (with $\delta X_0 \to 0$). After one iteration we have

$$\tilde{X}_1 = X_1 + \delta X_1 = f(X_0 + \delta X_0) = f(X_0) + f'(X_0)\delta X_0,$$

f' being the derivative of f with respect to X. The deviation is then given by

$$\delta X_1 = f'(X_0)\delta X_0.$$

After a second iteration, by the rule of chained differentiation, we get

$$\delta X_2 = f'(X_1)\delta X_1 = f'(X_1)f'(X_0)\delta X_0,$$

and at the nth step

$$\delta X_n = \left(\prod_{m=0}^{n-1} f'(X_m)\right)\delta X_0. \tag{2}$$

7. Characterization of Temporal Chaos

The evolution of the distance between the two trajectories is obtained after taking the absolute value of this product. Expecting an exponential convergence/divergence of trajectories, we assume $|\delta X_n| \sim (\gamma_{\text{eff}})^n |\delta X_0|$, where γ_{eff} is an effective rate *per* iteration step obtained from

$$\gamma_{\text{eff}} = \lim_{n\to\infty} \left(\left|\frac{\delta X_n}{\delta X_0}\right|\right)^{1/n} = \left(\prod_{m=0}^{n-1} |f'(X_m)|\right)^{1/n}, \quad (3)$$

which, once logarithms are taken, gives

$$\lambda = \lim_{n\to\infty} \frac{1}{n} \sum_{i=0}^{n-1} \log\left(|f'(X_i)|\right). \quad (4)$$

This limit, called the *Lyapunov exponent*, clearly presents itself as the time average of $\log(|f'|)$, the local divergence rate. After the transient has decayed, all physically relevant trajectories belonging to an attractor are expected to yield equivalent time averages. The so-defined Lyapunov exponent is thus expected to exist (the series converge) and to be *independent of the initial condition* taken in the basin of attraction of the attractor. Furthermore, it is easily seen to be *invariant under smooth changes of variables*. Indeed, assume that instead of measuring the state variable itself, we measure $Y = h(X)$ where h is a differomorphism, i.e., an invertible differentiable map. Then, from the time series of the observable Y, i.e.,

$$Y_{n+1} = g(Y_n) = (h \circ f \circ h^{-1})(Y_n),$$

using $(h^{-1})'(Y) = [h'(X)]^{-1}$, by the rule of chained differentiation we obtain easily

$$\left|\frac{\delta Y_{n+1}}{\delta Y_0}\right| = h'(X_{n+1}) \left(\prod_{i=0}^{n-1} |f'(X_i)|\right) (h^{-1})'(Y_0)$$

or, taking the logarithms

$$\lambda^{(Y)} = \lim_{n\to\infty} \left[\frac{1}{n} \sum_{i=0}^{n-1} \log(|f'(X_n)|)\right]$$
$$+ \lim_{n\to\infty} \frac{1}{n} \left(\log(|h'(X_n)|) + \log(|(h^{-1})'(Y_0)|)\right),$$

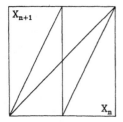 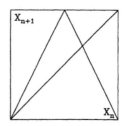 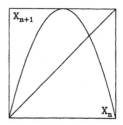

Fig. 1. Left: Diadic map. Middle: Tent map; Right: Logistic map.

where the "boundary terms" on the second line become negligible when n tends to infinity, so that we get $\lambda^{(Y)} = \lambda^{(X)}$.

Considering for example the diadic map, $X_{n+1} = 2X_n \pmod{1}$, Fig. 1a, we obtain $\lambda = \log(2)$ since $f' \equiv 2$. In fact, the same result is obtained for the tent map given by $X_{n+1} = 2X_n$ for $0 < X_n < 1/2$ and $X_{n+1} = 2(1 - X_n)$ for $1/2 < X_n < 1$ (Fig. 1b, $f' \equiv \pm 2 \to |f'| \equiv 2$) and also for the logistic map $X_{n+1} = 4X_n(1-X_n)$ (Fig. 1c) which can be transformed into the tent map through the change of variable $X_{\text{logistic}} = (2/\pi)\sin^{-1}(\sqrt{X_{\text{tent}}})$.

1.2. Generalization to d-Dimensional Maps

We examine first the case of a d-dimensional map and extend the above analysis to define the Lyapunov spectrum that measures the streching part of the tangent evolution. Writing the map in vector form as $\mathbf{X}_{n+1} = \mathbf{f}(\mathbf{X}_n)$, around a given iterate we have

$$\delta X_{i,n+1} = \sum_{j=1}^{d} \partial_j f_i(\mathbf{X}_n)\, \delta X_{j,n}$$

where $\partial_j f_i$ denotes the partial derivative of component f_i with respect to variable X_j. The $d \times d$-matrix $\mathbf{J}_n = [\partial_j f_i(\mathbf{X}_n)]$ is by definition the Jacobian matrix of $\mathbf{f}$ evaluated at $\mathbf{X}_n$ and, by analogy with (2), in vector notation we have

$$\delta \mathbf{X}_n = \left(\prod_{m=0}^{n-1} \mathbf{J}_m \right) \delta \mathbf{X}_0 \,,$$

7. Characterization of Temporal Chaos

where the product must remain time-ordered ... $\mathbf{J}_1 \mathbf{J}_0$ since the matrices do not commute in general. The stretching of the distance between two neighboring trajectories is given by the average evolution of the length $|\delta \mathbf{X}_n|$ or, preferably by $|\delta \mathbf{X}_n|^2$:

$$|\delta \mathbf{X}_n|^2 = \delta \mathbf{X}_n^t \delta \mathbf{X}_n = \left[\left(\prod_{m=0}^{n-1} \mathbf{J}_m\right) \delta \mathbf{X}_0\right]^t \left[\left(\prod_{m=0}^{n-1} \mathbf{J}_m\right) \delta \mathbf{X}_0\right]$$

$$= \delta \mathbf{X}_0^t \left[\mathbf{J}_0^t \mathbf{J}_1^t \ldots \mathbf{J}_{n-1}^t \mathbf{J}_{n-1} \ldots \mathbf{J}_1 \mathbf{J}_0\right] \delta \mathbf{X}_0,$$

where the superscript "t" denotes the transposition (remember that $(\mathbf{M}_2 \mathbf{M}_1)^t = \mathbf{M}_1^t \mathbf{M}_2^t$). The effective evolution rate is now defined by

$$|\delta \mathbf{X}_n|^2 = (\gamma_{\text{eff}})^{2n} |\delta \mathbf{X}_0|^2$$

and we have to study the limit

$$\gamma = \lim_{n \to \infty} \sqrt[2n]{\frac{\delta \mathbf{X}_0^t \left[\mathbf{J}_0^t \mathbf{J}_1^t \ldots \mathbf{J}_{n-1}^t \mathbf{J}_{n-1} \ldots \mathbf{J}_1 \mathbf{J}_0\right] \delta \mathbf{X}_0}{\delta \mathbf{X}_0^t \delta \mathbf{X}_0}} \quad (5)$$

or rather its logarithm $\lambda = \log(\gamma)$.

Staying at the formal level, we suppose first that the Jacobian matrix is constant and equal to $\mathbf{J}$ throughout the phase space so that $\delta \mathbf{X}_n = \mathbf{J}^n \delta \mathbf{X}_0$, where $\delta \mathbf{X}_0$ is an arbitrary initial vector. For n large enough, $\delta \mathbf{X}_n$ becomes aligned with the eigenvector corresponding to $\gamma_{\max}$, $\mathbf{J}$'s largest eigenvalue, and furthermore $|\delta \mathbf{X}_n|$ diverges geometrically as $(\gamma_{\max})^n$. This is the well-known *method of powers* for extracting the largest eigenvalue and the corresponding eigenvector of a given matrix. Generically, $\mathbf{J}$ is not symmetric and the spectrum is complex-valued so that the eigenvalue with largest modulus, i.e., the most unstable, is extracted by this method. The rotation in phase space associated with the imaginary part of the eigenvalue is eliminated by taking the scalar product.

Since $\mathbf{J}$ is not constant in general, we expect the $\mathbf{J}_m$ to fluctuate along the trajectory and the divergence rate to be obtained only after the limit defined by (5) is taken. λ is again understood as the time average of the instantaneous growth rate of small fluctuations around a given trajectory. The random initial fluctuation is not expected to have a strictly vanishing component along the most unstable direction; the procedure must therefore converge to give the *largest Lyapunov exponent*.

In contrast with the case of a constant matrix, the significance of $\delta \mathbf{X}_n$ is not obvious. Indeed, this vector neither tends to a fixed direction in phase space (simple real eigenvalue) nor rotates at some uniform rate (complex eigenvalue) but rather fluctuates owing to the variations of the $\mathbf{J}_m$ along the trajectory. At any time the *Lyapunov vector* corresponding to largest Lyapunov exponent points in the direction of space maximizing the divergence rate of trajectories in the long term ("time averaged matrix" $\sqrt[2n]{\mathbf{J}_0^t \mathbf{J}_1^t \ldots \mathbf{J}_1 \mathbf{J}_0}$). This vector is obviously different from the eigenvector of the most unstable eigenvalue of the "instantaneous" matrix $\mathbf{J}_n$ which rather relates to the short term evolution.

The method of powers should allow the determination of the whole spectrum of a given matrix recursively. Indeed, once the largest eigenvalue and its eigenvector have been extracted, the next-largest eigenvalues and its eigenvector can be obtained by applying the same method for vectors in the orthogonal subspace, then the next-next-largest eigenvalue and its eigenvector using vectors orthogonal to the two first eigenvectors, etc. In practice, this method is highly unstable from a numerical viewpoint but it works in principle. Transposed to the present context, this approach suggests that the tangent space can be split into a direct product of orthogonal subspaces ordered by decreasing values of the corresponding Lyapunov exponents. *Oseledec's theorem* implies the existence of a whole *Lyapunov spectrum*.

The procedure suggested above for the extraction of the Lyapunov spectrum can be carried out by studying the asymptotic growth rate of infinitesimal elements of increasing dimension in phase space. The key remark is that the largest Lyapunov exponent, hereafter denoted as λ_1, measures the divergence rate of the length of a infinitesimal one-dimensional element in phase space, i.e., the line

7. Characterization of Temporal Chaos

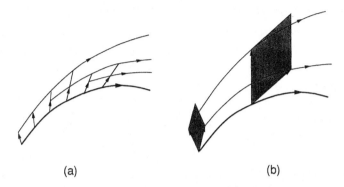

(a) (b)

Fig. 2. Lyapunov exponents can be determined by monitoring the evolution of a set of neighboring trajectories; a) the largest Lyapunov exponent λ_1; b) the next-largest exponent λ_2 is obtained indirectly from the growth rate of the surface of the parallelogram constructed on two perturbed trajectories.

segment joining two points, one on the reference trajectory, the other on the perturbed trajectory, see Fig. 2a. Considering two perturbed trajectories with linearly independent initial conditions $\delta \mathbf{X}_0^{(1)}$ and $\delta \mathbf{X}_0^{(2)}$, at any time the two vectors $\delta \mathbf{X}_n^{(1)}$ and $\delta \mathbf{X}_n^{(2)}$ define a parallelogram in tangent space. The evolution rate of the length of these two vectors is still given by the largest Lyapunov exponent λ_1 since they both have a nonvanishing projection onto the most unstable Lyapunov direction. However besides this "longitudinal" deformation, the parallelogram also experiences a "transverse" deformation at a rate governed by the next-largest Lyapunov exponent. After one iteration its surface is multiplied by a factor that tends asymptotically to $\gamma_1 \gamma_2$, so that, taking the logarithms to get the evolution rate of a 2-dimensional element μ_2, we obtain $\mu_2 = \lambda_1 + \lambda_2$. More generally, for a p-dimensional parallelepipedic element ($p \leq d$, the dimension of space) we have

$$\mu_p = \sum_{i=1}^{p} \lambda_i. \qquad (6)$$

In practice, to avoid the occurrence of angles becoming too small to be computed with accuracy (remember that the surface of a parallelogram is given by the product of the length of the sides by the sine of the angle) a procedure akin to Gram-Schmidt orthogonalization has to be developed, see Fig. 2b. Only the logarithms of the rescaling factors need to be cumulated for the determination of Lyapunov exponents.

1.3. Generalization to Differential Systems

We now examine the case of general autonomous systems. Assuming first that the attractor is a fixed point we see immediately that Lyapunov spectrum is given by the decreasingly ordered sequence of real parts of the eigenvalues of the tangent matrix at the fixed point.

When the attractor is not a fixed point and no reduction to a Poincaré map is available, we can still come back to the case studied in the previous section by constructing the so-called *time-one map*. This map applies the whole phase space onto itself by an integration of the motion between regularly spaced times and thus remains d-dimensional in contradistinction with the Poincaré map which is $(d-1)$-dimensional. However, it is important to note that the Lyapunov spectrum then trivially contains a vanishing exponent. This property is linked to the time-translational invariance displayed by any autonomous system; the distance between trajectories starting on the same orbit is bounded both from below and from above as soon as it is not asymptotic to a fixed point at finite distance (fixed point attractor) or at infinity (unbounded orbit). Indeed, assuming that the reference trajectory starts at $\mathbf{X}_0$ at time $t = 0$ and that the perturbed trajectory starts at the same point but at time δt then $\delta \mathbf{X}(0) = \mathbf{f}(\mathbf{X}(0))\, \delta t$. At time t, the distance between points on the two trajectories is simply given by $|\mathbf{f}(\mathbf{X}(t))|\, \delta t$ and cannot diverge to infinity nor converge to zero since the trajectory does not approach a singular point by assumption. The corresponding growth rate thus averages to zero. We are then left with the determination of $(d-1)$ nontrivial exponents (notice that the present analysis contains Floquet stability analysis as a special case and that the trivial Lyapunov exponent found here is the counterpart of the trivial exponent in the Floquet spectrum).

7. Characterization of Temporal Chaos

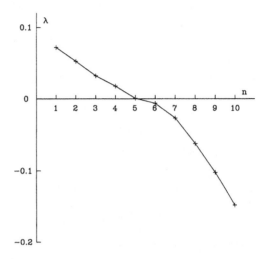

Fig. 3. Lyapunov spectrum for the Kuramoto-Sivashinsky equation with $\ell = 50$; here $\lambda_5 \approx 0$.

A last extension deals with infinite-dimensional systems usually stated in terms of partial differential equations (the case of ordinary differential equations with delays is examined in Appendix 2). As long as boundary conditions are set at finite distance the situation is not basically different from that already treated owing to the elimination of slaved modes that leaves us with an effective low-dimensional problem. Formally, the spectrum extends down to $-\infty$ since there is an infinity of slaved stable modes but we are not interested in the tail of the spectrum, only in its head that may contain a small number of positive exponents. To be treated numerically, the partial differential equation has to be reduced to a finite-dimensional problem by projection/truncation (spectral method) or by discretization of the differential operators (finite difference methods). Figure 3 displays the head of the Lyapunov spectrum $\{\lambda\}$ obtained by the method explained above for the so-called Kuramoto-Sivashinsky equation, $\partial_t w = -\partial_{x^2} w - \partial_{x^4} w - w\partial_x w$, simulated on an interval of length $\ell = 50$ using finite differences.

The real problem comes with partial differential equations defined on an unbounded domain. In this case a density of exponents by unit volume in physical space has been predicted to exist by Ruelle (1982). Figure 4 displays the number of Lyapunov exponents

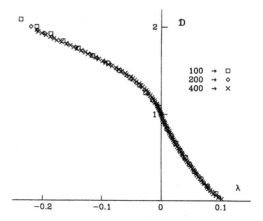

Fig. 4. Distribution of Lyapunov exponents rescaled by the number of nonnegative exponents for the Kuramoto-Sivashinsky equation with $\ell = 100, 200, 400$.

larger than a given λ (rescaled by an extensive quantity) as a function of λ itself, for the Kuramoto-Sivashinsky equation with $\ell = 100$, 200, and 400, which shows that such a limiting density can obtained numerically (here, the convergence toward the limiting distribution is faster when rescaling by the number of nonnegative exponents, $\nu_{nn} \simeq 0.131 \times \ell - 1.3$, than when taking ℓ itself).

1.4. Lyapunov Signature of Temporal Behavior

The characterization of temporal behavior by Lyapunov exponents is summarized in the table below

Regime	Continuous Time	Discrete Time
Steady	$--\ldots$	$--\ldots$
Periodic	$0----\ldots$	$----\ldots$
2-periodic	$00---\ldots$	$0---\ldots$
n-periodic	$0\ldots0--\ldots$ (n zeros)	$0\ldots0-\ldots$ ($n-1$ zeros)
Chaotic	at least $1+$ e.g., $+0---\ldots$ or $++00-\ldots$	at least $1+$ e.g., $+---\ldots$ or $++0-\ldots$

where the spectrum is ordered from the left to the right with $(+/0/-)$ meaning positive, vanishing, and negative numbers, respec-

7. Characterization of Temporal Chaos

tively. The correspondence between continuous and discrete time assumes implicitly that either a Poincaré section can be taken, or that the trivial vanishing Lyapunov exponent has been removed, so that the effective dimension of the problem is lowered.

2. Probabilistic Approach

2.1. Entropy

Instead of concentrating our attention on one single trajectory and fluctuations around it—a local approach—we now turn to a more global viewpoint and look for a way to obtain statistical information on the system without *a priori* knowledge of specific trajectories. In phase space, probability is usually introduced by attributing weights to volume elements covering the phase space. Let ϵ be the resolution of this covering. Owing to the sensitivity to initial conditions, two typical trajectories of a chaotic system with positive Lyapunov exponents, starting at different points inside the same small ball ϵ^d, become inevitably distinguishable at the considered resolution after a certain time spent at a distance smaller than ϵ. The first global quantity that can be defined, called the *topological entropy* of the system, is related to the rate of "birth" of "new trajectories." Of course, these trajectories are not new in any sense; only the fact that they started at different initial conditions is brought to our knowledge.

To be more specific, let us consider the case of the diadic map $X_{n+1} = 2X_n$ (mod 1) defined on the unit interval $[0,1]$, and consider a partition of this interval into M subintervals ϵ_i of length $\epsilon = 1/M$. Trajectories starting in two different subintervals ϵ_i and ϵ_j are clearly different so that, at this level, we have M different states ($\mathcal{N}(\epsilon, 0) = M$, where 0 means no iteration). But two different trajectories starting in the same ϵ_i will appear different only after a certain number of iterations. After one iteration, the difference between two neighboring trajectories is doubled, and the two halves of a given subinterval ϵ_i map on two different subintervals, ϵ_{i1} and

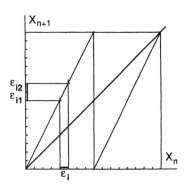

Fig. 5. For the diadic map, the number of ϵ-separated trajectories grows exponentially at a rate $H = \log(2)$.

ϵ_{i2}, see Fig. 5. The number of trajectories that have separated after one iteration is therefore $\mathcal{N}(\epsilon, 1) = 2M$. Similarly, after n iterations this number becomes $\mathcal{N}(\epsilon, n) = 2^n M = M \exp(n \log(2))$ so that the growth rate H of the number of new trajectories is just $\log(2)$. Things are particularly simple here because the expansion rate of the map is uniform. Usually, two successive limits are involved in the definition of the *topological entropy*

$$H = \lim_{\epsilon \to 0} \limsup_{n \to \infty} \frac{1}{n} \log\left(\mathcal{N}(\epsilon, n)\right),$$

where, as suggested heuristically above, $\mathcal{N}$ is the number of ϵ-separated trajectories after at most n iterations (more generally, n units of time).

This new approach marks a shift towards *information theory* concepts since the partition of phase space can be viewed as an alphabet, the outcome of an evolution step as a letter in this alphabet, a whole trajectory as a text, and H something related to the gain of information on the initial conditions as time evolves.

Let us make this more precise and consider first "messages" written with some M-letter alphabet. The probability of a random one-letter message is of course $p_i = 1/M$ and there are M possible different messages. The corresponding elementary entropy reads

$$K_1 = \sum_i p_i \log(1/p_i) = M \times (1/M) \times \log(M) = \log(M).$$

7. Characterization of Temporal Chaos

The probability of a given n-letter random message is $p_i = (1/M)^n$, and there are M^n possible messages so that the elementary entropy of an n-letter message reads

$$K_n = \sum_i p_i \log(1/p_i) = M^n \times (1/M)^n \log(M^n) = n \log(M).$$

The information gain when passing from n to $n+1$ letters is therefore

$$K = K_{n+1} - K_n = \log(M).$$

For a fully predictable message, i.e., a message that is entirely known once the first letter is known, whatever its length, there are M different possible possible first letters and therefore only M possible messages with probability $1/M$ so that $K_{n+1} = K_n = \log(M)$ and therefore $K = 0$. Clearly, interesting messages are neither fully deterministic nor fully random, so that we expect $0 < K < \log(M)$. The average gain of information will therefore be defined as the time average of the instantaneous gains:

$$K = \lim_{n \to \infty} \frac{1}{n} \sum_{m=1}^{n} (K_{m+1} - K_m).$$

To go from messages written with some finite alphabet to trajectories of dynamical systems we need only consider the covering of phase space introduced earlier. For the diadic map with its partition into M subintervals of length $\epsilon = 1/M$, we have $K_1 = \log(M)$ and every "one-letter message" can be at the origin of two different but equally probable "two-letter messages" so that

$$K_2 = 2 \times \big((1/2M) \log(2M)\big) = \log(2M) = \log(2) + \log(M)$$

and further, 2^n equally probable distinct "n-letter messages," so that

$$K_n = 2^n \times (1/2M)^n \log\big((2M)^n\big) = \log(M) + n \log(2)$$

which obviously leads to

$$K = \log(2).$$

Here the limit $n \to \infty$ is trivial since $K_{n+1} - K_n$ is independent of n and the resolution of the partition does not appear explicitly. The fact that K is equal to H defined earlier is only a consequence of the particularly simple *ergodic properties* of the diadic map giving an equal weight to all phase space elements. In contrast to H for which no notion of probability is introduced, K takes into account the probability of occurrence of given sequences. This quantity measuring the statistical amount of chaos is called the *Kolmogorov–Sinai entropy* or the *K-entropy*.

2.2. Invariant Measures

We now have to discuss how probability weights can be given to elements of the partition that serves to define the "messages" since measurements on a given system usually take the form of time series of observables, i.e., functions of the state of the system. Let G be such an observable; the outcome of a given experiment is then a series $\{G_m = G(X_m); m = 0, 1, \ldots\}$, where $\{X_m; m = 0, 1, \ldots\}$ is the specific trajectory followed by the system. The time average $\langle G \rangle_t$ of this series is defined as

$$\langle G \rangle_t = \lim_{n \to \infty} \frac{1}{n} \sum_{m=0}^{n-1} G_m,$$

but higher order averages can also be considered, e.g., correlation functions.

This passive attitude that consists in taking measurements and making averages is not satisfactory from a theoretical viewpoint more concerned with "predictions." The procedure, which was already alluded to earlier, is to consider *ensemble averages*, that is to say averages over an ensemble of systems all prepared with the same external conditions, i.e., the same constraints and no specific request about the initial conditions, and having all reached their asymptotic regime. The problem is then to determine the probability distribution $dm(X)$ characterizing this ensemble of systems, and then to replace of time average $\langle G \rangle_t$ by an *ensemble average* $\langle G \rangle_e$ defined as

7. Characterization of Temporal Chaos

an integral over the phase space:

$$\langle G \rangle_e = \int G(X)\, dm(X),$$

where the weight $dm(X)$ attributed to point X is interpreted as the fraction of systems from the ensemble which can be found "at" X. At least formally, $dm(X)$ can be expressed using a density: $dm(X) = \mu(X)\, dX$, where dX is the ordinary volume element in phase space (Lebesgue measure) and $\mu(X)$ is a positive integrable function normalized to unity.

Writing

$$G(X_i) = \int G(X)\, \delta(X - X_i)\, dX,$$

we check that $\mu(x)$ can be expressed in terms of a time average (Bowen, Ruelle):

$$\mu(x) = \lim_{n \to \infty} \frac{1}{n} \sum_{m=0}^{n-1} \delta(X - X_m),$$

where $\{X_m; m = 0, 1, \ldots\}$ is some trajectory taken on the attractor.

A possible way to determine the probability density in phase space is therefore by using the ϵ-partition defined earlier and not only detecting the presence of the system in one given phase space element but also counting the number of times the system visit the elements covering the attractor during a long typical trajectory after the decay of the transient. Unfortunately, this *box counting* method is impractical (except in one or two dimensions) and remains still passive. From a theoretical viewpoint, we therefore need to replace this *a posteriori* determination by some *a priori* argument deriving from a functional definition of the weight $dm(X)$.

Instead of introducing concepts from *measure theory* necessary to set this presentation on a rigorous basis, we will keep arguing at a heuristic level and understand $dm(X)$ as the infinitesimal form of the mass $m(A)$ of a subset A taken on the attractor around point X, i.e., something proportional to the number of visits of a single long trajectory in the vicinity of that point. Assuming for simplicity that we have a discrete time system $X_{n+1} = f(X_n)$, we see that, after a single iteration, though the mass of $f(A)$ may well be scattered over the whole attractor, it must remain constant since, by assumption, all the systems in the considered ensemble remain on the attractor. The asymptotic measure must be *invariant*, explicitly

$$m(A) = \int_A \mu(X)dX = \int_{f(A)} \mu(X)\,dX = m(f(A)), \qquad (7)$$

which clearly expresses the stationary character of the probability density on the attractor. In practice, as will be seen later, it is preferable to work with counterimages and write, e.g., the differential form of (7) as

$$dm(X) = dm(f^{-1}(X)). \qquad (8)$$

Now, μ defined above can be understood as a fixed point in the space of densities. Indeed, taking an arbitrary density μ_0, i.e., no longer restricted to take its values on the attractor, one can determine its transform μ_1 under f, e.g., by letting f operate on some approximation defined from a partition, define μ_2 from μ_1, and iterate the process up to convergence toward μ. The operator insuring the correspondence between two successive densities is called the *Frobenius-Perron operator*. It is implicitly defined *via* the relation

$$\int_B \mu_{n+1}(X)\,dX = \int_{f^{-1}(B)} \mu_n(X)\,dX, \qquad (9)$$

where B is an arbitrary subset not necessarily bound to be in the attractor.

Since dissipation concentrates an arbitrary distribution of initial conditions on a set with vanishing Lebesgue measure usually continuous in some directions and transversally fractal, the density that emerges from the iterative procedure sketched above is highly singular, being strictly zero everywhere except on the fractally distributed sheets of the attractor. However, when the system is sufficiently contracting, the problem can often be reduced to the determination

7. Characterization of Temporal Chaos

of some effective density probability along these continuous sheets, which justifies the study of ergodic properties of one-dimensional maps developed below.

2.3. Invariant Measures for One-Dimensional Maps

Reduced iterations of a single real variable involved in the description of chaotic phenomena are not one-to-one so that some care is required when writing the basic invariance condition for the probability density. Actually, it is the main reason why counterimages have to be introduced. Here we consider only the case where f is strictly expanding, i.e., $|f'| > 1$ everywhere. A theorem due to Lasota and Yorke (1973) then insures the existence of a continuous invariant measure that we now attempt to obtain analytically.

Denoting by $\{X_i; i = 1, \ldots q\}$ the q counterimages of a given point X, i.e., $f(X_i) = X$ for $i = 1, \ldots q$, we can write relation (8) as:

$$dm(X) = \mu(X)\, dX = dm(f^{-1}(X))$$
$$= \sum_{i=1}^{q} dm(X_i) = \sum_{i=1}^{q} \mu(X_i)\, dX_i,$$

which yields

$$\mu(X) = \sum_{i=1}^{q} \mu(X_i) \frac{dX_i}{dX} = \sum_{i=1}^{q} \frac{\mu(X_i)}{|f'(X_i)|} \qquad (10)$$

(the presence of the absolute value on the r.h.s. is required to avoid the interchange of bounds in the calculation of integrals since f, which expresses the change of variable from X_i to X, may be decreasing in some intervals; in the same way, changes of variables in d-dimensional integrals involve the inverse of the absolute value of the Jacobian).

Let us apply this first to the case of the tent map (Fig. 1b); every point X has two counterimages and $f' \equiv 2$ for $X_1 \in [0, 1/2]$, whereas $f' \equiv -2$ for $X_2 \in [1/2, 1]$, so that, for all X, we get $|f'| = 2$ and $\mu(X) = \frac{1}{2}(\mu(X_1) + \mu(X_2))$ (the same functional relation obviously

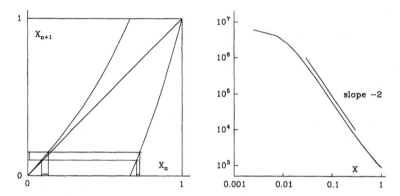

Fig. 6. Left: one dimensional map modeling type-III intermittency; a given X has two counterimages X_1 and X_2. Right: non-normalized invariant measure obtained by box counting for $\epsilon = 10^{-4}$ and 2^7 iterations (log–log scale); the $[0,1]$ interval is covered with 200 sub-intervals and the number of iterates falling in each sub-interval is counted; notice the X^{-2} behavior at large X where the nonlinear term controls the dynamics.

holds for the diadic map, Fig. 1a). The evident solution $\mu \equiv 1$ is the only one to be absolutely continuous with respect to the Lebesgue measure. As a by-product we also get the invariant measure for the logistic map at the crisis point $r = 1$, $X_{n+1} = 4X_n(1 - X_n)$, Fig. 1c, by performing the change of variables that maps it onto the tent map.

Next, let us consider the map:

$$X_{n+1} = f(X_n) = (1 + \epsilon)X_n + X_n^3 \quad (\text{mod } 1)$$

arising in the modeling of type III intermittency and illustrated in Fig. 6a. The fixed point at the origin is unstable for $\epsilon \geq 0$. Moreover, for $\epsilon > 0$, $f'(X) = 1 + \epsilon + 3X^2 > 1$, so that the map is everywhere expanding. The corresponding invariant measure (obtained by box counting) is displayed in Fig. 6b which shows that it tends become singular close to the origin.

This fact can be understood easily from the remark that, X_1 being the counterimage closest to the origin, $X_1 \sim X$ implies $\mu(X) \sim \mu(X_1)$, while dX_2/dX remains large. Equation (10) can thus be

7. Characterization of Temporal Chaos

fulfilled only if $\mu(X_2) \ll \mu(X_1) \sim \mu(X)$, i.e., if $\mu(X)$ becomes large for X close to the origin.

Denoting $\delta(X_1) = X - X_1$, and assuming that $\mu(X)$ is differentiable, we obtain

$$\mu(X) \simeq \mu(X_1) + \mu'(X_1)\delta(X_1),$$

but since $\mu(X)$ is not expected to be singular far from the origin, we can suppose

$$\mu(X_2)\frac{dX_2}{dX} = \frac{\mu(X_2)}{f'(X_2)} \sim c,$$

where c is some constant to be adjusted for normalization. Inserting these expressions and $dX/dX_1 \simeq 1 + \delta'(X_1)$ into (10), we obtain

$$(\mu + \mu'\delta)(1 + \delta') = \mu + c$$

or

$$\mu\delta' + \mu'\delta = (\mu\delta)' = c \quad \Rightarrow \quad \mu\delta = cx$$

(δ going to zero at $X = 0$ and μ remaining presumably finite, the integration constant must vanish). Finally, we get

$$\mu = \frac{cX}{\delta} = \frac{cX}{\epsilon X + X^3} = \frac{c}{\epsilon + X^2}.$$

As long as ϵ remains strictly positive, this expression can be normalized to unity, but at threshold, $\epsilon = 0$, a power law distribution is found that cannot be normalized ($\mu \sim 1/X^2$; note that since $\delta \sim X^3$, $\mu\delta \sim X \to 0$ so that the integration constant still vanishes). In fact, this means that the system is unsteady; time averages do not converge because any experimental trajectory happens to visit the neighborhood of the origin where the repulsion rate goes to zero so that the fraction of time spent close to the origin can become arbitrarily long. By contrast, when $\epsilon > 0$ iterates are expelled at a finite rate from this neighborhood and time averages converge since they involve a large number of independent and (relatively) short visits to $X = 0$.

2.4. Natural Measures

A given map has in general an infinite number of invariant measures, but not all of them are physically relevant. Considering again the diadic map, we see that there is a countable infinity of periodic points corresponding to rationals, e.g., $1/3, 2/3, 4/3 \equiv 1/3 \pmod 1$. Thus, there is an infinite number of invariant measures, one for each rational number. They are all discrete with equally weighted delta functions peaked at the points of the corresponding periodic orbit. However, these measures are physically irrelevant since these periodic orbits are all unstable. On the other hand, the continuous measure associated with the uniform density $\mu \equiv 1$ correctly describes the mixing character of the diadic map among the uncountable set of irrational numbers; furthermore, since slightly perturbing an irrational number yields generically another irrational number this invariant measure is stable against fluctuations. Such a measure is called *natural*. Natural measures are expected to be absolutely continuous along noncontracting directions (nonnegative Lyapunov exponents), stable against small perturbations and therefore representative of the statistical properties of a chaotic attractor (Ruelle, 1981).

The knowledge of the invariant measures allows us to determine observables using ensemble averages. For example the Lyapunov exponent of a one-dimensional map will be given by $\lambda = \int \log\left(|(f'(X)|\right) \mu(X)\, dX$. Now, returning to the argument used for the definition of the entropy, we see that we can interpret $\log(|f'(X)|)$ as the local growth rate of the number of "new trajectories." Averaging this observable over the phase space using the natural measure yields the K-entropy which is therefore equal to the Lyapunov exponent. Here we have implicitly assumed a chaotic regime $\lambda > 0$. To include the predictable case ($K = 0, \lambda \leq 0$) in a single formula, we can write

$$K = \sup(0, \lambda).$$

For higher dimensional systems, every unstable direction ($\lambda_i > 0$) is expected to contribute to the growth rate of the number of new trajectories as measured by the K-entropy, so that we expect (Piesin,

1976)
$$K = \sum_{i, \lambda_i \geq 0} \lambda_i.$$

For more details and related results, see Eckmann and Ruelle (1985).

3. Chaos and Dimensions

3.1. Introduction

Dimensions and codimensions are notions related to the number of conditions needed to specify sets of points. For example, in a d-dimensional space, a single point with dimension 0, is a set of codimension $d - 0 = d$. Its location requires d coordinates and can be seen as one of the (generically isolated) roots of a system of d equations with d unknowns. In the same way, a line (dimension 1) has codimension $d - 1$, since $d - 1$ equations with d unknowns have a one-parameter continuous family of solutions. A surface in the usual sense has dimension 2 and codimension $d - 2$, etc. The dimension introduced here is called the *topological dimension* of the set and will be denoted as d_t.

Simple nonchaotic attractors are expected to have a well-defined topological dimension since they "live" on ordinary manifolds, typically d_t-dimensional tori. But what for strange attractors? Clearly the topological dimension is insufficient to characterize the fractal "transverse" structure which is a global feature while it remains useful locally for describing the "longitudinal" structure. As a result, strange attractors seem to occupy space more than their topological dimensions suggest. In this section, we present some of the tools used to account for this fact first from a purely geometrical viewpoint and then including more dynamical aspects.

3.2. Fractal Geometry

The first concept to introduce is that of *internal similarity*. Let us consider for example the classical *triadic Cantor set* and recall that it is obtained from the (closed) unit interval [0, 1] by removing the (open) middle third]1/3, 2/3[, repeating the process on each remaining (closed) intervals [0, 1/3] and [2/3, 1], which yields four intervals [0, 1/9], [2/9, 1/3], [2/3, 7/9], and [8/9, 1], and doing it again and again indefinitely. In practice, the resulting fractal structure can be observed down to some small *inner scale* below which it is blurred by fluctuations, i.e., here truncation errors. Towards large scales, the construction proceeds by pasting two copies of the obtained set on both sides of a "blank" interval of length 1 obtaining a set of length 3, then repeating the process to get a set of length 9, etc., up to some *outer scale*, here given by the actual width of the display system. The internal similarity of the ideal indefinite structure is obvious from the construction rule; at each step, the pattern is reproduced $N = 2$ times and its size has increased by a factor $r = 3$. For a set that would behave "normally" we would have: $N = r^{d_t}$, which gives $d_t = \log(N)/\log(r)$. By analogy, we define the *similarity dimension* as

$$d_s = \frac{\log(N)}{\log(r)}.$$

For the triadic Cantor set we get $d_s = \log(2)/\log(3) = 0.631\ldots$. Other self-similar Cantor sets with arbitrary dimensions between 0 and 1 can be built simply by keeping the construction rule but varying N and r, "sparse" or "dense" sets having d_s close to 0 or 1, respectively. Other fractal sets can be constructed by applying analogous rules, e.g., *von Koch's snow flake* which is a fractal line with similarity dimension larger than 1 (see Bibliographic Notes).

When the construction rule is not known, factors N and r have to be determined empirically. Covering the set with balls of diameter ϵ, counting the number of balls $\mathcal{N}(\epsilon)$, and considering the limit

$$d_f = \lim_{\epsilon \to 0} \frac{\log(\mathcal{N}(\epsilon))}{\log(1/\epsilon)} \tag{11}$$

defines a quantity called the *capacity* of the set (Kolmogorov), also called the *fractal dimension* (Mandelbrot). It is easily seen that $d_f = d_t$ whenever the set is "continuous." To see how this definition works in the fractal case, we go back to the Cantor set;

7. Characterization of Temporal Chaos

at step n we take $\epsilon = (1/3)^n$ and get $\mathcal{N}(\epsilon) = 2^n$, which yields $d_f = \log(2^n)/\log(3^n) = \log(2)/\log(3) = d_s$. The built-in internal similarity no longer appears explicitly but results in the fact that n drops out from the calculation so that there is no need to take the limit. The introduction of the similarity dimension is only a matter of convenience; the only quantity of practical interest is the fractal dimension defined by (11).

Interpreting the fractal dimension in a slightly different way leads to the definition of the *Hausdorff dimension*. In a d-dimensional space the mass of a compact ball with radius ϵ varies as ϵ^d. Furthermore, a given ϵ-covering of a set can be given a mass $\mathcal{M}_d \sim \mathcal{N}(\epsilon) \times \epsilon^d$ and, since $\mathcal{N}(\epsilon) \sim \epsilon^{-d_f}$ by definition of d_f, we get $\mathcal{M}_d \sim \epsilon^{-d_f} \times \epsilon^d = \epsilon^{d-d_f}$ when $\epsilon \to 0$, so that for a fractal set with dimension $d_f < d$, $\mathcal{M}_d \to 0$ when $\epsilon \to 0$ (as expected for attractors of dissipative dynamical systems). Now, let $d' \leq d$ be a test-dimension and cover the set with d'-dimensional balls. Measuring $\mathcal{M}_{d'}$ we get, similarly, $\mathcal{M}_{d'} \sim \epsilon^{d'-d_f}$ so that when $\epsilon \to 0$ we have $\mathcal{M}_{d'} \to 0$ if $d' > d_f$, $\mathcal{M}_{d'} \to \infty$ if $d' < d_f$, and $\mathcal{M}_{d'} \sim$ Cst when $d' = d_f$. The dimension d' such that the d'-mass neither vanishes nor diverges at the limit $\epsilon = 0$ is therefore just the fractal dimension of the set. The Hausdorff dimension d_H is then obtained by optimizing the determination of the mass of the covering. Relaxing the condition on the radius of the balls, i.e., covering the set with balls of radius ρ not larger than ϵ, we define its *Hausdorff measure* by

$$\mathcal{M}_{d'}(\epsilon) = \inf \left(\sum \rho^{d'}(i) \right) \qquad \text{with} \quad \rho \leq \epsilon,$$

where the sum runs over the optimal covering of the set. We then define d_H as the value of d' such that $\mathcal{M}_{d'} \to \infty$ for $d' < d_\mathrm{H}$ and $\mathcal{M}_{d'} \to 0$ for $d' > d_\mathrm{H}$.

Noticing that $\inf(\Sigma \cdots) \leq \mathcal{N}(\epsilon)\epsilon^{d'}$, we see that $d_\mathrm{H} \leq d_f$. In practice, d_f is often easier to determine than d_H but mathematical results are usually expressed using d_H. However, both quantities characterize only the geometry of the attractor since all regions contributes to $\mathcal{N}(\epsilon)$ or to $\mathcal{M}_{d'}$ with equal weight; for strange attractors, this is clearly insufficient since frequently visited parts have a greater dynamical significance than those rarely visited. As already done for the entropy when passing from the topological entropy of the map to the K-entropy, we have to turn to dimensions relative to the natural measure.

3.3. Probabilistic Viewpoint

A first quantity easy to obtain is the so-called *pointwise dimension*. Its definition relies on a still different approach to the fractal dimension: In a d–dimensional space, the measure of a portion of a non-fractal set (line, surface, ...) with topological dimension d_t included in a sphere of radius R centered at a given point of that manifold varies as R^{d_t}, which extends as R^{d_f} for a fractal set with fractal dimension d_f. Clearly, in the present context, we can obtain the probabilistic weight (the natural measure) of the fraction of the attractor comprised in a sphere of radius R centered at one of its points X_0 by counting the number of points $\mathcal{N}_1(R, X_0)$ of a single long trajectory falling in that ball after the transient has decayed. Therefore, we define

$$d_p(X_0) = \lim_{R \to 0} \frac{\log\left(\mathcal{N}_1(R, X_0)\right)}{\log(R)}.$$

At this stage, it is however not completely clear whether d_p characterizes the attractor as a whole or depends on the point X_0 chosen on the attractor. In principle, d_p is independent of X_0 since the trajectory is supposed to explore the whole attractor densely so that a change of point comes to a time shift. In practice, some averaging over the attractor may have to be performed.

As already pointed out, the natural measure is fractal. As such, it must be characterized by a whole family of *generalized dimensions* (Renyi, 1970). At least formally, we can assume that the fractal measure is specified by the occupation probabilities p_i of cells i belonging to a partition with resolution ϵ. Then we define

$$d_q = \lim_{\epsilon \to 0} \frac{1}{q-1} \frac{\log\left(\sum_i p_i^q\right)}{\log(1/\epsilon)}$$

7. Characterization of Temporal Chaos

and notice immediately that, for $q = 0$,

$$\log\left(\sum_i p_i^0\right) = \log\left(\sum_i 1\right) = \log\left(\mathcal{N}(\epsilon)\right),$$

$\mathcal{N}(\epsilon)$ being the number introduced in the definition of the fractal dimension, so that we obtain $d_0 = d_f$.

Besides d_0, two other dimensions have a special interest: d_1 and d_2. As seen from the presence of $q-1$ in the denominator of d_q, the evaluation of d_1 requires a little care. We have

$$\sum_i (p_i)^q = \sum_i p_i\,(p_i)^{q-1} = \sum_i p_i \exp\left((q-1)\log(p_i)\right)$$
$$= \sum_i p_i\left(1 + (q-1)\log(p_i)\right) = \sum_i p_i + \sum_i (q-1)p_i \log(p_i)$$
$$= 1 + \sum_i (q-1)p_i \log(p_i),$$

and, using $\log(1+u) \sim u$,

$$d_{q\to 1} = \lim_{\epsilon \to 0} \lim_{q \to 1} \frac{1}{q-1} \frac{\log\left(1 + \sum_i (q-1)p_i \log(p_i)\right)}{\log(\epsilon)}$$
$$\sim \lim_{\epsilon \to 0} \frac{1}{q-1} \frac{\sum_i (q-1)p_i \log(p_i))}{\log(\epsilon)},$$

which shows that the limit $q \to 1$ is well behaved. After simplification, we get

$$d_1 = \lim_{\epsilon \to 0} \frac{\sum_i p_i \log(p_i)}{\log(1/\epsilon)}.$$

This quantity is known as the *information dimension* and hence often denoted as d_I ($I(\epsilon) = \sum_i p_i \log(1/p_i)$ was defined as the entropy of the weighted partition). When the distribution is uniform, $p_i \equiv$

$p = 1/\mathcal{N}(\epsilon)$, so that $d_\mathrm{I} = d_\mathrm{f}$, but in general $I(\epsilon) \le \log(1/\mathcal{N}(\epsilon))$, so that $d_\mathrm{I} \le d_\mathrm{f}$.

To interpret d_2 we need only notice that p_i^2 is the probability of finding two points of the attractor in cell i weighted by p_i. This relates to the two-point correlation function of the attractor in phase space which can be obtained from the distribution of the distances between pairs of points on the attractor (Grassberger and Proccacia, 1983):

$$C(R) = \lim_{N \to \infty} \frac{1}{N^2} \sum_{\{X_i, X_j\}} Y(R - d(X_i, X_j)),$$

where $d(X_1, X_2)$ is the distance between two points taken on the attractor and Y the Heavyside function ($Y(u) = 0$ for $u < 0$ and $Y(u) = 1$ for $u > 0$). The correlation dimension is then defined by

$$\nu = \lim_{R \to 0} \frac{\log(C(R))}{\log(R)}$$

and since, at a given resolution ϵ of the partition, the probability of finding two points in the same cell should be proportional to $C(R = \epsilon)$, we have $d_2 = \nu$. Moreover, since the correlation function C presents itself as an average of $\mathcal{N}_1(R, X)$ over the attractor, we should have $d_\mathrm{p} = d_2$. Higher order correlation functions and higher order dimensions could be considered as well.

The inequality $d_\mathrm{I} \le d_\mathrm{f}$ quoted above is a special case of the more general inequality

$$d_q \le d_{q-1}.$$

If the probability distribution is uniform then $p_i \equiv p$ so that all d_q are equal to d_f, but in general the attractor is not homogeneous and the function d_q is not a Dirac peak at d_f but a monotonously decreasing function. The spreading of d_q as a function of q gives an idea of the inhomogeneity of the distribution of points on the attractor. Increasing q gives more weight to the frequently visited cells and decreasing q favors the rarely visited cells. In particular, for $q \to +\infty$ only the cells with largest probability remain in the sum since $(p_i/p_\mathrm{max})^q \to 0$ for all cells except the most probable. Not

7. Characterization of Temporal Chaos

only positive integer values of q can be considered but more generally any real value, positive or negative. One can understand d_q as the fractal dimension of subsets of the attractor with a given level of probability. A lot of work has been devoted to distinguish between attractors using d_q or quantities related to it (see Bibliographic Notes for references).

3.4. Dimensions and Lyapunov Exponents

Up to now, the dynamical origin of the fractal measure has not been taken into consideration. As in the case of the entropy, we should be able to relate dimensions and Lyapunov exponents since the fractal structure of the attractor is the combined result of the contraction associated with dissipation and the stretching due to the divergence of trajectories.

The first and most reliable quantity that emerges from the determination of Lyapunov exponents is the *number of nonnegative exponents*, here denoted as κ_{nn}, giving the number of expanding or neutral directions, i.e., the topological dimension of the continuous component of the attractor. This is obviously the case for simple nonchaotic attractors, e.g., a limit cycle (two-torus) with only one (two) nonnegative exponent(s), but likely also for chaotic attractors. As conjectured by Ruelle, some smoothing is expected that along these directions owing to the sensitive dependence on initial conditions (for the Kuramoto–Sivashinsky equation with $\ell = 50$, we have $\kappa_{nn} = 5$, see Fig. 7).

The second quantity that can be extracted from the Lyapunov spectrum is called the *Lyapunov dimension* or the *Kaplan–Yorke dimension*, after Kaplan and Yorke (1978). Its significance is best understood from a definition in terms of the cumulated Lyapunov exponents μ_k:

$$d_L = \nu_{nn} + \frac{\mu_{\nu_{nn}}}{\mu_{\nu_{nn}} - \mu_{\nu_{nn}+1}} = \nu_{nn} + \frac{1}{|\lambda_{\nu_{nn}+1}|} \left(\sum_{i=1}^{\nu_{nn}} \lambda_i \right)$$

where ν_{nn} is the number of nonnegative μ_k. The μ_ks give the dilation rate of k-dimensional volume elements in phase space. The spectrum

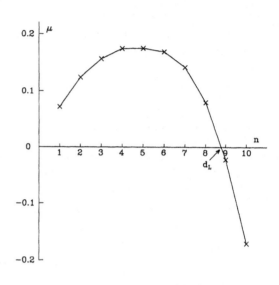

Fig. 7. μ-spectrum for the Kuramoto-Sivashinsky equation with $\ell = 50$; there are $\kappa_{nn} = 5$ nonnegative λs ($\lambda_5 = \mu_5 - \mu_4 \approx 0$), and $\nu_{nn} = 8$ non-negative μs; the Lyapunov dimension is defined by linear interpolation as the dimension of volume elements that would neither expand ($\mu > 0$) nor contract ($\mu < 0$); $d_L \simeq 8.8$.

of μ_k for the Kuramoto-Sivashinsky equation is given in Fig. 7; for $k < \nu_{nn}$, k-dimensional parallelepipeds expand along their κ_{nn} noncontracting directions more than they shrink along their $k - \kappa_{nn}$ supplementary directions, so that their k-volume grows on average. For $k > \nu_{nn}$, they shrink more than they expand. The Lyapunov dimension defined above by linear interpolation is therefore the dimension for which the volume of a parallelepiped neither expands nor contracts on average, which should be related to some dimension of the invariant measure. More specifically, since a relation already exists between the entropy and the Lyapunov exponents, it has been conjectured that the Lyapunov dimension was equal to the information dimension: $d_L = d_I$. Even though their relation with the quantities introduced before is not yet well established, κ_{nn} and ν_{nn}, or better d_L, are interesting characteristics of a chaotic attractor.

7. Characterization of Temporal Chaos 275

4. Experimental Approach

4.1. The Method of Time Delays

In all the preceding sections, it was tacitly assumed that the dynamical system was known explicitly, especially the number of degrees of freedom (the dimension of phase space), the structure of their couplings (the map or the differential system). When dealing with experimental data, we never have direct access to the degrees of freedom but acquire information on the dynamics by *sampling observables*. Before using the main tools developed up to now, Poincaré maps, Lyapunov exponents and dimensions, we have to face the problem of extracting a faithful representation of the system in some *pseudo-phase space* from the time-series of observables accessible experimentally. Moreover, since the physical system is usually a continuous medium with (formally) an infinite number of degrees of freedom, we have to be sure that the reconstruction obtained accounts adequately for the reduction to a small number of effective degrees of freedom (note that this problem is not restricted to laboratory experiments but also arise in computer simulations).

Let us assume that we have a discrete time system $X_{n+1} = f(X_n)$, where X is d-dimensional array (d and f unknown), and that the dynamics has to be reconstructed from the time series of some scalar observable $Y = Y(X)$. Ideally, we want to invert the relation between Y and the primitive variables X_i ($i = 1, \ldots, d$) from the knowledge of the series $\{Y_n, n = 0, 1, \ldots\}$. Of course, a single measurement Y_0, i.e., $Y_0 = Y(X_0)$ cannot be sufficient to define a point in phase since this requires d independent relations. A second measurement $Y_1 = Y(X_1) = Y(f(X_0))$ (since we believe in the existence of a deterministic underlying dynamics accounted for by f) gives a second relation between the d components of X_0. Going further we see that d successive measurements $Y_0, Y_1, \ldots, Y_{d-1} = Y(X_{d-1}) = Y(f^{d-1}(X_0))$ gives us d relations involving the d unknown components of X_0, which defines X_0 in principle. Repeating the argument with the series $Y_1, \ldots, Y_d$ then yields point X_1, etc.

Let us call $\mathbf{Y}_n$ the array formed by d consecutive values of the observable Y starting at Y_n. This defines a point in a pseudo-phase space and we expect to be able to monitor the whole forward orbit in the actual phase space from its representation in this pseudo-phase space. In general, much more than d successive measurements are necessary to obtain a faithful description of the system.

In the case of a continuous-time dynamical system formally given by $dX/dt = f(X)$, the reduction to a map is always possible but has an intrinsic significance only if measurements correspond to a stroboscopic analysis or to a Poincaré section of the dynamics. In all other cases, this is merely a "time-one map" accounting for the evolution of the system during some time lapse, i.e., it has an extrinsic significance (e.g., sampling time in laboratory experiments, time stepping in numerical simulations of differential systems). Working with a scalar observable, we get a function $Y(t) = Y(X(t))$ and we can understand time delays as the discrete analogue of derivatives with respect to time: $Y(t); Y'(t) = dY/dt; Y''(t) = d^2Y/dt^2; \ldots$. Using time derivatives could have seemed more natural in this context but the numerical differentiation of an experimental signal amplifies the noise so that it is preferable to keep the series $Y_0, Y_1, Y_2, \ldots$ rather than to try to evaluate $Y_0, Y_0', Y_0'', \ldots$.

4.2. Embedding

The next problem is to find a way to choose the dimension of the arrays just sufficient to avoid both ambiguities and redundancy in the representation of the dynamics, i.e., in the definition of points belonging to the attractor. The kind of problems to be dealt with can be understood from the consideration of a continuous closed loop which is seen as a "figure eight" in some representation. Whether the observed crossing is real or an "effect of perspective" linked to the given representation is not known *a priori*. To solve this question, one must be able to look at this object from different directions in space, i.e., we have to embed the loop in a sufficiently high-dimensional space.

When the considered set is a non-fractal manifold with topolog-

7. Characterization of Temporal Chaos

ical dimension d_t, a simple argument shows that embedding it in a space with at least $d_e = 2d_t + 1$ dimensions sufficient to resolve ambiguities (Whitney, 1936). Indeed, in a d-dimensional space the generic point of this manifold is defined by $d - d_t$ equations with d unknowns (codimension $d - d_t$). A double point requires generically $2(d - d_t)$ conditions. As long as $2(d - d_t) < d$, i.e., $d < 2d_t$, the number of unknowns is larger than the number of equations so that ambiguities are not resolved. On the contrary, if $2(d - d_t) > d$, the system of equations is incompatible in general. If it still has solutions, the corresponding points are real. In order to resolve ambiguities, we must therefore embed the set in a space with dimension $d = d_e \geq 2d_t + 1$.

For a fractal set, typically a strange attractor, taking into account the topological dimension of the continuous component of the set is not sufficient since this neglects the fuzziness introduced by the Cantorian transverse structure. It can then be shown (Mañe, 1981) that the embedding dimension has to be increased up to $2d_f + 1$, where d_f is the fractal dimension of the set.

In practice, neither d_t nor d_f are known *a priori* so that d_e has to be increased progressively up to a value where results become independent of d_e, which is the sign that additional data are redundant as will be seen in the applications below.

4.3. Practical Problems

The method of time delays has been widely used by experimentalists to determine the amount of chaos present in a system at a qualitative, semi-quantitative, and even quantitative level. The main difficulty is related with the determination of the best sampling rate. If it is too fast, one gets a huge amount of data that becomes difficult to handle. If it is too slow, important features of the deterministic dynamics may escape the analysis, e.g., in the case of period doubling. The most important parameter to fix correctly is the width of the gliding window translated on the data, i.e., the product $d_e \times \delta t$, where d_e is the embedding dimension and δt the sampling time interval. It is usually thought that, if a characteristic period is present in the signal, the window should have a width at least twice that period, in analogy with Nyquist's criterion which states that the Fourier representation of a signal is statistically faithful up to frequencies corresponding to half the sampling frequency; but more refined techniques have been designed (see Bibliographic Notes). Several attempts may be necessary to get reliable results and the stability of the procedure against changes of the sampling frequency and the embedding dimension has to be checked carefully.

The method has been used mainly for estimates of the correlation dimension but more recently also to extract the head of the Lyapunov spectrum.

4.4. Dimension Estimates

It is customary to plot the logarithm of the number of pairs of points $\mathcal{N}_2$ falling at a distance smaller than R as a function $\log(R)$, not taking all recorded points but only a large enough subset of them in order to limit the number of distances to determine, while keeping the estimates statistically reliable. The norm $\sum_{i=1}^{d_e} |(\cdots)|$ is usually preferred to the Euclidian norm $(\sum_{i=1}^{d_e} (\cdots)^2)^{1/2}$ to avoid the calculation of square roots. Furthermore, since integer arithmetic is faster than floating-point arithmetic, the data is often converted to integers first.

A particularly well-behaved system is illustrated in Fig. 8; slopes are easy to measure and the corresponding correlation dimension converges rapidly as the embedding dimension is increased. Notice that the raw curves giving the distribution of distances $\log(C(\log(R))$ often display "bends," which may obscure the determination of the correlation dimension more and cast a doubt on the value obtained. If we follow definitions given in Section 3 strictly, we have to take the limit $R \to 0$ to extract the value of the dimension. However, when R is small a spurious slope corresponding to $d_2 = d_e$ generally appears, which corresponds to the fact that, owing to the presence of extrinsic noise, the data is essentially random at that scale. This will be the case particularly if the sampling rate is too low. On the contrary, if the sampling rate is too high, a spurious part with slope 1 can appear

7. Characterization of Temporal Chaos

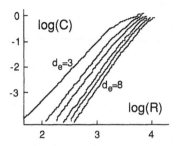
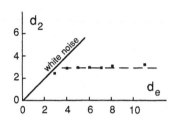

Fig. 8. Correlation dimension of a chaotic Bénard convection regime observed in the laboratory (after Malraison et al. 1983); left: variation of $\log(C)$ as a function of $\log(R)$; right: the correlation dimension saturates at $d_2 \simeq 2.8$ as soon as the embedding dimension d_e gets larger than 5.

from the contributions of successive points on the same trajectory which dominate at small distances. Obviously, this difficulty can be suppressed by imposing a minimum time lapse between two points (e.g., nonoverlapping windows). When R is large, $C(R)$ is expected to saturate since for R larger than the diameter of the attractor, C is simply the number of pairs of points retained in the statistics. Generally speaking, the part of the $\log(C)$ function actually relevant for the determination of d_2 corresponds to intermediate values of $\log(R)$.

4.5. Determination of Lyapunov Exponents

The largest Lyapunov exponents can been determined from the reconstructed dynamics in the pseudo-phase space in basically two ways. The first one directly follows from the definition in terms of the divergence of neighboring trajectories effectively followed by the system, the second one works with statistical approximations to the tangent operator.

In practice, the first method can be used only when the largest exponent is required; an initial point is taken on the reference orbit and its evolution is followed for a while; data is then scanned to find another piece of the trajectory close enough to the reference trajectory. The divergence rate of the two portions of the orbit is then measured. If the trajectories fly too far apart, a rescaling procedure has to be used. The global picture is therefore the same as that depicted in Fig. 2, except that instead of being exact in the theoretical case, it is now approximate. Indeed, when some rescaling is necessary the data is scanned again to find in roughly the direction of the perturbation another portion of orbit sufficiently close to the reference trajectory so that the calculation of the divergence rate can be continued with a reasonable accuracy.

Clearly, the rescaling step becomes cumbersome if more than one exponent is required since several adequate neighboring trajectories need to be found. The second method may then be more practical. It consists in looking at a cloud of neighboring portions of trajectories taken in a small sphere centered at a given point and then to follow their evolution during a short time interval to determine an approximation of the tangent operator (Sano and Sawada, 1985). From a practical viewpoint, the embedding dimension d_e being chosen once for all, the $d_e \times d_e$ matrix corresponding to the tangent operator is determined by least square fitting for a sufficiently large number of trajectory segments. As already stated, the calculation of the product of successive matrices that yields the Lyapunov spectrum requires some care, owing to the exponential divergence of the order of magnitude of the terms involved (typically as $(\lambda_1/\lambda_2)^n$ if n is the number of calculations). The trick used to get out of this problem follows directly from the analogy between the determination of Lyapunov exponents and that of eigenvalues of a matrix. Like in this latter case, it is found preferable (Eckmann and Ruelle, 1985) to replace the method of powers joined to Gram-Schmidt orthogonalization by a QR-algorithm that amounts to decomposing the given matrix into a product of two matrices $\mathbf{Q}$ and $\mathbf{R}$, $\mathbf{Q}$ orthogonal (i.e., $\mathbf{Q}^{-1} = \mathbf{Q}^t$) and $\mathbf{R}$ upper triangular. The algorithm goes as follows (Conte and Dubois, 1988): At the first step, $\mathbf{J}_0$ is decomposed into $\mathbf{Q}_0\mathbf{R}_0$; then, computed from $\mathbf{J}_1$, the product $\mathbf{J}_1\mathbf{Q}_0$ is further decomposed into $\mathbf{Q}_1\mathbf{R}_1$ and at step m, $\mathbf{Q}_m\mathbf{R}_m = \mathbf{J}_m\mathbf{Q}_{m-1}$. It is easily seen that

$$\mathbf{J}_{n-1}\mathbf{J}_{n-2}\ldots\mathbf{J}_1\mathbf{J}_0 = \mathbf{Q}_{n-1}\mathbf{R}_{n-1}\mathbf{R}_{n-2}\ldots\mathbf{R}_1\mathbf{R}_0,$$

7. Characterization of Temporal Chaos

so that the product of the general matrices $\mathbf{J}$ is replaced by a product of upper triangular matrices $\mathbf{R}$. The Lyapunov spectrum is obtained from the matrix product

$$\left(\mathbf{J}_{n-1}\ldots\mathbf{J}_0\right)^t\left(\mathbf{J}_{n-1}\ldots\mathbf{J}_0\right) = \mathbf{R}_0{}^t\ldots\mathbf{R}_{n-1}{}^t\,\mathbf{R}_{n-1}\ldots\mathbf{R}_0\,.$$

The orthogonal matrices $\mathbf{Q}$ have disappeared since $\mathbf{Q}^t\mathbf{Q} = \mathbf{Q}^{-1}\mathbf{Q} = \mathbf{I}$, the d_e-dimensional identity matrix. The exponents are then given by:

$$\lambda_i = \lim_{n\to\infty}\frac{1}{n}\log\left(|(\mathbf{R}_{n-1}\ldots\mathbf{R}_0)_{ii}|\right),$$

where ii denotes the ith diagonal element of the matrix product. Owing to the triangular structure of the matrices involved, these diagonal elements are easily computed as $\prod_{m=0}^{n-1}(\mathbf{R}_m)_{ii}$, which avoids the addition of several terms of very different orders of magnitude, as would be the case when using the primitive matrices $\mathbf{J}$.

4.6. Final Remark

In conclusion, a good understanding of the transition to chaos can be obtained by a combination of traditional signal processing, e.g., Fourier analysis, and a reconstruction of dynamics from time delays. Moreover, the amount of disorder can even be determined quantitatively by estimating dimensions and Lyapunov exponents. However, all this works well when a small number of effective degrees of freedom is implied and one runs into difficulties when more than a single observable is required to characterize the state of the system, especially when the spatial structure is insufficiently frozen by confinement effects, a situation to be studied in the forthcoming chapters.

5. Bibliographical Notes

A general reference for this chapter is:

[1] J. P. Eckmann and D. Ruelle, "Ergodic theory of chaos and strange attractors," Rev. Mod. Phys. **57**, 617 (1985).

For early discussions of the characterization of temporal chaos, see:

[2] D. Ruelle, "Sensitive dependence on initial condition and turbulent behavior of dynamical systems," Ann. N.Y. Acad. Sc. **316**, 408 (1979).

Lyapunov exponents and their properties are discussed by:

[3] G. Benettin and L. Galgani, "Ljapunov characteristic exponents and stochasticity," in:

[4] G. Laval and D. Gresillon, eds., *Intrinsic stochasticity in plasmas* (Editions de Physique, 1979).

that also contains several interesting contributions, notably an article on "Strange attractors" by Y. Pomeau. Another interesting early reference is:

[5] I. Shimada and T. Nagashima, "A numerical approach to ergodic problem of dissipative dynamical systems," Prog. Theor. Phys. **61**, 1605 (1979).

The existence of a density of Lyapunov exponents for infinite dimensional systems described by partial differential equations has been first discussed in:

[6] D. Ruelle, "Large volume limit of the distribution of characteristic exponents," Comm. Math. Phys. **87**, 287 (1982).

The experimental confirmation on the Kuramoto-Sivashinsky equation comes from:

[7] P. Manneville, "Liapounov exponents for the Kuramoto-Sivashinsky model," in *Macroscopic modeling of turbulent flows*, O. Pironneau, ed., Lect. Notes in Physics **230** (Springer-Verlag, 1985), p. 319.

A brief and elementary but illuminating introduction to the probabilistic description of strange attractors is given by

[8] D. Ruelle, "What are the measures describing turbulence ?," Supplement of the Prog. Theor. Phys. **64**, 339 (1978).

For a more thorough mathematical survey, consult, e.g.,:

[9] A. Lasota and M. C. Mackey, *Probabilistic properties of deterministic systems* (Cambridge University Press, 1985).

For fractal aspects of strange attractors, of course begin with:

[10] B. Mandelbrot, *The fractal geometry of nature* (Freeman, 1983).

7. Characterization of Temporal Chaos

A review of mathematical results has been given by:

[11] D. Farmer, E. Ott, and J. Yorke, "The dimension of strange attractors," at the Los Alamos Conference, BN 6 [10], p. 153.

The correlation dimension and the algorithm to get it are discussed in:

[12] P. Grassberger and I. Procaccia, "Measuring the strangeness of strange attractors," Physica **9D**, 189 (1983).

while a more elaborate characterization of fractal measures is developed in:

[13] H. G. E. Hentschel and I. Procaccia, "The infinite number of generalized dimensions of fractals and strange attractors," Physica **8D**, 435 (1983).

Further developments can be found in the proceedings of the conference:

[14] *Dimensions and entropies in chaotic systems—quantification of complex behavior*, G. Mayer–Kress, ed., Springer Series in Synergetics Vol. 32 (Springer-Verlag, 1986).

(see Procaccia's contribution: "The characterization of fractal measures as interwoven set of singularities: global universality at the onset of chaos," on p. 8).

The mathematical foundation of the method of delays is discussed by:

[15] F. Takens, "Detecting strange attractors in turbulence," Lect. Notes in Mathematics **898** (Springer-Verlag, 1981), p. 366.

The implementation of the method and some concrete applications to experimental data with optimization criteria are presented in:

[16] N. Gershenfeld, "An experimentalist's introduction to the observation of dynamical systems" in Hao Bai–lin's series, BN 6 [19], p. 310.

[17] G. Mayer–Kress, "Application of dimension algorithms to experimental chaos" in [18] below, p. 122.

[18] Hao Bai–lin, ed., *Directions in chaos*, Vol. 1 (World Scientific, Singapore, 1987).

The example given here is adapted from:

[19] B. Malraison, P. Atten, P. Bergé, and M. Dubois, "Dimension d'attracteurs étranges: une détermination expérimentale en régime chaotique de deux systèmes convectifs," C. R. Acad. Sc. Paris **297**, Série II, 209 (1983).

Limitations of the reconstruction procedure when the dimension increases have been analyzed by:

[20] J. G. Caputo, B. Malraison, and P. Atten, "Determination of attractor dimensions and entropy of different flows," in [14] above.

The direct calculation of largest Lyapunov exponents from time series is presented in:

[21] A. Wolf, J. B. Swift, H. L. Swinney, and J. A. Vastano, "Determining Lyapunov exponents from a time series," Physica **16D** (1985) 285.

whereas the method resting on a least square fit of the tangent operator has been proposed independently by:

[22] M. Sano and Y. Sawada, "Measurement of the Lyapunov spectrum from a chaotic time series," Phys. Rev. Lett. **55**, 1082 (1985).

and by Eckmann and Ruelle, see [1] above.

Chapter 8

Basics of Pattern Formation in Weakly Confined Systems

Whereas the approach in terms of dissipative dynamical systems developed in previous chapters is fully adapted when the spatial structure remains frozen, the case of weakly confined systems is much less clear. In this case the number of interacting modes is linked to confinement effects measured by *aspect ratios* rather than to the number of independent physical processes. The geometry then allows a large number of equivalent configurations to arise, and disorder gains an irreducible spatial meaning linked to the specific position/orientation degeneracy.

In this chapter we recall the most important features of the *linear* behavior close to the threshold associated with the notion of aspect ratio (Section 1). Then we illustrate the phenomenology of nonlinear evolution above threshold in the weak-confinement limit by presenting simulation results on a simplified model of convection (Section 2), as a prelude to the analytical formulation developed in the two remaining sections.

Nonlinear solutions close to the threshold will be obtained in two steps first by searching small amplitude uniform solutions for a laterally unbounded system (Section 3), and then by adding long wavelength modulations (Section 4). In the vicinity of the threshold, the resulting envelope equations will yield a self-contained description of pattern selection at lowest order to be developed in Chapter 9, with more delicate problems related to the "nature" of *weak turbulence* being deferred to Chapter 10.

1. Instabilities, Confinement, and Aspect Ratios

Usual situations of macroscopic physics involve a large number of atoms or molecules so that the limit of an infinite medium is a good approximation. In the field of dissipative structures, the situation is less favorable: lateral confinement is expected to play a nontrivial role since most often the wavelength of the unstable mode, i.e., the width of the elementary cells, is not infinitesimally small when compared with the size of the experimental set-up.

The forthcoming discussion will be restricted to the case of stationary cellular instabilities. Writing the growth rate of Fourier modes as $s(k,r) = \sigma(k,r) + i\omega(k,r)$, we recall that the the threshold r_c corresponds to the absolute minimum of the marginal stability condition $\sigma(k,r) = 0$ as a function of k. The corresponding wavevector $k = k_c$ is called the critical wavevector, and the instability is *cellular* when $k_c \neq 0$ and *stationary* when $\omega_c = \omega(k_c, r_c) = 0$. The extension of the discussion to the case of waves in a closed container ($k_c \neq 0$ and $\omega_c \neq 0$) and *a fortiori* to the case of waves in open flows requires some care. The coherence of the structure is directly related to the curvature of the marginal stability curve at $k = k_c$. Let us write the marginal stability condition in the form

$$r = \xi_0^2 (k - k_c)^2,$$

where, for notational simplicity, r now stands for the relative distance to the threshold, i.e., we perform the change $(r - r_c)/r_c \to r$ which makes r dimensionless. On the right hand side, ξ_0, which has by definition the dimensions of a length (the inverse of a wavevector), is called the *coherence length* (see later Fig. 2). It is specific to the instability mechanism and, for dimensional reasons, we expect $\xi_0 \sim 1/k_c$ (see Chapter 4, Section 2.3, for plain Rayleigh–Bénard convection).

In order to evaluate the number of degrees of freedom, one can think of counting just the number of elementary cells, assuming tacitly that they behave independently. With the direction called "vertical" being that along which the instability mechanism operates, ℓ

8. Basics of Pattern Formation

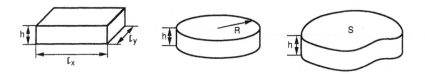

Fig. 1. Definition of aspect ratios; h is the "height" of the container; for a rectangular box one can define $\Gamma_{x,y} = \ell_{x,y}/h$, where ℓ_x and ℓ_y are the "transverse" dimensions; for a cylinder one can choose $\Gamma = 2R/h$, where $2R$ is the diameter; for a container with arbitrary shape of surface S one could take $\Gamma = \sqrt{S}/h$.

being the typical size of the system in the transverse "horizontal" direction, and the width of a cell being given by $\lambda_c/2 = \pi/k_c$, the number of cells reads

$$\Gamma = \frac{\ell k_c}{\pi}.$$

This defines the intrinsic aspect ratio of the system, intrinsic in the sense that it makes reference to the specific instability considered through k_c. Since k_c usually scales as the inverse of the height h of the system along the vertical, it may be more convenient to define an extrinsic aspect ratio making exclusive reference to the geometry of the system (see Fig. 1):

$$\Gamma = \frac{\ell}{h}.$$

In usual cases, the proportionality factor that relates the two definitions is expected to be of order unity (e.g., $\lambda_c \simeq 2h$ for Rayleigh–Bénard convection).

Close to the instability threshold, the "bare" form factor Γ overestimates the actual number of degrees of freedom by a large amount. Indeed, if we think of the cells as small bricks that can move around, we understand easily that there cannot be as many degrees of freedom as cells since the macroscopic coherence inherent in the instability mechanism implies highly correlated motions. Let the linear evolution rate s of a given mode k be given by

$$\tau_0 \sigma = r - \xi_0^2 (k - k_c)^2, \tag{1}$$

Fig. 2. Typical marginal stability curve of a cellular instability; at large aspect ratios, the eigenmodes form a quasi-continuum with spacing $2\pi/\ell$. Above threshold, unstable modes belong to a band of width $\mathcal{O}(\sqrt{r})$.

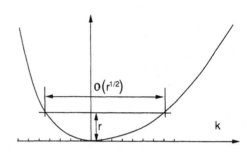

where τ_0 is the natural *relaxation time* of a typical unstable mode in the system. Assume first only one relevant horizontal direction and periodic boundary conditions at a distance ℓ. Allowed wavevectors are then multiples of $\delta k = 2\pi/\ell$ (see Fig. 2). From (1) the width of the domain of unstable wavevectors is given by $\Delta k = 2\sqrt{r}/\xi_0$ and the number Δn of easily excitable modes by $2 \times \Delta k/\delta k$ (for each value of k we have two modes, either $\pm$ or odd/even). Assuming $\xi_0 \sim 1/k_c$ as stated earlier, we get

$$\Delta n = 2\frac{\ell}{\pi\xi_0}\sqrt{r} = 2\frac{\ell k_c}{\pi}\sqrt{r} = 2\Gamma\sqrt{r}.$$

At the linear stage, the number of degrees of freedom is therefore reduced by a factor of order $\sqrt{r}$ with respect to the strictly geometrical estimate.

On general grounds, a solution built with a finite width wavepacket is better interpreted as a *modulation* to some ideal periodic solution at wavevector k_c. Going back from Fourier space to physical space, we see that the typical length-scale of this modulation is $\mathcal{O}(1/\sqrt{r})$. The divergence of this length-scale at threshold makes the complete coherence of the most unstable mode fully explicit. Further above the threshold, more modes can take part in the dynamics; the coherence length is reduced to

$$\xi = \xi_0/\sqrt{r}, \qquad (2)$$

so that modulations can get steeper.

8. Basics of Pattern Formation

Before discussing further the large aspect ratio linear dynamics, let us insist first on some implications of relation (2) regarding the nature of the nonlinear regime. Three situations can occur:

$$1)\ \xi \gg \ell, \quad 2)\ \ell > \xi \gg \lambda_c, \quad 3)\ \ell \gg \xi > \lambda_c.$$

Assuming first that r is so small that $\xi \gg \ell$, we see that the limit of an infinite medium is irrelevant and a solution of the linearized problem is required, taking full account of the finite size of the system. In fact, from (1) it is not difficult to guess that finite size effects usually imply a shift of the actual instability threshold to a value $r_c \sim 1/\ell^2$ since the presence of lateral boundaries imposes a modulation at a scale $\sim \ell$. Moreover, the distance between neighboring eigenmodes is also expected to vary as $1/\ell^2$ (all this can be derived asymptotically from the case study developed at the beginning of Chapter 5). For $r \sim r_c$ the description of the system rather refers to the theory of low-dimensional systems. This first nonlinear stage, which can be called the *elementary bifurcation* regime, breaks down when the lowest lying mode can no longer be considered as isolated from its next neighbors, which therefore happens when $r - r_c \sim 1/\ell^2$. For ℓ large enough, this parameter range may well be inaccessible experimentally and we expect the elementary bifurcation regime to be mostly irrelevant to the understanding of the dynamics of weakly confined systems.

As soon as $\ell > \xi$ the reference to periodic pattern that would be modulated by an *envelope* begins to make sense. The envelope equations at lowest order, supposed to give an adequate description of the dynamics in the weakly nonlinear regime ($\ell > \xi \gg \lambda_c$), will be seen to explain many features of the short-term/small-distance dynamics in the strongly nonlinear regime ($\ell \gg \xi$), whereas a higher order approach will prove necessary to account for its long-term/large-distance behavior.

Up until now, we have assumed that everything could be discussed in terms of a single direction x transverse to the reference direction z, i.e., that modulations in the other transverse direction y were forbidden in some way. This can happen if the spatial structure

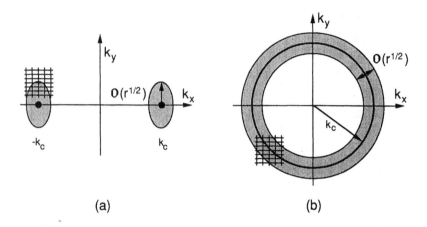

Fig. 3. Domain of unstable modes for horizontally anisotropic (a) and isotropic (b) instabilities. In the presence of lateral boundaries, modes are in fact at the nodes of a lattice with spacing $(2\pi/\ell_x, 2\pi/\ell_y)$.

in the y direction is frozen by strong confinement effects, that is to say if $\ell_y \sim h \ll \ell_x$. When $\ell_y \gg h$, modulations along y cannot be excluded for geometrical reasons as above but they can be restricted by anisotropy effects intrinsic to the instability mechanism (e.g., for convective instabilities in nematic liquid crystals with planar anchoring along the x direction). The expansion of the dispersion relation then reads

$$\tau_0 s = r - \xi_0^2 (k_x - k_c)^2 - \xi_0'^2 k_y^2,$$

where a second coherence length ξ_0' accounts for the additional damping experienced by modes with $k_y \neq 0$. The unstable modes now belong to elliptical domains centered at $k_x = \pm k_c$, $k_y = 0$ with typical sizes $(\sqrt{r}) \times (\xi_0^{-1}, \xi_0'^{-1})$ as illustrated in Fig. 3a.

When the mechanism is horizontally isotropic, e.g., for Rayleigh–Bénard convection in ordinary liquids, directions x and y are equivalent and only the modulus of the wavevector is fixed, not its orientation. At lowest order, this implies

$$\tau_0 s = r - \xi_0^2 (|\mathbf{k}| - k_c)^2, \tag{3}$$

so that, for $r > r_c$, unstable modes are seen to belong to a ring of width $\mathcal{O}(\sqrt{r})$ with average radius k_c, as illustrated in Fig. 3b.

8. Basics of Pattern Formation   291

In the large aspect ratio limit, the number of degrees of freedom is expected to vary as $\Gamma_x\Gamma_y$ times the surface of the unstable domain in Fourier space. This number can readily become very large, and we must raise the question of the number of modes in *effective interaction*, i.e., the problem of *nonlinear pattern selection*, especially when the orientational degeneracy has not been removed at the linear stage, which leads to disordered patterns also called *textures*.

2. Pattern Formation

2.1. Modeling of Weakly Confined Systems

In order to clarify the role of nonlinearities, we will report results of numerical experiments on specially designed simplified models of convection (standard numerical techniques are briefly reviewed in Appendix 3). As discussed previously, only the horizontal dependence of the fields is directly relevant to the nontrivial dynamics above threshold in the large aspect-ratio limit. This makes the consideration of two-dimensional models particularly rewarding. Indeed, the amount of data to be stored and the time required for an evolution step are drastically reduced whereas the visualization of the structures obtained is much easier.

Such simplified models can be derived from the primitive (e.g., Boussinesq) equations, at least heuristically. Important qualitative features that they should include are as follows:

- appropriate global translational and rotational symmetries;
- generation of calibrated cells with given wavelengths around some critical value $\lambda_c \neq 0$ (linear part of the evolution equation);
- appropriate type of bifurcation, e.g., supercritical, and saturation of the unstable modes with nonlinear interactions favoring the expected type of cells (rolls, squares, hexagons, ...) above the threshold;
- among supplementary requirements: no variational structure

since this is not generic for systems far from equilibrium and, especially for fluid system modeling, presence of the specific coupling between global advection by large scale flows and long wavelength distortions.

Here we will consider different variants of the Swift–Hohenberg model (1977) sharing some of the properties stated above. This model reads

$$\tau_0 \partial_t w = \left(r - \xi^4 (\nabla^2 + k_c^2)^2 \right) w - g(w), \qquad (4)$$

where the field w accounts for perturbations to the basic state "$w \equiv 0$"; τ_0 and ξ are natural time and space scales introduced for dimensional consistency.

In the context of convection, the field w has to be viewed as the local amplitude of the main convective variable (adequate mixture of the temperature and velocity field as discussed in Chapter 3, Section 1). The linear term is easily seen to be the translation in physical space of expression (3) valid in Fourier space. It can however be shown to derive from the Boussinesq equations close to the threshold. In the stress-free case, the calculation is straightforward and yields the appropriate values of τ_0, ξ, and k_c. Semi-realistic models can be obtained from various Galerkin approximations.

The growth rate s of an infinitesimal fluctuation with wavevector $\mathbf{k}$ is given by

$$\tau_0 s(\mathbf{k}) = r - \xi^4 (\mathbf{k}^2 - k_c^2)^2 \,.$$

It displays a maximum for $|\mathbf{k}| = k_c$, positive when r is positive. In the following, we choose a length unit such that $k_c = 1$, i.e., $\lambda_c = 2\pi/k_c = 2\pi$. We also change the time unit and rescale r so that $\tau_0 = 1$ and $\xi = 1$, which is always possible. At $r = 0$ modes with $|\mathbf{k}| = 1$ become unstable and for r slightly larger, a ring of width $\mathcal{O}(\sqrt{r})$ with radius 1 is destabilized.

Nonlinear interactions described by $g(w)$ ensure the saturation of the unstable modes. The global dynamics of the original model with $g(w) \propto w^3$ is easily seen to derive from a potential

8. Basics of Pattern Formation

($G = \frac{1}{2}(1-r)w^2 + \frac{1}{2}(\nabla^2 w)^2 - (\nabla w)^2 + \frac{1}{4}w^4$), so that the asymptotic time dependence remains trivial, but this does not exclude the existence of complicated metastable textures. Though this may be adequate for modeling convection in high Prandtl number fluids, extensions are clearly required to get a more realistic behavior at intermediate or small Prandtl numbers, for which nonpotential contributions to $g(w)$ and the coupling to large scale secondary flows induced by curvature effects become essential.

2.2. Pattern Formation in Two Dimensions

Here, we consider model (4) with:

$$g(w) = (w^2 + (\nabla w)^2)w \qquad (5)$$

and boundary conditions $w = \nabla_n w = 0$, where ∇_n is the normal gradient to the boundary. The value of the bare coherence length ξ_0 is easily obtained by expanding the marginal stability curve $r = (k^2 - 1)^2$ in powers of $\delta k = k - 1$, which immediately yields $\xi_0 = 2$, i.e., about $\lambda_c/3$. Simulations are performed on a disc of radius $R = 25$ for $r = 0.1$ starting with low amplitude random initial conditions, which places the system well inside the strongly nonlinear domain of interest (from (2), $\xi = \xi_0/\sqrt{r} \simeq 6 \sim \lambda_c \ll 2R \sim 8\lambda_c$). (A finite differences implicit/explicit Euler scheme was used, second order in space with $\delta x = \delta y = 0.5$, first order in time with $\delta t = 0.01$; the implicit part was solved by a conjugate gradient algorithm, see Appendix 3 for explanations if necessary.)

During the experiment we monitor

$$w_1(t) = \max_{\{\mathcal{D}\}}\left(|w(x,y;t)|\right) \quad \text{and} \quad w_2(t) = \left(\int_\mathcal{D} w^2(x,y;t)\,dx\,dy\right)^{1/2}$$

and display some instantaneous pictures of the pattern, the structure being visualized by some level lines of w (most often only the level line $w = 0$).

Three different stages are clearly visible in Fig. 4, which displays the evolution of w_1 and w_2 as a function of time: 1) exponential growth; 2) saturation; 3) slow adjustment at roughly constant amplitude.

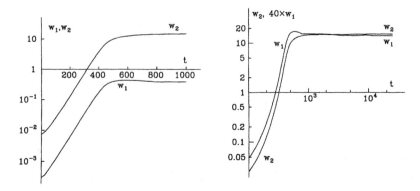

Fig. 4. Modified Swift–Hohenberg model on a disc: evolutions of w_1 and w_2. Left: initial linear stage of exponential growth (lin–log scales). Right: nonlinear evolution at constant "amplitude" in the long term (log–log scales).

Figure 5 displays pictures of the pattern during the exponential growth stage. As long as the amplitudes of the modes present in the solution remain "sufficiently small," mutual interactions are negligible. Linear considerations developed above imply that all modes are damped except those with unstable wavevectors in the ring of radius $k_c = 1$ and width $\mathcal{O}(\sqrt{r})$, which are exponentially amplified. The small-scale structure present in the low level random initial conditions is not yet completely eliminated at $t = 0.5$ but at $t = 5$, with $w_1 = 3.187 \times 10^{-4}$ and $w_2 = 7.693 \times 10^{-3}$, this elimination has been completed and we easily identify locally periodic structures with wavelengths of the order of λ_c. Comparing the solution a $t = 50$ with $w_1 = 5.419 \times 10^{-4}$ and $w_2 = 1.239 \times 10^{-2}$, and $t = 350$ with $w_1 = 8.214 \times 10^{-2}$ and $w_2 = 1.686$, we see that the patterns are roughly similar, whereas the amplitude of the solution, measured by either w_1 or w_2, has increased by a factor of order 150. Modes taking part in the solution evolve at an effective rate $s_{\text{eff}} \simeq \log(150)/300 = 1.67 \times 10^{-2}$ about six times smaller than the maximum growth rate $s_{\max} = r = 10^{-1}$ for a pure mode with $k = k_c$, which is not too surprising owing to the incoherence of the *mode superposition* over the whole system.

8. Basics of Pattern Formation 295

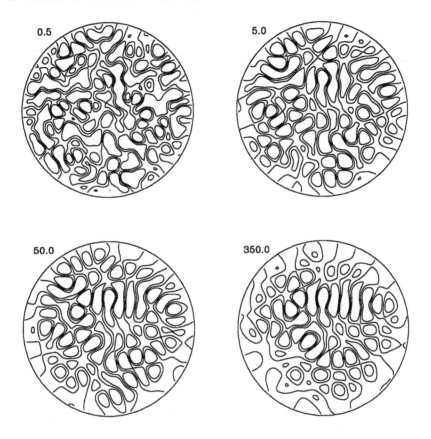

Fig. 5. Linear selection stage: $t = 0.5, 5, 50, 350$.

The *linear selection* process ends when the surviving unstable modes reach a finite level such that *saturation* takes place. An estimate of the duration of the phase controlled by the linear dynamics is easily obtained from the previous calculation giving the average growth rate: A being the amplitude of a representative unstable mode, A_0 its initial value, and A_s the typical value at which nonlinear interactions become significant, we define the saturation time t_s, often called the *onset time*, by

$$A(t_s) = A_s = A_0 \exp(s_{\text{eff}} t_s) \quad \Rightarrow \quad t_s \sim \frac{1}{s_{\text{eff}}} \log\left(\frac{A_s}{A_0}\right).$$

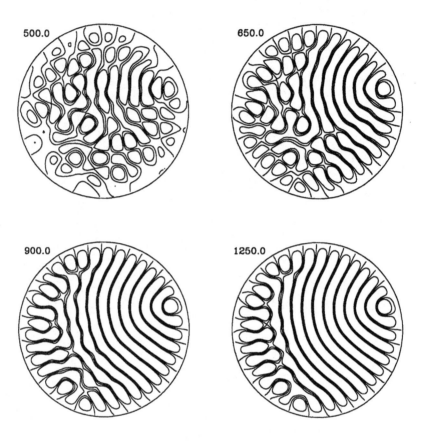

Fig. 6. Short-term nonlinear selection: $t = 500, 650, 900, 1250$.

Therefore, we see that the duration of this phase is controlled mainly by the time scale of the unstable modes $(1/s_{\text{eff}} \sim 1/r)$ and only *logarithmically* by their initial amplitudes, which is important since the onset time then remains reasonable even if the level of the background fluctuations is reduced to that of the unavoidable microscopic thermal fluctuations (notice however that, in open flows with a very reduced level of residual turbulence, the length ℓ of the channel may be insufficient to let the unstable modes reach an observable level, i.e., when $\ell < Ut_s$, where U is the average velocity downstream).

The subsequent nonlinear evolution displayed in Fig. 6 is more delicate to understand. At the end of the linear stage, a large number

8. Basics of Pattern Formation

of modes remain, yielding a highly modulated solution. In dissipative dynamical systems with a small number of degrees of freedom, the asymptotic dynamics remains confined to a "small" subset of the phase space. At the large aspect-ratio limit, by analogy, we expect the long-time evolution to be governed by a restricted set of unstable modes taken among those allowed by the symmetries of the problem and present at the end of the linear selection stage. Amplified modes have no preferential orientation and their superposition can yield polygonal level lines at a scale of order λ_c as, e.g., in the "south" region of the pattern obtained at $t = 500$ ($w_1 = 0.3983$, $w_2 = 9.791$), whereas in the "north-east" quarter the pattern is made of well-aligned rolls.

In fact, interactions between modes are expected to depend on the angle made by the corresponding wavevectors, which in turn affects the saturation process. Certain wavevector combinations are favored at a local scale and the corresponding, supposedly uniform, patterns are expected to be stable at a global scale. This results in a first *nonlinear selection criterion*: selected dissipative structures are formed with those regular superpositions of Fourier modes that are linearly stable (small perturbations decay since they correspond to less favored configurations).

For model (4–5), calculations to be developed in Section 4.5 show that a roll pattern made of a superposition of a single pair of opposite wavevectors is selected according to this criterion. This explains that the messy area in the "west" and "south" of the system recedes to the benefit of the well-ordered roll pattern in the "north-east" region (see Fig. 6, $t = 650$, 900, and 1250). w_1 reaches its maximum of 0.4367 at $t \simeq 595$ and then decreases slightly, whereas w_2 still slowly increases.

The nature of the emerging pattern depends on the instability considered. Rolls are obtained for plain Rayleigh–Bénard convection; squares, i.e., superposition of two pairs of opposite wavevectors at right angles, for convection between insulating horizontal plates; hexagons, i.e., three pairs are 120°, for convection in the absence of top–bottom symmetry, especially Bénard–Marangoni

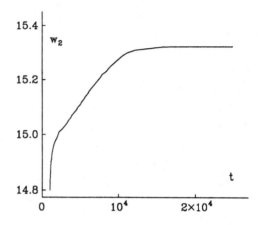

Fig. 7. Long-term variation of w_2: steady growth for $2000 < t < 10000$ and final saturation for $t > 12000$.

convection,.... The number of cell types is limited and one can speak of a *dissipative crystallography* by analogy with the crystallography of atomic edifices.

The disordered system resulting from the linear selection stage can be decomposed into several subsystems, more or less independent but sufficiently coherent to be described by partial envelopes. Corresponding envelope equations at lowest order will account for the intense competition between modes that characterize this *early nonlinear selection* stage (see Section 4.5).

The locally regular pattern that emerges has a much simpler global topology but it still continues to evolve as seen in Fig. 7, which displays w_2 as a function of time for $1000 < t < 25000$ (in the context of convection, this quantity would be interpreted as the square root of the convective part of the heat flux $\int w\theta\, dx\, dy$ since at lowest order $w \sim \theta$ and $w_2 = (\int w^2\, dx\, dy)^{1/2}$).

As is easily understood from the evolution of the pattern depicted in Fig. 8, the selection criterion just stated concerns the *local structure* and holds in the *short term* but does not tell much about the final pattern. In the *long term*, the solution is seen to converge slowly towards a very symmetrical time-independent *global texture*, which can be analyzed in terms of two *grains*, i.e., two domains with nearly uniform structure but well-defined different orientations, separated by a narrow region called a *grain boundary*. In each grain,

8. Basics of Pattern Formation

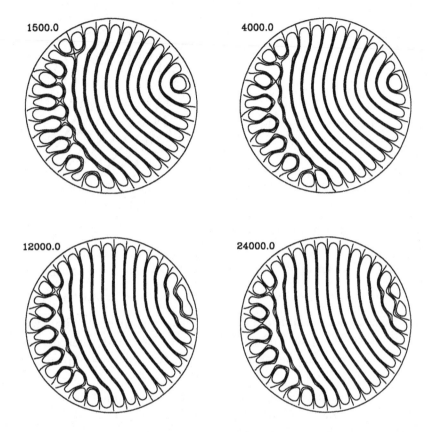

Fig. 8. Long-term evolution: $t = 1500, 4000, 12000, 24000$.

a nearly uniform orientation field can be defined, Fig. 9. The grain boundary is a *structural defect* of this orientation field.

The time scales controlling in this last evolution stage are no longer given by the growth rate of the primary instability (linear stage) or the decay rate of unfavored wavevector combinations (short term nonlinear selection). The presence of lateral boundaries, which was irrelevant to the determination of these time scales, now becomes essential. Inhomogeneities imposed by the presence of the lateral boundaries and structural defects are seen to propagate inside the nearly uniform domains, which yields a much slower evolution since processes involved in this last stage are nearly neutral, mainly

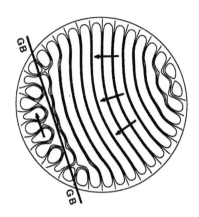

Fig. 9. The orientation field is defined from the direction of the local wavevector. Here the pattern is made of two grains of rolls at right angles separated by a grain boundary (GB).

displacement of rolls and defect motions.

Choosing the x axis along the symmetry axis of the pattern, we can describe the final evolution in the main grain by writing

$$w(x, y; t) = a \cos \left(k_c (x - x_0) \right) \quad \text{with } x_0 = x_0(x, y; t), \quad (6)$$

where a is the amplitude of the rolls and x_0 gives the absolute position of the pattern with respect to an ideal reference structure. Indeed, the level lines "$w = 0$" are given from (6) by $x = x_0(x, y; t) + (n + \frac{1}{2})\pi/k_c$. The quantity x_0 or else the phase $\phi = -k_c x_0$ of the solution is supposed to be a slowly varying function of x, y, and t. For example, the phase describing the curved rolls close to the x axis in Fig. 7 is of the form $\phi = -\alpha_0 - \alpha_1 x - \beta y^2$, where $\alpha_0(t)$ accounts for the translation of the rolls along the x axis depicted in Fig. 10a, $\alpha_1 = k - k_c$ gives the (here mostly undetectable) small wavevector shift away from k_c, and $\beta(t)$ accounts for the varying curvature. Here, the final evolution, therefore, amounts to a relaxation of the phase of the rolls.

The problem of finding the equations governing the *phase dynamics* will be examined in Chapter 10. Processes involved in these dynamics are related to perturbations that are neutral at the limit of the laterally infinite medium. Neutral modes are those induced by the invariance properties of the system, translation in all cases, rotation in the case of isotropic instability mechanisms, and possibly Galilean invariance in some special cases. Here, the relaxation

8. Basics of Pattern Formation

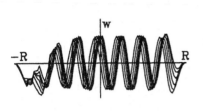
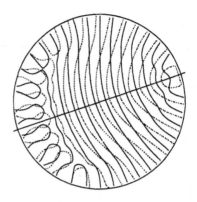

Fig. 10. Left: profiles of the solution along the symmetry axis of the final pattern for $t = 4000, 6000, 8000, 10000, 12000$, and 24000 showing an eastward drift of the order of $\lambda_c/3$. Right: relaxation of the curvature and associated translation of the grain boundary explaining this drift: superposition of patterns for $t = 4000$ (dotted line) and $t = 24000$ (solid line).

of curvature evidenced in Fig. 10b by superimposing solutions for $t = 4000$ and 24000 is a nearly neutral process related to the rotational invariance built in the model.

The resulting pattern selection is global in the sense that the bulk of the texture feels the geometry of the lateral boundaries. Here, we note that rolls arrive mostly perpendicular to them everywhere, which justifies the presence of the grain boundary. Other patterns are possible depending on the details of the preliminary linear and early nonlinear selection stages. In fact, owing to these geometrical constraints, the system usually does not succeed in finding its optimum with defect free straight rolls.

Before turning to the analytical description of nonlinear solutions adapted to the large aspect-ratio limit, let us conclude by noting that long-lived imperfect structures or *textures* are characterized by the overall topology of the pattern and the precise value of the selected average wavelength. They are usually obtained by *quenching* unstable modes above the threshold. When such textures may persist indefinitely in a metastable way, chaos remains purely spatial.

But more complicated situations can happen, especially when the feedback coupling to large scale flows generated by the curvature of the rolls is added, which may result in sustained chaotic space-time evolution (weak turbulence) close to the threshold, as observed in convection at low Prandtl numbers. This may be due to the incompatibility between different selection processes, so that the nontrivial dynamics may be attributed to a *frustration* of the system which cannot reach its "optimum" for reasons linked to the geometry of the boundaries and/or the kinematics of large scale flows.

3. Uniform Nonlinear Solutions

3.1. General Setting

Briefly stated, our problem is to find nonlinear solutions to

$$\frac{dV}{dt} = F_r(V) = L_r V + N(V, V)$$

slightly beyond a point in parameter space where a stationary instability mode becomes marginal. At such a point, say r_c, the linearized problem $L_r V = 0$ has a nontrivial solution $L_c V = 0$.

This *singular perturbation problem* is best understood as a term-by-term search for the Taylor expansion of an implicit function relating the amplitude of the bifurcated state to the control parameter. To make this expansion explicit, a formal parameter ε is introduced so that the solution *and* the control parameter are written as

$$V = \varepsilon V_1 + \varepsilon^2 V_2 + \ldots,$$
$$r = r_c + \varepsilon r_1 + \varepsilon^2 r_2 + \ldots,$$

where $V_1, V_2, \ldots$ define the structure of the nonlinear solution perturbatively and ε relates to its amplitude relative to the normalization chosen for V_1. We can notice immediately that if $r_1 \neq 0$ the lowest order bifurcated solution reads

$$V \simeq \varepsilon V_1 \quad \text{with } \varepsilon \simeq \frac{r - r_c}{r_1},$$

8. Basics of Pattern Formation

so that the bifurcation is *two-sided* or *transcritical*, whereas if $r_1 = 0$ and $r_2 \neq 0$, we have

$$V \simeq \varepsilon V_1 \quad \text{with } \varepsilon \simeq \sqrt{\frac{r - r_c}{r_2}},$$

so that the bifurcation is *one-sided*, either *supercritical* or *subcritical* according to whether r_2 is positive or negative.

Usually the control parameter enters the expression of L_r in such a way that one can write

$$L_r = L_c + (r - r_c)M.$$

Assuming in addition that the nonlinearities are formally quadratic and inserting these *ansatz* into the equations, at steady state we obtain:

$$\begin{aligned}
L_r V + N(V, V) &= 0 \\
&= \varepsilon L_c V_1 + \varepsilon^2 \Big(L_c V_2 + r_1 M V_1 + N(V_1, V_1) \Big) \\
&\quad + \varepsilon^3 \Big(L_c V_3 + r_1 M V_2 + r_2 M V_1 \\
&\quad\quad + N(V_1, V_2) + N(V_2, V_1) \Big) \\
&\quad + \ldots .
\end{aligned} \tag{7}$$

Different orders in ε are then isolated, which yields a series of linear problems:

$$L_c V_1 = 0, \tag{8a}$$
$$L_c V_2 = -r_1 M V_1 - N(V_1, V_1), \tag{8b}$$
$$L_c V_3 = -r_1 M V_2 - r_2 M V_1 - N(V_1, V_2) - N(V_2, V_1), \tag{8c}$$
$$\ldots = \ldots .$$

All problems are inhomogeneous except the first one that defines the critical point (r_c, V_1). Moreover, since the l.h.s. of these inhomogeneous problems have a nontrivial kernel, namely the critical solution

V_1, they have no solution in general except when their r.h.s. are orthogonal to the kernel of L_c^+, the operator adjoint to L_c (see Appendix 2 for an introductory presentation of algebraic properties of differential problems and especially this classical *compatibility condition*, also called the *Fredholm alternative*).

Writing problems $(8b, c\ldots)$ formally as:

$$L_c V_n = F_n,$$

denoting the appropriate scalar product as $\langle \cdot | \cdot \rangle$ and $\tilde{V}$ the solutions of $L_c^+ \tilde{V} = 0$, we obtain the compatibility condition in the form

$$\langle \tilde{V} | F_n \rangle = 0. \tag{9}$$

When applied to Equation (7b), condition (9) determines the value of r_1, the only parameter left free on the r.h.s. since V_1 is supposed to be known from the solution of the first-order problem. When the compatibility condition is fulfilled, the solution exists but is not unique. To remove the indeterminacy, we can add a supplementary condition, e.g., V_2 orthogonal to V_1, i.e.,

$$\langle V_1 | V_2 \rangle = 0, \tag{10}$$

so that the second-order solution appears as a true correction to the first-order solution in the sense of the scalar product. Carrying the expansion at higher orders, we see that the free parameters $r_2, \ldots$ introduced in the expansion of r are determined by compatibility conditions (9) relative to $(8c, \ldots)$ at order $3, \ldots$ involving the complete solution determined at lower orders.

This formal approach will be implemented first on the simplified models of convection used in the previous section, introducing the notion of *spatial resonance* as a consequence of compatibility conditions in a very transparent way. Then, we will consider Rayleigh–Bénard convection but restrict ourselves to the case of roll structures. The two classical cases of horizontal boundary conditions, stress-free and no-slip, will be presented and, skipping without damage the horizontal contribution to the compatibility conditions, we will see the

8. Basics of Pattern Formation

generic role of the (nonuniversal) spatial dependence along the vertical.

3.2. Steady Solutions of Two-Dimensional Models

The variant of the Swift–Hohenberg model (4–5) is rewritten here as

$$\tau_0 \, \partial_t w = \left(r - \xi^4(\nabla^2 + k_c^2)^2\right)w - g\left[\left(\frac{1}{k_c}\nabla w\right)^2 + w^2\right]w, \quad (11)$$

with time τ_0, length ξ, and wavevector k_c reintroduced explicitly to make the homogeneity of the model more apparent. With these notations, $(1/k_c)\nabla$ is dimensionless so that g is homogeneous to $1/w^2$.

The nonlinear solution is usually sought around the critical point, here $k = k_c$ and $r_c(k_c) = 0$. Accordingly, we assume

$$w = \varepsilon \, w_1(\mathbf{x}) + \varepsilon^2 w_2(\mathbf{x}) + \dots ,$$
$$r = \varepsilon \, r_1 + \varepsilon^2 r_2 + \dots .$$

At order ε, the problem simply reads

$$L_c w_1 = 0 ,$$

with $L_c \equiv -\xi^4(\partial_{x^2} + k_c^2)^2$. This yields

$$w_1(\mathbf{x}) = \sum_{\mathbf{k}_j} W_{\mathbf{k}_j} \exp(i\mathbf{k}_j \cdot \mathbf{x}) , \quad (12)$$

with $|\mathbf{k}_j| = k_c$. Owing to the condition $\bar{w}_1 = w_1$, the superposition of Fourier modes must involve pairs of opposite wavevectors $\pm \mathbf{k}_j$ with complex conjugate amplitudes:

$$W_{-\mathbf{k}_j} = \bar{W}_{\mathbf{k}_j} . \quad (13)$$

Let us begin with a single pair of wavevectors leading to a spatially periodic solution along the x axis: $\mathbf{k}_j = \pm k_c \hat{\mathbf{x}}$; then we have

$$w_1(x) = W_1 \exp(ik_c x) + \bar{W}_1 \exp(-ik_c x)$$

or, equivalently,
$$w_1 = W_s \sin(k_c x) + W_c \cos(k_c x).$$

Here L_c is self-adjoint and its kernel is spanned by $\exp(\pm i k_c x)$ or $(\sin(k_c x), \cos(k_c x))$ (this corresponds only to periodic functions with period $\lambda_c = 2\pi/k_c$ since the kernel is in fact four-dimensional with two supplementary unbounded eigenfunctions $x \exp(\pm i k_c x)$ or $(x\sin(k_c x), x\cos(k_c x))$ irrelevant in the present context).

Choosing for simplicity a phase and a normalization such that $w_1(k_c x) = \sin(k_c x)$, at order ε^2 we obtain
$$L_c w_2 = f_2 = -r_1 w_1 = -r_1 \sin(k_c x).$$

The compatibility condition reads
$$\frac{1}{\lambda_c}\int_0^{\lambda_c} w_1 f_2(k_c x)\,dx = 0 = -r_1 \frac{1}{\lambda_c}\int_0^{\lambda_c} \sin^2(k_c x)\,dx = -r_1/2, \tag{14}$$
i.e., $r_1 = 0$ (the orthogonality to the other component of the kernel is automatically satisfied). Since the problem at order ε^2 reduces itself to $L_c w_2 = 0$, with solution $w_2 \propto w_1$, by virtue of the supplementary orthogonality condition (10) we can directly set $w_2 \equiv 0$.

At order ε^3 we obtain
$$L_c w_3 = f_3 = -r_2 w_1 + g\left(\frac{1}{k_c^2}(\partial_x w_1)^2 + w_1^2\right) w_1$$

or, expanding w_1,
$$L_c w_3 = (-r_2 + g)\sin(k_c x)$$

so that, by projection, we get $r_2 = g$. We notice that the value of r_2 depends on the normalization chosen for w_1 but is independent of its phase: taking, e.g., $w_1 = \cos(k_c x)$ gives the same result. As before, w_3 can be taken to vanish identically. Therefore, up to terms of order higher than ε^3, the solution reads
$$w(k_c x) = \varepsilon \sin(k_c x), \quad \text{with } \varepsilon = \sqrt{r/g}.$$

8. Basics of Pattern Formation

Let us now consider a solution with $k \neq k_c$. From the marginal stability condition, we get

$$r_c = \xi^4 (k^2 - k_c)^2 \quad \Rightarrow \quad L_c \equiv r_c - \xi^4 (\partial_{x^2} + k_c^2)^2 \,.$$

Taking $w_1 = \sin(kx)$, we still have $r_1 = 0$ and $w_2 \equiv 0$. At third order, we obtain

$$L_c w_3 = \left[-r_2 + \frac{g}{4} \left(\frac{k^2}{k_c^2} + 3 \right) \right] \sin(kx) + \frac{g}{4} \left(\frac{k^2}{k_c^2} - 1 \right) \sin(3kx) \,. \quad (15)$$

The third harmonic is nonresonant and the compatibility condition now gives

$$r_2 = \frac{g}{4} \left(\frac{k^2}{k_c^2} + 3 \right) = g_{\text{eff}}(k)$$

($g_{\text{eff}} = g$ is recovered when $k = k_c$). The solution up to terms of order higher than ε^3 now contains a third harmonic term obtained by identification. Inserting $w_3 = W_3 \cos(3kx)$ in (15) above, assuming $k \simeq k_c$ and $r_c \simeq 0$ so that $L_c(3k) = r_c - \xi^4(-(3k)^2 + k_c^2)^2 \simeq -\xi^4(-9 + 1)^2 k_c^4 = -64(\xi k_c)^4$, we obtain

$$-64(\xi k_c)^4 W_3 = \frac{g}{4} \left(\frac{k^2}{k_c^2} - 1 \right) \,,$$

which vanishes as expected for $k = k_c$ (the undetermined contribution of the homogeneous problem is once again omitted). The complete solution then reads

$$w = \varepsilon \cos(kx) + \varepsilon^3 W_3 \cos(3kx) \,, \quad \text{with } \varepsilon \simeq \sqrt{(r - r_c)/g_{\text{eff}}(k)} \,.$$

Let us now examine the case of a pattern with several pairs of degenerate wavevectors, starting with superposition (12) where all the wavevectors have the same length $k = k_c$ but different orientations and opposite wavevectors have complex conjugate amplitudes, condition (13).

At order ε^2, we simply get

$$L_c w_2 = -r_1 w_1 = -r_1 \sum_j W_{\mathbf{k}_j} \exp(i \mathbf{k}_j \cdot \mathbf{x})$$

since there are no quadratic terms in the model in order to preserve the supercritical character of the bifurcation more easily. Beyond this specific case, on general grounds we would expect

$$L_c w_2 = -r_1 \sum_{\mathbf{k}_j} W_{\mathbf{k}_j} \exp(i\mathbf{k}_j \cdot \mathbf{x}) \\ + \sum_{\mathbf{k}_j, \mathbf{k}_{j'}} h(\alpha_{\mathbf{k}_j \mathbf{k}_{j'}}) W_{\mathbf{k}_j} W_{\mathbf{k}_{j'}} \exp\big(-(\mathbf{k}_j + \mathbf{k}_{j'}) \cdot \mathbf{x}\big),$$
(16)

where h is an interaction coefficient function of the angle $\alpha_{\mathbf{k}_j \mathbf{k}_{j'}}$ formed by $\mathbf{k}_j$ and $\mathbf{k}_{j'}$.

When a single pair of opposite wavevectors is involved in the superposition, the definition of the horizontal part of the scalar products expressing the compatibility condition is obvious (see, e.g., (14)). When two pairs or more are present, things are more complicated, since the functions are in general multiperiodic in several directions of space. In addition, we see that from a basic set $\{\mathbf{k}_j\}$ nonlinearities generate a larger set $\{\mathbf{k}_j + \mathbf{k}_{j'}\}$, where j and j' run on the whole set of indices.

Keeping in mind that we have to solve Equation (16), we must look for compatibility conditions associated with all possible critical wavevector combinations appearing in it. Let us suppose that we have only two wavevectors $\mathbf{k}_1$ and $\mathbf{k}_2$. As pictured in Fig. 11, as long as $\mathbf{k} = \mathbf{k}_1 + \mathbf{k}_2$ does not fall on the critical circle (Fig. 11a), the corresponding second-order problem can be solved without restriction. On the contrary, when $|\mathbf{k}| = k_c$, a compatibility condition must be fulfilled.

If the interaction coefficient between $\mathbf{k}_1$ and $\mathbf{k}_2$ is nonvanishing, then $\mathbf{k} = \mathbf{k}_1 + \mathbf{k}_2$ falling right on the critical circle ($|\mathbf{k}| = k_c$, Fig. 11b) must be present in the superposition and a nontrivial compatibility condition is obtained. This obviously implies an initial superposition containing at least one "star" of wavevectors made of three pairs at 120° (Fig. 11c) and therefore a triangular or hexagonal pattern.

Since in that case $r_1 \neq 0$, the corresponding bifurcation is expected to be transcritical. Notice however that this happens only if

8. Basics of Pattern Formation

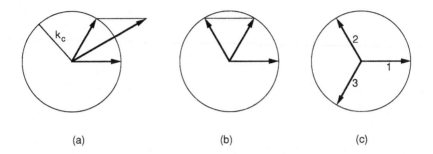

Fig. 11. Spatial resonance conditions at second order: a) non-resonant combination of two wavevectors $\mathbf{k} = \mathbf{k}_1 + \mathbf{k}_2$ with $|\mathbf{k}| \neq k_c$; b) resonant combinations involve wavevectors making an angle $\alpha_{\mathbf{k}_1 \mathbf{k}_2} = 2\pi/3$; c) regular superposition of three pairs of wavevectors at 120°.

the interaction coefficient h does not vanish for some symmetry reason. For example, with symmetrical boundary conditions at top and bottom, at the Boussinesq approximation, Rayleigh–Bénard convection remains supercritical ($r_1 = 0$), while non-Boussinesq corrections and unsymmetrical boundary conditions yield a transcritical bifurcation.

Returning to our simple example, we see that the solution at second order can be kept identically vanishing (which is not generic).

The problem at third order:

$$L_c w_3 = - r_2 \sum_j W_{\mathbf{k}_j} \exp(i\mathbf{k}_j \cdot \mathbf{x})$$

$$+ g \sum_{j,j'j''} \left(1 - \frac{\mathbf{k}_j \cdot \mathbf{k}_{j'}}{k_c^2}\right) W_{\mathbf{k}_j} W_{\mathbf{k}_{j'}} W_{\mathbf{k}_{j''}}$$

$$\times \exp(i(\mathbf{k}_j + \mathbf{k}_{j'} + \mathbf{k}_{j''}) \cdot \mathbf{x})$$

is approached along parallel lines. The result for critical combinations contributing to compatibility conditions is sketched in Fig. 12. The resonance condition obviously leads to

$$\mathbf{k} = \mathbf{k}_j + \mathbf{k}_{j'} + \mathbf{k}_{j''},$$

which means that $\mathbf{k}$, $\mathbf{k}_j$, $\mathbf{k}_{j'}$, and $\mathbf{k}_{j''}$ form a parallelogram, possibly degenerate.

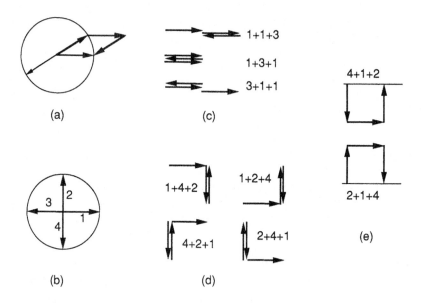

Fig. 12. Resonance condition at third order: a) example of resonant superposition; b) regular superposition of four wavevectors yielding a square pattern; c) the three resonant combinations involving a single pair of wavevectors $\pm \mathbf{k}$; d) the four combinations forming a degenerate parallelogram; e) the two remaining nontrivial combinations.

At this stage, one usually restricts oneself to the consideration of regular patterns obtained by superposition of N pairs of wavevectors and making angles multiples of π/N with each other. When $N = 1$, the case of rolls is recovered; $N = 2$ leads to square patterns (level lines of some observable look like a chess-board); $N = 3$ gives triangles or hexagons; more complicated higher order superpositions yielding multiperiodic patterns are not forbidden.

For regular superpositions, all directions defined by specific pairs of wavevectors play symmetrical roles. Taking one of them as a reference, say $j = 0$, we can write down the corresponding compatibility condition as

$$r_2 W_{\mathbf{k}_0} = g \sum_{\mathbf{k}_1 \mathbf{k}_2 \mathbf{k}_3} \left(1 - \frac{\mathbf{k}_1 . \mathbf{k}_2}{k_c^2}\right) W_{\mathbf{k}_1} W_{\mathbf{k}_2} W_{\mathbf{k}_3} \, \delta(\mathbf{k}_1 + \mathbf{k}_2 + \mathbf{k}_3 - \mathbf{k}_0).$$

(17)

8. Basics of Pattern Formation

For each specific regular pattern considered, from Fig. 2 we see that the reference vector is necessarily present in any resonant combination so that $W_{\mathbf{k}_0}$ can be dropped out from (17). Moreover, since the two remaining wavevectors are opposite, whatever their orientation, the phase of their complex amplitudes W never appears and only the square of modulus remains. Again the value of r_2 depends on the normalization chosen for w_1. It is convenient to take $|W| = \frac{1}{2}N$, which allows us to recover the choice made earlier for rolls. Note that the phase is left undetermined at this stage and that the aspect of the pattern may be different for different phase choices when $N > 2$. Finally, we have to evaluate

$$r_2^{(N)} = \frac{g}{4N^2} \sum_{\mathbf{k}_1 \mathbf{k}_2 \mathbf{k}_3} \left(1 - \frac{\mathbf{k}_1 \cdot \mathbf{k}_2}{k_c^2}\right) \delta(\mathbf{k}_1 + \mathbf{k}_2 + \mathbf{k}_3 - \mathbf{k}_0),$$

which can be decomposed into two contributions, one from the pair of wavevectors aligned along the reference direction (Fig. 12a), the other involving off-aligned pairs (Fig. 12b, c). Weights $(1 - \cos(\alpha))$ depend on the relevant angles: $\alpha = 0$ for configuration 2; $\alpha = \pi$ for configurations 1, 3, 6, and 9; $\alpha = p\pi/N$ for 4 and 5; $\alpha = (N-p)\pi/N$ for 7 and 8.

- For $N = 1$, only configurations 1, 2, and 3 remain and we have:

$$r_2^{(1)} = \frac{g}{4 \times 1^2} \Big(2 \times (1 - \cos(\pi)) + 1 \times (1 - \cos(0))\Big) = g$$

  as expected from the previous calculation.
- For $N = 2$, with only one off-aligned pair, we get

$$r_2^{(2)} = \frac{g}{4 \times 2^2} \Big(4 \times (1 - \cos(\pi)) + 1 \times (1 - \cos(0))$$
$$+ 2 \times 2 \times (1 - \cos(\pi/2))\Big) = \frac{12g}{16}$$
$$\Rightarrow \quad r_2^{(2)} = \frac{3g}{4}.$$

- For $N = 3$, with two off-aligned pairs at angles $\pi/3$ and $2\pi/3$, we have

$$r_2^{(3)} = \frac{g}{4 \times 3^2}\Big(6 \times \big(1 - \cos(\pi)\big) + 1 \times \big(1 - \cos(0)\big)$$
$$+ 2 \times \big(2 \times \big(1 - \cos(\pi/3)\big) + 2 \times \big(1 - \cos(2\pi/3)\big)\big)\Big)$$
$$= \frac{20g}{36} \quad \Rightarrow \quad r_2^{(3)} = \frac{5g}{9}$$

and so on. Corresponding solutions can be calculated, which then serve as a starting point for the analytic study of secondary instabilities.

3.3. Rayleigh–Bénard Convection

Original calculations of Malkus and Veronis for stress-free boundary conditions, as well as those of Schlüter, Lortz, and Busse in the no-slip case, were dealing with superpositions of several wavevectors pairs $\pm\mathbf{k}_i$, $i = 1, N$ with $N \geq 1$. As such, they were complicated but they both concluded that for plain Rayleigh–Bénard convection, roll structures should be preferred in a sense to be discussed in the next section. We will therefore concentrate on the determination of the solution describing steady straight rolls.

Assuming that the rolls are aligned along the y axis (x axis parallel to the single pair of wavevectors $\pm\mathbf{k}$), we can derive from the general Boussinesq equations of Chapter 4 the following set of time-independent equations:

$$(\partial_{x^2} + \partial_{z^2})^2 v_z + \partial_{x^2}\theta = \frac{1}{P}\Big(\partial_{x^2}(v_x\partial_x + v_z\partial_z)v_z$$
$$- \partial_{xz}(v_x\partial_x + v_z\partial_z)v_x\Big),$$
$$(\partial_{x^2} + \partial_{z^2})\theta + Rv_z = (v_x\partial_x + v_z\partial_z)\theta,$$
$$\partial_x v_x + \partial_z v_z = 0$$

(all fields depend on x and z only, the pressure has been eliminated, v_y vanishes identically; for notations, return to Chapter 4). The exact analytical solutions to the linearized problem given by neglecting

8. Basics of Pattern Formation

the r.h.s. have been obtained for good conducting and both stress-free and no-slip top/bottom boundary conditions. Here we are interested in obtaining the bifurcated solutions at finite distance from the marginal stability curve of the lowest lying mode $R^{(1)}$. The adjoint linearized problem reads

$$(\partial_{x^2} + \partial_{z^2})^2 \tilde{v}_z + R^{(1)}\tilde{\theta} = 0,$$
$$(\partial_{x^2} + \partial_{z^2})\tilde{\theta} + \partial_{x^2}\tilde{v}_z = 0,$$

where adjoint functions $\tilde{v}_z$ and $\tilde{\theta}$ can be shown to fulfill the same boundary conditions as v_z and θ (the general procedure used to determine the adjoint problem, i.e., the adjoint equations plus the boundary conditions on the adjoint functions is examined in Appendix 2).

Dropping the horizontal contribution to the compatibility conditions, which amounts to a supplementary integration $(1/\lambda) \int_0^\lambda dx\,(\cdots)$, we have to evaluate scalar products of the form

$$\int_{z_b}^{z_t} (\tilde{w}f + \tilde{\theta}g)\,dz = 0,$$

where subscripts "t" and "b" mean "top" and "bottom" and where f and g denote the the functions on the r.h.s. of the equations for v_z and θ.

3.3.1. Stress-Free Solution (Malkus and Veronis, 1958)

Taking $z_b = 0$, $z_t = 1$, and choosing the x origin so that $v_z = 0$ at $x = 0$, we can write the lowest lying solution of the linearized problem with horizontal wavevector k as

$$v_1(x, z) = \sin(kx)\sin(\pi z), \quad \theta_1(x, z) = \Theta_{11} v_1(x, z)$$

with $\Theta_{11} = (k^2 + \pi^2)^2/k^2$. In the following, indices will serve to distinguish between contributions from different orders in ε, first index, and from different horizontal harmonics, second index; furthermore, as exemplified previously, lowercase variables will denote functions at a given order (with only one index), and uppercase variables will correspond to numerical coefficients with two indices in general (the subscript of v_z has been dropped to simplify notations).

The marginal stability condition $R_1(k) = (k^2 + \pi^2)^3/k^2$ reaches its minimum at $k_c = \pi/\sqrt{2}$ and $R_c = 27\pi^4/4$. Owing to its simplicity, the expansion will be developed for arbitrary k, i.e., not necessarily at k_c as done in the general case. The solution of the adjoint problem reads

$$\tilde{v}(x,z) = v_1(x,z), \quad \tilde{\theta}(x,z) = \tilde{\Theta} v_1(x,z), \quad \text{with } \tilde{\Theta} = -\frac{k^2}{k^2 + \pi^2}.$$

At order ε^2, we obtain

$$(\partial_{x^2} + \partial_{z^2})^2 v_2 - \partial_{x^2}\theta_2 = f_2 = 0,$$

$$(\partial_{x^2} + \partial_{z^2})\theta_2 + R_1 v_2 = g_2 = -r_1 \sin(kx)\sin(\pi z) - \frac{\pi\Theta}{2}\sin(2\pi z).$$

The vanishing of f_2 is a nongeneric feature of the stress-free boundary conditions. In the second equation, g_2 contains two contributions. The first one involving r_1 is resonant with the kernel of the critical linear problem. The second one is independent of x (i.e., $k = 0$), and therefore non-resonant. The compatibility condition (9) simply reads $-r_1 \int_0^1 v_1 \tilde{\theta}\, dz = 0$, which imposes immediately $r_1 = 0$. The solution of the second-order problem is then obtained from

$$\partial_{z^2}\theta_2(z) = -\frac{\pi\Theta_{11}}{2}\sin(2\pi z) \quad \Rightarrow \quad \theta_{20}(z) = \Theta_{20}\sin(2\pi z),$$

with $\Theta_{20} = -\Theta_{11}/8\pi$.

The general solution of the corresponding homogeneous problem, $\alpha z + \beta$, must vanish identically in order to fulfill the boundary conditions $\theta(0) = \theta(1) = 0$. A supplementary correction proportional to the critical solution is set to zero owing to condition (10) used to remove the indeterminacy.

At order ε^3, we obtain

$$(\partial_{x^2} + \partial_{z^2})^2 v_3 - k^2 \theta_3 = f_3 = 0,$$
$$(\partial_{x^2} + \partial_{z^2})\theta_3 + R_1 v_3 = g_3 = -r_2 v_1 + v_1 \partial_z \theta_2 \qquad (18)$$
$$= -\sin(kx)\sin(\pi z)\left(r_2 + \frac{\Theta}{4}\cos(2\pi z)\right).$$

8. Basics of Pattern Formation

Since there is no velocity correction at order ε^2, we have $f_2 = 0$, which is again nongeneric. Expanding the compatibility condition reduced to $\int_0^1 \tilde{\theta} g_3 \, dz = 0$, since $f_3 = 0$, we obtain

$$\int_0^1 \tilde{\Theta} \sin(\pi z) \times \left(r_2 + \frac{\Theta_{11}}{4} \cos(2\pi z) \right) \sin(\pi z) \, dz$$

$$= \int_0^1 \tilde{\Theta} \frac{1}{2}(1 - \cos(2\pi z)) \left(r_2 + \frac{\Theta_{11}}{4} \cos(2\pi z) \right) dz$$

$$= \tilde{\Theta} \left(\frac{r_2}{2} - \frac{\Theta_{11}}{8} \int_0^1 \cos^2(2\pi z) \, dz \right) \quad \Rightarrow \quad r_2 = \frac{\Theta_{11}}{8}.$$

The solution at order ε^3 can be obtained from (18) by inserting this value of r_2:

$$(\partial_{x^2} + \partial_{z^2})^2 v_3 - k^2 \theta_3 = 0,$$
$$(\partial_{x^2} + \partial_{z^2}) \theta_3 + R_1 v_3 = -r_2 v_1(x,z) + v_1(x,z) \partial_z \theta_{20}(z)$$
$$= -\frac{\Theta_{11}}{8} \sin(kx) \sin(\pi z)(1 + 2\cos(2\pi z))$$
$$= -\frac{\Theta_{11}}{8} \sin(kx) \sin(3\pi z)$$

(the fully resonant term "$\sin(kx)\sin(\pi z)$" has disappeared, as expected, since the compatibility condition is fulfilled). The equations above give a correction of the form

$$(w_3, \theta_3) = (V_{31}, \Theta_{31}) \sin(kx) \sin(3\pi z),$$

where V_{31} and Θ_{31} are the solution of

$$(k^2 + 9\pi^2)^2 V_{31} - k^2 \Theta_{31} = 0,$$
$$-(k^2 + 9\pi^2)\Theta_{31} + R_1 V_{31} = -\frac{\Theta_{11}}{8}.$$

Finally, the complete solution at order ε^3 reads

$$v(x,z) = \varepsilon \sin(kx) \left(\sin(\pi z) + \varepsilon^2 V_{31} \sin(3\pi z) \right),$$
$$\theta(x,z) = \varepsilon \sin(kx)(\Theta_{11} \sin(\pi z) + \varepsilon^2 \Theta_{31} \sin(3\pi z)) + \varepsilon^2 \Theta_{20} \sin(2\pi z),$$

where ε is obtained from

$$R = R_1 + r_2\varepsilon^2 \quad \Rightarrow \quad \varepsilon = \sqrt{8(R-R_1)/\Theta_{11}}\,.$$

Owing to the normalization chosen for v_1, ε is simply the peak-to-peak amplitude of the vertical velocity at lowest order. Note the presence of a nontrivial uniform (i.e., independent of x) second order correction to the temperature field.

3.3.2. No-Slip Solution (Schlüter, Lortz, and Busse, 1965)

Being free from nongeneric cancellations, the solution in the no-slip case is more representative of the general case. Here, we will give only a sketch of the derivation using compact notations, the expansion of which is straightforward but somewhat tedious. Moreover we will consider only the case $k = k_c$ (when $k \neq k_c$ the solution can be obtained by means of an expansion in powers of $k - k_c$). From the analysis developed in Chapter 4, we know that the critical solution can be written as

$$v_1(x,z) = \cos(k_c x) \sum_{m=0}^{2} V_m \cos(q_m z)$$

with analogous expressions for $\theta_1(x,z)$ (boundary conditions are taken at $z = \pm 1/2$). It will be convenient to abridge the notations by replacing the explicit sums with curly brackets $\{\cdots\}$ and to use the same conventions as above for the indices. At first order this yields

$$v_1(x,z) = \cos(k_c x)\{V_{11}\cos(q_1 z)\},$$
$$\theta_1(x,z) = \cos(k_c x)\{\Theta_{11}\cos(q_1 z)\}$$

(where q_1 now represents the set $q_m, m = 0, 2$). Calculating the inhomogeneous system at order ε^2, we find that

$$(\partial_{x^2} + \partial_{z^2})^2 w_2 - k^2 \theta_2 = f_2 = \cos(2k_c x)\frac{1}{P}\{F_{22}\sin(q_2 z)\}$$
$$(\partial_{x^2} + \partial_{z^2})\theta_2 + R_c w_2 = g_2 = -r_1\cos(k_c x)\{V_{11}\cos(q_1 z)\}$$
$$+ \cos(2k_c x)\{G_{22}\sin(q_2 z)\}$$

8. Basics of Pattern Formation

where F_{22} and G_{22} can be calculated from V_{11} and Θ_{11}, and where "q_2" denotes all the combinations of the form "$q_1 \pm q_1$" plus the three roots of the linear problem for $k = 2k_c$. The compatibility condition again leads to $r_1 = 0$ and we can write the second order solution in the form

$$v_2 = \cos(2k_c x)\{V_{22} \sin(q_2 z)\},$$
$$\theta_2 = \cos(2k_c x)\{\Theta_{22} \sin(q_2 z)\},$$

where each coefficient V_{22} and Θ_{22} is the sum of a term in $1/P$ involving F_{22} and a term independent of P involving G_{22}.

At order ε^3, we obtain

$$(\partial_{x^2} + \partial_{z^2})^2 v_3 + \partial_{x^2}\theta_3 = f_3 = \cos(k_c x)\frac{1}{P}\{F_{31}\sin(q_3 z)\}$$
$$+ \cos(3k_c x)\frac{1}{P}\{F_{33}\sin(q_3 z)\},$$
$$(\partial_{x^2} + \partial_{z^2})\theta_3 + R_c v_3 = g_3 = -r_2 \cos(k_c x)\{V_{11}\cos(q_1 z)\}$$
$$+ \cos(k_c x)\{G_{31}\sin(q_3 z)\}$$
$$+ \cos(3k_c x)\{G_{33}\sin(q_3 z)\},$$

where "q_3" stands for all possible combinations "$q_1 \pm q_2$" and $F_{31}, F_{33}, G_{31}, G_{33}$ all contain contributions in $1/P$ and contributions independent of P. The scalar product expressing the now complete compatibility condition $\int_{-1/2}^{1/2} (\tilde{v} f_3 + \tilde{\theta} g_3)\, dz = 0$ now reads

$$0 = \int_{-1/2}^{1/2} dz \left(\{\tilde{V}\cos(q_1 z)\}\frac{1}{P}\{F_{31}\sin(q_3 z)\}\right.$$
$$\left. + \{\tilde{\Theta}\cos(q_1 z)\}\bigl(-r_2\{V_{11}\cos(q_1 z)\} + \{G_{31}\sin(q_3 z)\}\bigr)\right),$$

which yields a nontrivial result for r_2 containing all generic contributions the form $1/P^n$ with $n = 0, 1, 2$. With the normalization chosen for v_1, we get

$$r_2(P) = 10.757 - 0.073/P + 0.128/P^2$$

and therefore, $\varepsilon = \sqrt{(R - R_c)/r_2(P)}$. Notice that since $q_2 = q_1 \pm q_1$, the set q_1 itself is included in the set q_3. Accordingly, secular terms are present in the expression of the solution at order ε^3, which formally reads

$$v_3 = \cos(k_c x)\Big(\{V_{31}\cos(q_3 z)\} + \{zV'_{31}\sin(q_1 z)\}\Big)$$
$$+ \cos(3k_c x)\{V_{33}\cos(q_3 z)\},$$

where the first term involves the whole set q_3 and the primed term the subset q_1 only.

4. Modulated Structures

Up until now, we have tacitly assumed that structures were spatially homogeneous and stationary so that amplitudes introduced are just *scalars*, independent of time and horizontal space coordinates. However, the relevance of such "pure states" is questionable for spatially extended systems beyond the instability threshold since, from a practical viewpoint, there is usually no operational procedure able to generate them experimentally. On the other hand, since "natural" initial conditions contain a whole continuous spectrum of wavevectors in the unstable domain of Fourier space pictured in Fig. 3, a *wavepacket* centered on the most unstable wavevectors is expected to emerge from the linear selection stage. Corresponding *modulated structures* are therefore of much greater interest. As already introduced heuristically in the first chapter, they will be described by slowly variable *functions* of time and space called *envelopes*.

In this section, we first develop the systematic expansion yielding the envelope equation governing modulations to a strictly one-dimensional stationary periodic pattern. Next, we extend the calculation to account for (horizontal) rotational invariance and then perform several phenomenological generalizations mainly based on resonance and symmetry considerations. The case of stationary "plain" two-dimensional patterns will be considered first, which will give insight in the problem of "early nonlinear selection." Then we will extend the description to the case of dissipative waves generated by instabilities.

8. Basics of Pattern Formation

4.1. Systematic Expansion

The most transparent case surely corresponds to that of stationary rolls for which the only nontrivial basic dependence takes place in a single direction of space. For simplicity, we suppose that we are dealing with systems from which the nonuniversal "vertical" dependence has been removed. As can be anticipated from the calculations developed in the previous section, difficulties may arise in extending the formalism to more realistic problems, but they should remain "purely technical."

Let us consider a perfect stationary roll pattern with wavevector **k** aligned along the x axis. The corresponding uniform solution can be written as $w(x) = \frac{1}{2}(A\exp(ik_c x) + \text{c.c.})$, where $(A, \bar{A})$ form a pair of complex conjugate *scalars*. In this preliminary approach, we assume in addition that the transverse space dependence (i.e., in y) is frozen so that the solution, slowly modulated along x only, can be described by a pair of complex conjugate *functions* of x and t only:

$$w(x,t) = \tfrac{1}{2}\left(A\left(x,t\right)\exp\left(ik_c x\right) + \text{c.c.}\right).$$

Of course, such an expression cannot be an exact solution, and corrections of the order of the gradients of A and $\bar{A}$ will have to be introduced. For the moment, let us examine the effect of differential operators on the product $A(x,t)\exp(ik_c x)$. We have

$$\partial_x\left(A\left(x,t\right)\exp(ik_c x)\right) = \left(\partial_x A + ik_c A\right)\exp(ik_c x),$$
$$\partial_t\left(A\left(x,t\right)\exp\left(ik_c x\right)\right) = \left(\partial_t A\right)\exp(ik_c x).$$

By assumption, modulations are slow in space and time, which means that $|\partial_x A| \ll k_c|A|$ and $|\partial_t A|$ "small."

It may be useful to make an explicit distinction between the "carrier" $\exp(\pm ik_c x)$ and the modulation $A(x,t)$ by introducing specific variables, therefore writing

$$w(x,t) = \tfrac{1}{2}\left(A(X,T)\exp(ik_c x) + \text{c.c.}\right) \qquad (19)$$

and making the substitution $\partial_x \to \partial_x + \partial_X$, $\partial_t \to \partial_t + \partial_T$ in the equations. Derivatives ∂_X, ∂_T act on A only and $\partial_x \equiv \pm ik_c$, $\partial_t \equiv 0$ when applied to the carrier.

Clearly, the uniform solutions obtained in the previous section correspond to A constant in space and time. Deviations from the exact solution should therefore be sought as an expansion in powers of the gradients expressed in terms of the slow variables, treating ∂_X and ∂_T as formal parameters. The main problem is then to ensure the consistency of the expansion, i.e., to determine the relative orders of magnitude of the derivatives acting on the amplitude. In fact, this can be derived easily from an examination of the Taylor expansion of the linearized evolution operator.

For specificity, let us consider another variant of the Swift–Hohenberg model already used at the beginning of Chapter 5:

$$\tau_0 \, \partial_t w = r\, w - \xi^4 \bigl(\partial_{x^2} + k_c^2\bigr)^2 w - g w\, \partial_x w \qquad (20)$$

(in this expression, g is homogeneous to ξ/w so that, if w represents a velocity, g has the dimension of a time so that $w\,\partial_x w$ corresponds to the self-advection contribution to the dynamics of w). Assuming $k = k_c + \delta k_x$, at lowest order in δk_x we obtain

$$\tau_0 s = r - \xi^4 \left((k_c + \delta k_x)^2 - k_c^2\right)^2 \simeq r - \xi_0^2\, \delta k_x^2$$

with $\xi_0^2 = 4 k_c^2 \xi^4$. Going back to physical space, from the correspondence $\partial_X \longleftrightarrow i\,\delta k_x$ and $\partial_T \longleftrightarrow s$, we deduce the lowest order consistency condition between space derivatives acting on the envelope

$$\partial_T A \sim r A \sim \partial_X^2 A.$$

Though the formal expansion can be performed in a straightforward way, it is traditional to choose a common expansion parameter ε and perform the explicit changes of variables

$$\tilde{T} = \varepsilon^2 T, \quad \tilde{X} = \varepsilon X, \quad \text{and} \quad r = \varepsilon^2 \qquad (21)$$

so that when $(\tilde{T}, \tilde{X})$ vary by a quantity of order unity, (T, X) vary by a large amount of order $(\varepsilon^{-2}, \varepsilon^{-1})$ for ε small enough (we discuss later the implications of the choice $r = \varepsilon^2$, which seems at odds with

8. Basics of Pattern Formation

the expansion developed in the previous section). The assumptions above imply the substitutions

$$\partial_t \to \varepsilon^2 \partial_{\tilde{T}} \quad \text{and} \quad \partial_x \to \partial_x + \varepsilon \partial_{\tilde{X}},$$

which are further inserted into the model to yield

$$\begin{aligned}
L - \partial_t &= L_c + \varepsilon L_1 + \varepsilon^2 L_2 + \varepsilon^3 L_3 + \varepsilon^4 L_4 \\
&= \left(-\xi^4 \left(\partial_{x^2} + k_c^2\right)^2\right) + \varepsilon \left(-4\xi^4 \left(\partial_{x^2} + k_c^2\right) \partial_x \partial_{\tilde{X}}\right) \\
&\quad + \varepsilon^2 \left(1 - 2\xi^4 \left(\partial_{x^2} + k_c^2\right) \partial_{\tilde{X}^2} - 4\xi^4 \partial_{x^2} \partial_{\tilde{X}^2} - \tau_0 \partial_{\tilde{T}}\right) \\
&\quad + \varepsilon^3 \left(-4\xi^4 \partial_x \partial_{\tilde{X}^3}\right) + \varepsilon^4 \left(-\xi^4 \partial_{\tilde{X}^4}\right)
\end{aligned} \quad (22)$$

($L_c = -\xi^4(\partial_{x^2} + k_c^2)^2$ as in Section 3.2.

Up until now, nothing has been said about the amplitude of the solution. However, from results obtained in the previous section, we expect the modulus of the envelope to be of the order of ε for a supercritical bifurcation, hence the expansion

$$w = \varepsilon w_1 + \varepsilon^2 w_2 + \varepsilon^3 w_3 \ldots.$$

The nonlinear term then expands as

$$\begin{aligned}
gw\,\partial_x w &= g(\varepsilon w_1 + \varepsilon^2 w_2 + \ldots)(\partial_x + \varepsilon\,\partial_X)(\varepsilon w_1 + \varepsilon^2 w_2 + \ldots) \\
&= g\Big(\varepsilon^2(w_1\,\partial_x w_1) + \varepsilon^3(w_1\,\partial_x w_2 + w_2\,\partial_x w_1 + w_1\,\partial_X w_1) \\
&\quad + \varepsilon^4\,(w_1\,\partial_x w_3 + w_2\,\partial_x w_2 + w_3\,\partial_x w_1 + w_1\,\partial_X w_2 \\
&\quad + w_2\,\partial_X w_1) + \ldots\Big).
\end{aligned}$$

At order ε, we simply have

$$L_c w_1 = 0 \quad \Rightarrow \quad w_1(x) = w_{11}(x) = \tfrac{1}{2}\big(A_{11} \exp(ik_c x) + \text{c.c.}\big), \quad (23)$$

where the yet undetermined complex amplitude A_{11} is the lowest order contribution to the general amplitude A introduced in (19), with the convention that the first index corresponds to the order and the second index to the rank of the spatial harmonic as before.

At order ε^2, we obtain

$$L_c w_2 + L_1 w_1 = g w_1 \, \partial_x w_1,$$

but it is readily checked that $L_1 w_1 \equiv 0$; indeed, we have

$$L_1 w_1 = -2\xi^4 \left(2\, \partial_x \partial_{\tilde{X}}\right) \left(\partial_{x^2} + k_c^2\right) \tfrac{1}{2} \left(A_{11} \exp(ik_c x) + \text{c.c.}\right),$$

which vanishes identically since $(\partial_{x^2} + k_c^2) \exp(\pm ik_c x) \equiv 0$. This property is not accidental but can be shown formally to derive from the fact that we look for modulations around a perfect structure with $k = k_c$, the critical wavevector that defines the threshold as the minimum of the marginal stability curve.

The problem at order ε^2 reduces itself to

$$\begin{aligned} L_c w_2 &= \frac{g}{4} \left(A_{11} \exp(ik_c x) + \text{c.c.}\right) \left(ik_c A_{11} \exp(ik_c x) + \text{c.c.}\right) \\ &= \frac{g}{4} \left(ik_c A_{11}^2 \exp(2ik_c x) + \text{c.c.}\right). \end{aligned}$$

Since the r.h.s. contains nothing in resonance with $\exp(\pm ik_c x)$, the solution at order ε^2 is the superposition of the general solution of the homogeneous problem, which introduces a second unknown amplitude A_{21} and a special solution of the inhomogeneous problem A_{22}:

$$\begin{aligned} w_2(x) &= w_{21}(x) + w_{22}(x) \\ &= \tfrac{1}{2} \left(A_{21} \exp(ik_c x) + A_{22} \exp(2ik_c x) + \text{c.c.}\right), \end{aligned}$$

where the amplitude A_{22} of the forced response is easily obtained by identification. With

$$-\xi^4 (\partial_{x^2} + k_c^2)^2 w_{22} = -\xi^4 \left((\pm 2ik_c)^2 + k_c^2\right)^2 w_{22} = -9\xi^4 k_c^4 w_{22}$$

we obtain immediately

$$A_{22} = -\frac{2ik_c g}{9\zeta} A_{11}^2, \qquad (24)$$

8. Basics of Pattern Formation

where $\zeta = 4\xi^4 k_c^4 = \xi_0^2 k_c^2$ is a dimensionless parameter of the problem, expected to be of the order of 1 (see Chapter 4 for the case of convection).

At order ε^3, we obtain

$$L_c w_3 = -L_1 w_2 - L_2 w_1 + g\left(w_1 \partial_x w_2 + w_2 \partial_x w_1 + w_1 \partial_X w_1\right). \quad (25)$$

For the same reason as before, w_{21} being in the kernel of L_1, we have $L_1 w_{21} \equiv 0$, so that we obtain

$$-L_1 w_2 = -L_1 w_{22} = -\left[-2\xi^4 \left(2\partial_x \partial_{\tilde{X}}\right)\left(\partial_{x^2} + k_c^2\right) w_{22}\right]$$

$$= -\frac{4}{3} g A_{11} \partial_X A_{11} \exp(2ik_c x) + \text{c.c.}$$

which, involving only $\exp(\pm 2ik_c x)$, turns out to be nonresonant. The term $-2\xi^4 \left(\partial_{x^2} + k_c^2\right) \partial_{\tilde{X}^2} w_1$ in $L_2 w_1$ also vanishes identically, so that we get

$$-L_2 w_1 = -\left[1 - \xi^4 (2\partial_x \partial_{\tilde{X}})^2 - \tau_0 \partial_{\tilde{T}}\right] w_{11}$$

$$= -\left[1 + \xi_0^2 \partial_{\tilde{X}^2} - \tau_0 \partial_{\tilde{T}}\right] \frac{1}{2} A_{11} \exp(ik_c x) + \text{c.c.}$$

with $\xi_0^2 = 4k_c^2 \xi^4$ as defined earlier. This term is obviously resonant.

Turning to the nonlinear term on the r.h.s. of (25), we obtain:

$$\partial_x \left(w_{11}(w_{21} + w_{22})\right) = \tfrac{1}{4}\left(ik_c \bar{A}_{11} A_{22} \exp(ik_c x) + \text{c.c.}\right)$$
$$+ \tfrac{1}{4}\left(2ik_c A_{11} A_{21} \exp(2ik_c x) + \text{c.c.}\right)$$
$$+ \tfrac{1}{4}\left(3ik_c A_{11} A_{22} \exp(3ik_c x) + \text{c.c.}\right),$$

$$w_{11} \partial_X w_{11} = \tfrac{1}{4}\left(A_{11} \partial_X \bar{A}_{11} + \text{c.c.}\right)$$
$$+ \tfrac{1}{4}\left(A_{11} \partial_X A_{11} \exp(2ik_c x) + \text{c.c.}\right).$$

None of these contributions are resonant with $\exp(ik_c x)$, except the first one with $\bar{A}_{11} A_{22} \sim \bar{A}_{11} A_{11}^2$ from (24).

The compatibility condition associated with Equation (25) reduced to the resonant terms,

$$L_c w_3 = -\left[\left(1 + \xi_0^2 \partial_{\tilde{X}^2} - \tau_0 \partial_{\tilde{T}}\right)\frac{1}{2} A_{11} + \frac{g}{4}\left(ik_c \bar{A}_{11} A_{22}\right)\right] \exp(ik_c x)$$
$$+ \text{c.c.},$$

can be written as

$$\tau_0 \partial_{\tilde{T}} A_{11} = A_{11} + \xi_0^2 \partial_{\tilde{X}^2} A_{11} - g_{\text{eff}} |A_{11}|^2 A_{11},$$

with $g_{\text{eff}} = g^2 k_c^2 / 9\zeta$.

The solution at order ε^3 will be the superposition of the general solution of the homogeneous problem and of special solutions of the inhomogeneous problems relative to the nonresonant harmonics $(0, 1, 2) \times k_c$. We obtain

$$w_3(x) = w_{30}(x) + w_{31}(x) + w_{32}(x) + w_{33}(x)$$

with

$$w_{3n} = \tfrac{1}{2} \left(A_{3n} \exp(ink_c x) + \text{c.c.} \right) \quad \text{for } n = 1, 2, 3,$$

where A_{31} is a third unknown amplitude and A_{32} and A_{33} slowly varying amplitudes, functions of A_{11} and A_{21}:

$$A_{32} = \frac{26g}{27\zeta} A_{11} \partial_{\tilde{X}} A_{11} - \frac{4ik_c g}{9\zeta} A_{11} A_{21},$$

$$A_{33} = -\frac{g^2 k_c^2}{48\zeta^2} A_{11}^3.$$

The contribution at $k = 0$, w_{30}, is given by

$$w_{30} = -\frac{g}{\zeta} \partial_{\tilde{X}} |A_{11}|^2.$$

At order ε^4, we have

$$L_c w_4 = -L_1 w_3 - L_2 w_2 - L_3 w_1 + g \Big[\partial_x (w_1 w_3) + w_2 \partial_x w_2 + \partial_{\tilde{X}} (w_1 w_2) \Big].$$

Expanding the r.h.s. and keeping only resonant terms, we obtain the compatibility condition governing A_{21}:

$$\tau_0 \partial_{\tilde{T}} A_{21} = A_{21} + \xi_0^2 \partial_{\tilde{X}^2} A_{21} - i \frac{\xi_0^2}{k_c} \partial \tilde{X}^3 A_{11}$$
$$- g_{\text{eff}} \Big[2|A_{11}|^2 A_{21} + A_{11}^2 \bar{A}_{21}$$
$$- \frac{10i}{3k_c} \left(3 A_{11}^2 \partial_{\tilde{X}} \bar{A}_{11} + 2|A_{11}|^2 \partial_{\tilde{X}} A_{11} \right) \Big].$$

8. Basics of Pattern Formation

The same procedure can be repeated at higher orders, with each step introducing a new amplitude and fixing the amplitude introduced two steps before.

At this stage, the difference between the calculation performed in Section 3 and the present one appears clearly. The existence of the solution at a given order requires the adjustment of free parameters, either in the expansion of the solution or in the expansion of the control parameter. In Section 3, the control parameter was expanded in powers of ε, which introduced unknown coefficients $r_1, r_2, \ldots$ determined by the compatibility condition (9) with the orthogonality condition (10) removing the indeterminacy linked to the presence of the solution of the homogeneous solution at all orders. Here, the freedom is introduced at the level of the amplitudes A_{11}, A_{21}, ... while the control parameter is rigidly fixed to ε by the condition $r = \varepsilon^2$, which implicitly takes for granted the supercritical character of the bifurcation *via* the subsequent assumption $w \sim \varepsilon w_1$ with w_1 nontrivial. The compatibility conditions bear on the rapidly varying part of the solution (the carrier) and yield equations for the slowly varying part (the envelope). In concrete applications, computations are strictly the same, only the interpretation of the results differs slightly.

Now, it is interesting to gather the contributions at successive orders A_{11}, $A_{21}, \ldots$ to reconstruct a general amplitude A defined as

$$A = \varepsilon A_{11} + \varepsilon^2 A_{21} + \ldots . \tag{26}$$

Multiplying the different equations obtained at successive orders by the relevant factors $\varepsilon^3, \varepsilon^4 \ldots$, summing them side to side and returning to the independent variables in physical units (X, T) by performing the inverse of transform (21), we obtain

$$\tau_0 \partial_T A = rA + \xi_0^2 \left(1 - \frac{i}{k_c}\partial_X + \ldots\right)\partial_{X^2} A$$
$$- g_{\text{eff}}\left[|A|^2 A - \frac{10i}{3k_c}\left(3A^2 \partial_X \bar{A} + 2|A|^2 \partial_X A\right) + \ldots\right].$$

Fig. 13. $\mathcal{O}(r^{1/2}) \times \mathcal{O}(r^{1/4})$ domain of wavevectors allowed to take part in modulations of a one-dimensional roll pattern for a horizontally isotropic system.

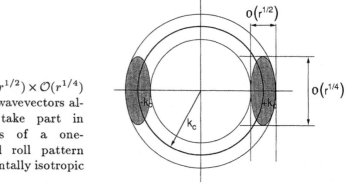

4.2. Roll Modulations in a Rotationally Invariant System

Model (20) cannot be extended straightforwardly to describe rolls generated by a horizontally isotropic instability mechanism. Therefore, to account for modulations taking place along the axis of the rolls (coordinate y) we return to model (11) used in Section 3.2. Assuming $\mathbf{k} = (k_c + \delta k_x)\hat{\mathbf{x}} + \delta k_y \hat{\mathbf{y}}$, the expansion of the linear operator $L = r - \xi^4(\nabla^2 + k_c^2)^2$ now reads

$$\begin{aligned}L(\mathbf{k}) &= r - \xi^4\left((k_c + \delta k_x)^2 + (\delta k_y)^2 - k_c^2\right)^2 \\ &= r - \xi^4\left(2k_c\,\delta k_x + \delta k_x^2 + \delta k_y^2\right)^2.\end{aligned} \quad (27)$$

Comparing independent lowest order terms in the parentheses, we see that

$$\delta k_x \sim \delta k_y^2,$$

as illustrated in Fig. 13.

Introducing specific independent variables for the slow modulations $A \to A(x,y,t)$ with $|\partial_y A| \ll |k_c A|$, i.e., $A(X,Y,T)$ as before and going back to physical space, we first note that ∂_y is merely equivalent to ∂_Y since the rapid space dependence is independent of y for a well-aligned periodic structure. This yields

$$\partial_T A \sim rA \sim \partial_X^2 A \sim \partial_Y^4 A. \quad (28)$$

8. Basics of Pattern Formation

In addition, we see from (28) that we can group $2k_c\,\delta k_x$ and δk_y^2, i.e., we can perform the replacement

$$2\,\partial_x\,\partial_X \longleftrightarrow (2\,\partial_x\,\partial_X + \partial_{Y^2})$$

everywhere in the expansion of $L - \partial_t$ written above (Equation (22)). Notice that from (28), gradients in the y direction can be larger than in the x direction.

The ε expansion proceeds as before with in addition $\tilde{Y} = \varepsilon^{1/2}Y$, which leads to the replacement of ∂_Y by $\varepsilon^{1/2}\,\partial_{\tilde{Y}}$ in the model. In principle, it should be performed in powers of $\varepsilon^{1/2}$, which seems to be the leading order but, owing to the fact that L involves ∂_{y^2} only and that nonlinear terms are quadratic in $\partial_y w$, it is easily checked that the terms present are all proportional to an integer power of ε which therefore remains the actual expansion parameter. (However, in the case of convection, the contribution at order $\varepsilon^{1/2}$ is nontrivial for stress-free boundary conditions; overlooking this fact leads to ignoring the role of the vertical vorticity in an important respect as shown by Siggia and Zippelius, 1981.)

Here, the expansion of the nonlinear term in powers of ε obviously begins at order 3:

$$\left(\frac{1}{k_c^2}(\nabla w)^2 + w^2\right)w$$
$$= \varepsilon^3\left(\frac{1}{k_c^2}(\partial_x w_1)^2 + w_1^2\right)w_1 + \varepsilon^4\left\{\left[2\frac{1}{k_c^2}\partial_x w_1(\partial_x w_2 + \partial_{\tilde{X}} w_1)\right.\right.$$
$$\left.\left.+\frac{1}{k_c^2}(\partial_{\tilde{Y}} w_1)^2 + 2w_1 w_2\right]w_1 + \left(\frac{1}{k_c^2}(\partial_x w_1)^2 + w_1^2\right)w_2\right\}$$
$$+ \mathcal{O}\left(\varepsilon^5\right).$$

As seen in Section 3.2 solutions up to order ε^3 inclusive can all be taken of the form $w_n(x) = w_{n1}(x) = \frac{1}{2}(A_{n1}\exp(ik_c x)+\text{c.c.})$ without other spatial harmonics but, as discussed at the end of Section 4.2, amplitudes $A_{n1}, n = 1,2,3$ are unknown functions of X, Y, T that must not be suppressed by using the supplementary orthogonality condition (10).

A straightforward computation yields the compatibility condition at order ε^3 known as the *Newell–Whitehead equation*:

$$\tau_0\, \partial_{\tilde T} A_{11} = A_{11} + \xi_0^2 \left(\partial_{\tilde X} + \frac{1}{2ik_c} \partial_{\tilde Y^2} \right)^2 A_{11} - g|A_{11}|^2 A_{11}, \quad (29)$$

where $\xi_0^2 = 4k_c^2 \xi^4$ as defined previously. Pushing the expansion at order ε^4, for amplitude A_{21} we get

$$\begin{aligned}
\tau_0\, \partial_{\tilde T} A_{21} = & \; A_{21} + \xi_0^2 \left(\partial_{\tilde X} + \frac{1}{2ik_c} \partial_{\tilde Y^2} \right)^2 A_{21} \\
& - i\frac{\xi_0^2}{k_c} \left(\partial_{\tilde X} + \frac{1}{2ik_c} \partial_{\tilde Y^2} \right) \partial_{\tilde X^2} A_{11} \\
& - g\bigg[(A_{11}^2 \bar A_{21} + 2|A_{11}|^2 A_{21}) + \frac{1}{4k_c^2} (2ik_c A_{11}^2\, \partial_{\tilde X} \bar A_{11} \\
& + 2A_{11} |\partial_{\tilde Y} A_{11}|^2 + \bar A_{11} (\partial_{\tilde Y} A_{11})^2) \bigg]
\end{aligned}$$

and, resumming the series of equations, we obtain

$$\begin{aligned}
\tau_0\, \partial_{\tilde T} A = & \; rA + \xi_0^2 \left(\partial_X + \frac{1}{2ik_c} \partial_{Y^2} \right)^2 A \\
& - \frac{i\xi_0^2}{k_c} \left(\partial_X + \frac{1}{2ik_c} \partial_{Y^2} \right) \partial_{X^2} A \\
& - g\bigg(|A|^2 A + \frac{1}{4k_c^2} (2ik_c A^2\, \partial_X \bar A \\
& + 2A |\partial_Y A|^2 + \bar A (\partial_Y A)^2) + \ldots \bigg).
\end{aligned}$$

4.3. Extension to Two-Dimensional Patterns

In section Section 3, exact resonance conditions between modes were assumed. This restriction has been relaxed above for structures that can be described by a single pair of wavevectors. Here we combine

8. Basics of Pattern Formation

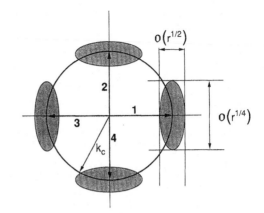

Fig. 14. Domains of wavevectors allowed for a modulated square pattern with two sets of modes at right angles.

the two approaches to treat the approximate resonance between several pairs of wavevectors accounting for more general modulated patterns. Therefore, we introduce one complex amplitude A_i for each wavepacket centered on a given pair of wavevectors $\pm \mathbf{k}_i$ present in the pattern, simply assuming no overlap in Fourier space, i.e., $\alpha_{\mathbf{k}_j \mathbf{k}_j} \gg r^{1/4}$ (Fig. 14 illustrates the case of two pairs at right angles).

Defining specific slow variables X_i, Y_i associated with the direction of a given wavevector pair, we obtain the linear part of the equation that governs the corresponding amplitude directly from the previous calculation. The determination of the nonlinear terms is easily adapted from Section 3 where only regular superpositions were considered since, at lowest nontrivial order, the approximate resonance is taken for exact. Therefore, relative to wavevector $\mathbf{k}_0$, we have

$$\tau_0 \partial_T A_0 = r A_0 + \xi_0^2 \left(\partial_{X_0} + \frac{1}{2ik_c} \partial_{Y_0^2} \right)^2 A_0 - N(A_1, A_2, A_3),$$

where

$$N(A_1, A_2, A_3) = \sum_{\mathbf{k}_1, \mathbf{k}_2, \mathbf{k}_3} g_{123} A_1 A_2 A_3 \, \delta(\mathbf{k}_1 + \mathbf{k}_2 + \mathbf{k}_3 - \mathbf{k}_0),$$

where g_{123} is a function of the angles between the wavevectors which depends on the problem considered. From Fig. 2, the resonance condition implies that we can treat pairs of wavevectors separately, which yields

$$N(A_1, A_2, A_3) = \left(g_0|A_0|^2 + \sum_{i \neq 0} g(\alpha_i)|A_i|^2\right) A_0$$

(note that there is one term for each direction $i \neq 0$ and that $g(\alpha) = g(-\alpha) = g(\alpha + \pi)$).

For model (11), we have

$$g_0 = g, \qquad g(\alpha) = 2g.$$

Replacing $g((1/k_c^2)(\nabla w)^2 + w^2)w$ by gw^3 (the original Swift–Hohenberg term), we obtain

$$g_0 = \frac{3g}{4}, \qquad g(\alpha) = \frac{3g}{2}.$$

A last nonlinear term, derived by Gertsberg and Sivashinsky (1981) for convection between poorly conducting horizontal boundaries, is also of interest. It reads

$$N(w) = \frac{g}{k_c^4}\Big(\left(3\,\partial_x w^2 + \partial_y w^2\right)\partial_{x^2}w \\ + \left(3\,\partial_y w^2 + \partial_x w^2\right)\partial_{y^2}w + 4\,\partial_x w\,\partial_y w\,\partial_{xy}w\Big)$$

and leads to

$$g_0 = \frac{3g}{4}, \qquad g(\alpha) = \frac{g}{2}(1 + 2\cos^2(\alpha)).$$

4.4. Short-Term Stability and Early Nonlinear Selection

Leaving applications of this formalism on truly modulated patterns to a forthcoming section, we consider here only the stability of uniform patterns with respect to uniform perturbations. At lowest order, the problem of the evolution of amplitudes A_i is easily seen

8. Basics of Pattern Formation

to reduce itself to the study of a dissipative dynamical system that derives from a potential. Indeed, we can write

$$\tau_0 \, \partial_T A_i = r A_i - \left(g_0 |A_i|^2 + \sum_{j \neq i} g(\alpha_{\mathbf{k}_i \mathbf{k}_j}) |A_j|^2 \right) A_i \, , \quad i, j = 1, 2, \ldots \, , \tag{30}$$

in the form

$$\tau_0 \, \partial_T A_i = -\frac{\partial G}{\partial \bar{A}_i}$$

with

$$G(\{A_i\}) = -r \sum_i |A_i|^2 + \frac{g_0}{2} \sum_i |A_i|^4 + \sum_{j \neq i} g(\alpha_{\mathbf{k}_i \mathbf{k}_j}) |A_i|^2 |A_j|^2 \, . \tag{31}$$

If modulations are taken into account, the potential has to be completed by adequate terms but the variational structure remains and only the partial derivatives need to be replaced by functional derivatives (see Chapter 9).

Patterns selected at this stage correspond to the linearly stable fixed points of flow (30), i.e., the local minima of potential (31). Here, we restrict ourselves to a two-mode case in order to keep calculations sufficiently simple. For the fixed point equations we obtain

$$(r - g_0 |A_1|^2 - g(\alpha) |A_2|^2) A_1 = 0 \, ,$$
$$(r - g_0 |A_2|^2 - g(\alpha) |A_1|^2) A_2 = 0 \, ,$$

that is to say:

$$A_1 = 0 \, , \quad A_2 = 0 \, , \tag{32a}$$
$$A_1 = 0 \, , \quad |A_2| = \sqrt{r/g_0} \, , \tag{32b}$$
$$A_2 = 0 \, , \quad |A_1| = \sqrt{r/g_0} \, , \tag{32c}$$
$$A_1 = A_2 = \sqrt{r/(g_0 + g(\alpha))} \, . \tag{32d}$$

For $r > 0$, solution (32a) is obviously unstable and we have to determine which solution is selected. Let us first consider solution (32c). The perturbation equations read

$$\tau_0 \, \partial_T \delta A_1 = r \, \delta A_1 - 2 g_0 |A_1|^2 \, \delta A_1 - g_0 A_1^2 \, \delta \bar{A} \, , \tag{33a}$$
$$\tau_0 \, \partial_T \delta A_2 = (r - g(\alpha) |A_1|^2) \, \delta A_2 = (g_0 - g(\alpha)) |A_1|^2 \, \delta A_2 \, , \tag{33b}$$

where (32c) has been used in (33b) to eliminate r. From this equation, it is easily understood that perturbation δA_2 is damped if $g_0 < g(\alpha)$. On the other hand, the growth rate s of perturbation δA_1 is given by

$$(\tau_0 s + r)\delta A_1 + r \exp(2i\phi)\, \delta \bar{A}_1 = 0,$$

where the phase ϕ of the envelope A_1 can be set equal to zero by a change of the origin on the x axis. The first root of this equation, $-2r/\tau_0$, is negative when the bifurcation is supercritical, as implicitly assumed. The corresponding stable mode is that already found in the real Landau model of Chapter 6, Section 2.1. The second root vanishes exactly, which corresponds to the marginal mode associated with the translational invariance of the physical system implying the recourse to complex variables.

Performing the same analysis for solution (32d) that corresponds to a truly two-dimensional pattern, after simplifications (and convenient phase choices), we find:

$$\tau_0\, \partial_T\, \delta A_1 = -\frac{r}{g_0 + g(\alpha)}\big(g_0(\delta A_1 + \delta\bar{A}_1) + g(\alpha)(\delta A_2 + \delta\bar{A}_2)\big)$$

plus a second equation where δA_1 and δA_2 are interchanged. In addition to the two neutral modes corresponding to the translational invariance properties, we have two modes whose evolution rates are given by

$$s_1 = -\frac{2r}{\tau_0}, \qquad s_2 = -\frac{2r}{\tau_0}\frac{g_0 - g(\alpha)}{g_0 + g(\alpha)}.$$

Therefore, we see that when $g(\alpha) > g_0$ the solution with a single amplitude is stable while the superposition of the two amplitudes is unstable (s_2 above is positive). This means that an initial state prepared with roughly equal amplitudes in both directions decays by damping of one or the other amplitude, depending on the position of the initial condition in the phase space spanned by A_1 and A_2, as illustrated in Fig. 15 (this can be generalized to patterns described by larger sets of amplitudes). Conversely, if $g(\alpha) < g_0$ and a solution

8. Basics of Pattern Formation

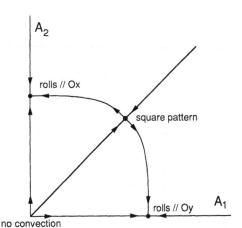

Fig. 15. The stability of a two-dimensional pattern can be decided at lowest order from the phase portrait of differential system (30).

with a single amplitude is prepared, the second one grows from an infinitesimal level. Among different possible perturbations, that with the largest growth rate, i.e., the smallest $g(\alpha)$ (from equation (27b)) dominates and a two-dimensional pattern results. When $\alpha = \pi/2$ this is called the *cross-roll instability*, otherwise an *oblique-roll instability* (return to Chapter 4, Section 3.2).

This stability analysis leads us to our first criterion of nonlinear selection. Indeed, starting an instability experiment usually amounts to taking an initial condition close to the origin in the phase space of all amplitudes. During the phase of linear growth, all modes allowed by the symmetries of the system are present, but with random amplitudes. Even if they happen to reach similar levels, some of them are damped as a result of their interactions with others. Only structures corresponding to the deepest potential wells can survive, as can be easily understood from the consideration of the phase portrait in Fig. 15.

Returning to the evaluation of nonlinear terms in the envelope equations for the modified Swift–Hohenberg model, we see that the first two models have $g_0 < g(\alpha)$ so that monoperiodic structures are expected. By contrast the third nonlinear term leads to $g(\alpha) < g_0$ for some $\pi/3 < \alpha < 2\pi/3$ so that a pattern periodic in two directions is expected. Moreover, $g(\alpha)$ is minimum for $\alpha = \pi/2$ so that the stable structure should correspond to a square pattern.

This selection criterion is very rough since it concerns only the type of the cells and not their wavelengths. Moreover, in spite of the fact that we have neglected modulations, it is valid only at a local scale: during the phase of *early nonlinear selection*, the type of pattern is selected but a very disordered *texture* emerges. The rotational invariance of the underlying problem is locally broken since the system condenses itself in a subset of modes, but it is globally restored at a statistical level when the average over the grain orientations is taken. This short-term behavior is controlled by the first nontrivial contribution to the envelope equations that derives from a potential in general (exception: stress-free Rayleigh–Bénard convection at finite Prandtl number). The subsequent evolution is much slower since it involves a diffusive regression of modulations getting slower as their wavelengths increase.

4.5. Phenomenological Extension: The Example of Waves

Up until now we have considered the envelope equations mostly from the viewpoint of their explicit derivation. This has the advantage of giving a specific value to the coefficients but also the drawback of becoming somewhat tedious. In fact, the *structure* of these equations rests on genericity considerations based on scale and symmetry arguments so that more complicated cases can be handled from a phenomenological viewpoint and the variety of possible behaviors explored systematically by freely varying the coefficients in the equations. This approach will be followed here in the case of dissipative waves generated by instability mechanisms, e.g., convection in binary mixtures introduced in Chapter 3, Section 3.

From Chapter 2, Section 3.1, we know that the generic form of the dispersion relation reads

$$s = \sigma(k) \pm i\omega(k),$$

and that the marginal stability condition is given by $\sigma(k) = 0$. In the following, we relax the condition $\omega(k) \equiv 0$ that leads to the stationary instabilities considered up until now, but we restrict ourselves to

8. Basics of Pattern Formation

a strictly one-dimensional case. The marginal stability condition has no reason to be changed and the real part of the dispersion relation σ may still be expanded as

$$\sigma = \frac{1}{\tau_0}(r - \xi_0^2 \, \delta k^2) + \ldots,$$

where δk is the distance to the critical wavevector k_c. In the same way, the frequency ω can be written as

$$\omega(k) = \omega_c + \omega_c' \delta k + \ldots,$$

where $\omega_c = \omega(k_c)$ and $\omega_c' = d\omega/dk$ at $k = k_c$.

The corresponding modes are of the form

$$w(x,t) = \frac{1}{2}\Big(A_{\mathrm{R/L}} \exp\big(i(k_c x \mp \omega_c t)\big) + \mathrm{c.c.}\Big),$$

i.e., describe waves propagating to the right ($-$ sign) or to the left ($+$ sign). The quantity ω_c/k_c is the phase velocity of the wave and ω_c' is its group velocity, i.e., the velocity at which wavepackets travel.

The dispersion relations of waves driven by external perturbations in equilibrium systems, e.g., elastic waves, are generally of the form $\omega = kc$, where c is the velocity of the waves and either k or ω is controlled from the outside. By contrast, waves appearing in out-of-equilibrium systems result from a spontaneous space-time symmetry breaking, k_c and ω_c being fixed by the instability mechanism.

Turning to modulated solutions, we replace scalars $A_{\mathrm{R/L}}$ by envelope functions $A_{\mathrm{R/L}}(x,t)$ and assume that, in addition to the condition of slow spatial modulations, $|\partial_x A_{\mathrm{R/L}}| \ll |k_c A_{\mathrm{R/L}}|$, we have $\partial_t \to \partial_t \mp i\omega_c$ when acting on the rapidly varying part of the solution and $|\partial_t A_{\mathrm{R/L}}| \ll |\omega_c A_{\mathrm{R/L}}|$ (ω_c is the internal frequency with which the slow evolution rate can be compared).

From a general point of view, introducing explicitly the slow variables X and T, we look for nonlinear solutions in the form

$$\begin{aligned}w(x,t) = \frac{1}{2}\Big(&A_{\mathrm{R}}(X,T) \exp\big(i(k_c x - \omega_c t)\big) \\ &+ A_{\mathrm{L}}(X,T) \exp\big(i(k_c x + \omega_c t)\big) + \mathrm{c.c.}\Big)\end{aligned}$$

with $A_{\text{R/L}}$ governed by envelope equations

$$\tau_0\, \partial_T A_{\text{R}} = F_{\text{R}}(r, A_{\text{R}}, A_{\text{L}}, \bar{A}_{\text{R}}, \bar{A}_{\text{L}}, \partial_X),$$
$$\tau_0\, \partial_T A_{\text{L}} = F_{\text{L}}(r, A_{\text{R}}, A_{\text{L}}, \bar{A}_{\text{R}}, \bar{A}_{\text{L}}, \partial_X)$$

whose structure is constrained by the following symmetry requirements:

(1) Invariance under space translation: changing the origin on the x-axis, i.e., replacing x by $x - x_0$ leads to uniform phase changes for $A_{\text{R/L}}$ induced by $\exp(i(k_c(x-x_0)\mp i\omega_c t)) = \exp(-i\phi)\exp(i(k_c x \mp i\omega_c t))$ with $\phi = k_c x_0$. This implies

$$A_{\text{R}} \to A_{\text{R}}\exp(-i\phi), \quad A_{\text{L}} \to A_{\text{L}}\exp(-i\phi), \quad \text{and c.c.}.$$

(2) Invariance under time translation: $t \to t - t_0$ similarly implies:

$$A_{\text{R}} \to A_{\text{R}}\exp(i\theta), \quad A_{\text{L}} \to A_{\text{L}}\exp(-i\theta), \quad \text{and c.c.},$$

with $\theta = \omega_c t_0$.

(3) Invariance under space reversal: $x \to -x$ leads to an interchange of A_{R} and A_{L}:

$$F_{\text{L}}(r, A_{\text{R}}, A_{\text{L}}, \bar{A}_{\text{R}}, \bar{A}_{\text{L}}, \partial_X) = F_{\text{R}}(r, A_{\text{L}}, A_{\text{R}}, \bar{A}_{\text{L}}, \bar{A}_{\text{R}}, -\partial_X).$$

At leading order, we obtain

$$\tau_0\, \partial_T A_{\text{R}} = r A_{\text{R}} - a\,\partial_X A_{\text{R}} + b\,\partial_{X^2} A_{\text{R}} - \left(g_0|A_{\text{R}}|^2 + g_1|A_{\text{L}}|^2\right) A_{\text{R}},$$
$$\tau_0\, \partial_T A_{\text{L}} = r A_{\text{L}} + a\,\partial_X A_{\text{L}} + b\,\partial_{X^2} A_{\text{L}} - \left(g_0|A_{\text{L}}|^2 + g_1|A_{\text{R}}|^2\right) A_{\text{L}},$$
(34)

where the coefficient a can easily be identified with the group velocity ω_c'. In the same way, the real part of b corresponds to ξ_0^2 as before, whereas its imaginary part is related to the second derivative of $\omega(k)$ with respect to k evaluated at k_c. The nonlinear coefficients g_0 and g_1 are complex in general and depend on the system considered. As in the case of the superposition of several stationary modes (Section 3.5 above), the stability analysis of uniform solutions ($\partial_X \equiv 0$) leads to

8. Basics of Pattern Formation

distinguishing between solutions involving a single amplitude, i.e., right or left *traveling waves*, and solutions involving an equal amount of the two amplitudes, i.e., *standing waves*.

As far as uniform solutions are concerned, the calculation is similar to that developed in the previous section except for the replacement of fixed points with limit cycles. For example, traveling waves (subscript "T") are obtained by assuming only one nonvanishing amplitude, e.g., $A_L = 0$ and $A_R = |A_T| \exp(i\Omega_T T)$. By identification, with $g_i = g'_i + ig''_i$ ($i = 0, 1$), we readily obtain

$$|A_T| = \sqrt{\frac{r}{g'_0}}, \qquad \Omega_T = -\frac{g''_0}{\tau_0}|A_T|^2 = -\frac{r}{g'_0}\frac{g''_0}{\tau_0}.$$

In the same way, for standing waves (subscript "S") we get

$$|A_S| = \sqrt{\frac{r}{g'_0 + g'_1}}, \qquad \Omega_S = -\frac{g''_0 + g''_1}{\tau_0}|A_S|^2 = -\frac{r}{g'_0 + g'_1}\frac{g''_0 + g''_1}{\tau_0}.$$

Both types of waves bifurcate supercritically if $g'_0 > 0$ and $g'_0 + g'_1 > 0$. In that case, the cubic nonlinearities are sufficient to ensure the saturation of the envelopes and Equations (34) make sense. A calculation parallel to that for stationary modes then shows that standing waves are selected when $g'_1 < g'_0$ and traveling waves when $g'_1 > g'_0$.

Finally, when traveling waves are preferred and one direction of propagation is selected, the envelope equation can be written in a frame moving at the group velocity, which removes the term $\partial_X A$. Indeed, for a right traveling wave, writing $\Xi = X - aT$ and $A(X, T) = \tilde{A}(\Xi, T)$ we have $\partial_T A = \partial_T \tilde{A} - a\partial_\Xi \tilde{A}$ and $\partial_X A = \partial_\Xi \tilde{A}$. Inserting these changes in the envelope equation for A suppresses the term with first-order space derivative. The resulting equation

$$\partial_t \tilde{A} = r\tilde{A} + b\partial_{\Xi^2} \tilde{A} - g_0|\tilde{A}|^2 \tilde{A}, \tag{35}$$

known as the *complex Ginzburg–Landau equation*, differs from that describing stationary rolls simply by the presence of complex coefficients.

In the following chapter, we develop some applications of this formalism to long wavelength secondary instabilities, lateral boundary effects, and structural defects, and analyze their consequences on nonlinear selection at lowest order.

5. Bibliographical Notes

The interest of two-dimensional models of convection was first recognized by:

[1] J. Swift and P. Hohenberg, "Hydrodynamic fluctuations at the convective instability," Phys. Rev. A **15**, 319 (1977).

who used the model they derived in an appendix to their paper to discuss the role of fluctuations close to the onset of convection. Later, detailed simulations were performed in rectangular geometry by:

[2] H. S. Greenside and W. M. Coughran Jr., "Nonlinear pattern formation near the onset of Rayleigh-Bénard convection," Phys. Rev. A **30**, 398 (1984).

In circular geometry, first but instructive simulation results, much less refined than those given here, have been presented in:

[3] P. Manneville, "A numerical simulation of convection in a cylindrical geometry," J. Physique **44**, 563 (1983).

The extension of the model including drift flows has been derived in:

[4] P. Manneville, "A two-dimensional model for three-dimensional convective patterns in wide containers," J. Physique **44**, 759 (1983).

The form of the nonlinear term adapted to nearly insulating boundary conditions with elongated cells was derived by:

[5] V. L. Gertsberg and G. I. Sivashinsky, "Large cells in nonlinear Rayleigh-Bénard convection," Prog. Theor. Phys. **66**, 1219 (1981).

For three-dimensional convection, original nonlinear calculations are developed in:

[6a] W. V. R. Malkus, G. Veronis, "Finite amplitude cellular convection," J. fluid Mech. **4**, 225 (1958).

and:

8. Basics of Pattern Formation

[6b] A. Schlüter, D. Lortz, and F. Busse, "On the stability of steady finite amplitude convection," J. Fluid Mech. **23**, 129 (1965).

In the context of convection, envelope equations have been introduced nearly simultaneously by:

[7a] L. A. Segel, "Distant side-walls cause slow amplitude modulation of cellular convection," J. Fluid Mech. **38**, 203 (1969).

and:

[7b] A. C. Newell and J. A. Whitehead, "Finite bandwidth, finite amplitude convection," J. Fluid Mech. **38**, 279 (1969).

The fact that their derivation overlooked the role of the large scale horizontal flow allowed by stress-free boundary conditions was pointed out by:

[8] E. D. Siggia and A. Zippelius, "Pattern selection in Rayleigh-Bénard convection near threshold," Phys. Rev. Lett. **47**, 835 (1981).

A derivation of the envelope equation for the modified Swift–Hohenberg model at high order (with applications to the wavelength selection problem by boundary effects) can be found in:

[9] M. C. Cross, P. G. Daniels, P. C. Hohenberg, and E. D. Siggia, "Phase-winding solutions in a finite container above the convective threshold," J. Fluid Mech. **127**, 155 (1983).

The systematic expansion at first nontrivial order for two-dimensional patterns with different types of cells was already considered by Newell and Whitehead [7b] and further reviewed by:

[10] M. C. Cross, "Derivation of the amplitude equation at the Rayleigh-Bénard instability," Phys. Fluids **23**, 1727 (1980).

who compiled the values of the coefficients adequate for plain convection no-slip boundary conditions. Imperfect thermal boundary condition were shown to lead to square patterns by:

[11] F. H. Busse and N. Riahi, "Nonlinear convection in a layer with nearly insulating boundaries," J. Fluid Mech. **96**, 243 (1980).

and:

[12] C. J. Chapman and M. R. E. Proctor, "Nonlinear Rayleigh–Bénard convection between poorly conducting boundaries," J. Fluid Mech. **101**, 759 (1980).

The phenomenological derivation of envelope equation for waves on the basis of symmetry considerations can be found in:

[13] P. Coullet, S. Fauve, and E. Tirapegui, "Large scale instability of nonlinear standing waves," J. Physique Lett. **46**, L-787 (1985).

The name *Ginzburg–Landau equation* comes from the theory of superconductivity. In the context of fluid dynamics one should probably prefer the name *Stewartson–Stuart equation* after:

[14] K. Stewartson and J. T. Stuart, "A nonlinear instability theory for a wave system in plane Poiseuille flow," J. Fluid Mech. **48**, 529 (1971).

Chapter 9

Applications of the Envelope Formalism

In the following, we will consider mainly modulations and defects appearing in stationary roll patterns and show that the formalism developed in the last part of the previous chapter already gives valuable hints on the selection processes at work in the nonlinear regime above the threshold of the instability that generates these patterns. We will first turn to the description of solutions with off-critical wavevectors within the envelope formalism (Section 1) and analyze their stability against long wavelength modes to which this formalism is specially adapted (Section 2). Next, dropping the reference to a uniform roll structure in a laterally unbounded medium, we will consider the effect of modulations imposed by horizontal confinement (Section 3) and defects (Section 4), which will yield a fairly good description of natural *textures* even at lowest nontrivial order (Section 5).

1. Phase Winding Solutions

Our starting point is Equation (29) of Chapter 8 that, forgetting the specific notation for slow variables, we rewrite as:

$$\tau_0 \, \partial_t A = rA + \xi_0^2 \left(\partial_x + \frac{1}{2ik_c} \partial_{y^2} \right)^2 A - g_{\text{eff}} |A|^2 A. \qquad (1)$$

Nonlinear solutions can have basically two kinds of modulations, either intensity variations (the modulus) at constant wavelength or wavelength modulations (the phase) at constant intensity. In fact, the two kinds of modulation are coupled through (1) but the distinction remains valuable. To make this more explicit, we insert $A = |A|\exp(i\phi)$ into (1). Neglecting space dependence along y for simplicity, after separation of real and imaginary parts, we obtain

$$\tau_0 \, \partial_t |A| = \left(r - \xi_0^2 (\partial_x \phi)^2\right)|A| + \xi_0^2 \, \partial_{x^2} |A| - g_{\text{eff}}|A|^3 , \qquad (2a)$$

$$\partial_t \phi = \frac{\xi_0^2}{\tau_0}\left(\partial_{x^2}\phi + 2\frac{\partial_x |A|}{|A|}\, \partial_x \phi\right). \qquad (2b)$$

From (2a), we see that the evolution of the modulus is mainly controlled by a balance between two contributions existing even in the absence of modulations ($\partial_x \equiv 0$): $r|A|$ and $g_{\text{eff}}|A|^3$. Spatial modulations of the modulus play a role in a second instance only, so that the relevant time scale is primarily linked to the distance to threshold r; this is precisely what we called *short-term nonlinear evolution* before. By contrast, the relaxation of the phase is mostly diffusive as seen from Equation (2b). The rate becomes very slow when the wavelength of the modulation is large and the corresponding diffusivity is given by ξ_0^2/τ_0. The formalism developed in the previous chapter is specially designed to treat these long wavelength fluctuations. The signature of instabilities will appear in a change of sign of the effective phase diffusivities that will be functions of the wavevector of the underlying pattern, an approach to be developed more extensively in Chapter 10.

An important class of exact time-independent solutions of Equations (2) is easily found by assuming a constant modulus $\partial_x |A| = 0$ and $\phi = \delta k\, x + \phi_0$. Indeed, this *phase winding solution* fulfills Equation (2b) with $\partial_t \equiv 0$ identically, whereas Equation (2a) reads:

$$0 = (r - \xi_0^2 \delta k^2)|A| - g_{\text{eff}}|A|^3 ,$$

which yields the nontrivial solution

$$|A| = \sqrt{(r - \xi_0^2\, \delta k^2)/g_{\text{eff}}} \qquad (3)$$

9. Applications of the Envelope Formalism

as long as
$$|\delta k| < \sqrt{r}/\xi_0. \tag{4}$$

Restoring for a while the short scale space dependence, we see that this describes a uniform solution periodic with wavevector $k = k_c + \delta k$. Indeed, inserting $A = |A|\exp(i\,\delta k\,x)$ into the definition of the complete solution, we obtain

$$w = \frac{1}{2}\left(|A|\exp\left(i(\delta k\,x + \phi_0)\right)\exp(ik_c x) + \text{c.c.}\right)$$
$$= \frac{1}{2}\left(|A|\exp(i\phi_0)\exp\left(i(k_c + \delta k)x\right) + \text{c.c.}\right).$$

Condition (4) obviously defines the domain of unstable wavevectors, whereas Equation (3) gives the amplitude of the corresponding nonlinear steady solution.

At this stage, it appears useful to simplify slightly the notations by performing scale changes that suppress phenomenological parameters. Taking ξ_0 and $\sqrt{\xi_0/2k_c}$ as length units along x and y respectively, τ_0 as time unit and $\sqrt{g_{\text{eff}}}$ as amplitude unit, we can rewrite (1) as

$$\partial_t A = rA + \left(\partial_x - i\partial_{y^2}\right)^2 A - |A|^2 A. \tag{5}$$

(Note that length units along x and y are different; r, which could have been eliminated since the equation is consistent at order $r^{3/2}$, has been kept to avoid the definition of space-time units depending on the distance to the threshold.)

In order to study the properties of a structure with a given phase winding δk, it is more convenient to rewrite (1) as an equation for a reduced envelope $\tilde{A}$ related to A by $A(x,y,t) = \tilde{A}(x,y,t)\exp(i\,\delta k\,x)$, which yields

$$\frac{d\tilde{A}}{dt} = (r - \delta k^2)\tilde{A} + 2i\,\delta k\left(\partial_x - i\,\partial_{y^2}\right)\tilde{A} + \left(\partial_x - i\,\partial_{y^2}\right)^2\tilde{A} - |\tilde{A}|^2\tilde{A}. \tag{6}$$

From the above introduction of phase winding solutions, we know that steady-state uniform solutions of Equation (6) read simply $\tilde{A}_0 = \sqrt{r - \delta k^2}$, up to a uniform phase choice ϕ_0.

2. Long Wavelength Instabilities

2.1. General Formulation

Starting with solution $\tilde{A}_0$ of Equation (6), we add an infinitesimal perturbation $a = u + iv$, and linearize the resulting] equation for a. After separation of real and imaginary parts, we obtain

$$\partial_t u = \left(-2(r - \delta k^2) + \partial_{x^2} + \delta k\, \partial_{y^2} - \partial_{y^4}\right)u - \left(2\,\delta k - \partial_{y^2}\right)\partial_x v,$$
$$\partial_t v = \left(2\,\delta k - \partial_{y^2}\right)\partial_x u + \left(\partial_{x^2} + \delta k\, \partial_{y^2} - \partial_{y^4}\right)v,$$

which we analyze using normal modes of the form

$$u = U \exp(st) \cos(q_x x) \cos(q_y y),$$
$$v = V \exp(st) \sin(q_x x) \cos(q_y y). \tag{7}$$

The resulting algebraic system reads

$$\left(s + 2(r - \delta k^2) + q_x^2 + q_y^2\, \delta k + q_y^4\right)U + q_x\left(2\,\delta k + q_y^2\right)V = 0,$$
$$q_x\left(2\,\delta k + q_y^2\right)U + \left(s + q_x^2 + q_y^2\, \delta k + q_y^4\right)V = 0,$$

which directly leads to the dispersion relation

$$\begin{aligned}0 =\,&s^2 + 2\left((r - \delta k^2) + q_x^2 + q_y^2\, \delta k + q_y^4\right)s \\ &+ \left(2(r - \delta k^2) + q_x^2 + q_y^2\, \delta k + q_y^4\right)\left(q_x^2 + q_y^2\, \delta k + q_y^4\right) \\ &- q_x^2(2\,\delta k + q_y^2)^2, \end{aligned} \tag{8}$$

whose roots $s_{(\pm)}$ are easily seen to be real. We obtain

$$s_{(\pm)} = -\left((r - \delta k^2) + q_x^2 + q_y^2\, \delta k + q_y^4\right) \\ \pm \sqrt{(r - \delta k^2)^2 + q_x^2(2\,\delta k + q_y^2)^2}.$$

Solution $s_{(-)}$ is obviously always negative, so that the corresponding mode is stable and rolls can be unstable only if $s_{(+)}$ is positive. Symmetry considerations help us to restrict the study of $s_{(+)}(q_x, q_y; \delta k, r)$ to a domain $(q_x \geq 0; q_y \geq 0)$.

9. Applications of the Envelope Formalism

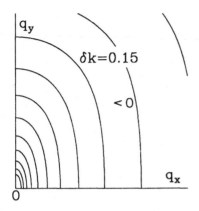
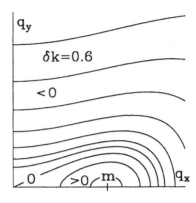
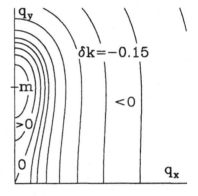

Fig. 1. Level curves of $s_{(+)}(q_x, q_y; \delta k)$: a) for $\delta k = 0.15$, the roll pattern is stable ($s_{(+)} \leq 0$ everywhere); b) for $\delta k = 0.6$, rolls are unstable against a longitudinal mode ($s_{(+)} > 0$ for $q_y = 0$ and $q_x \neq 0$); c) for $\delta k = -0.15$, rolls are unstable against a transverse mode ($s_{(+)} > 0$ for $q_x = 0$ and $q_y \neq 0$).

Level curves of $s_{(+)}(q_x, q_y; \delta k, r = 1)$ are given in Fig. 1 for three values of δk of interest. Maxima take place for either $q_x \equiv 0$ or $q_y \equiv 0$. When $q_y \equiv 0$ ($q_x \equiv 0$), the wavevector of the perturbation is parallel (perpendicular) to the wavevector of the roll pattern. The corresponding mode is called *longitudinal* (*transverse*).

2.2. Longitudinal Perturbations and Eckhaus Instability

Inserting $q_y = 0$ in (8), we obtain

$$s^2 + 2\big((r - \delta k^2) + q_x^2\big)s + q_x^2\big(2(r - 3\delta k^2) + q_x^2\big) = 0.$$

Since the roots are real and their sum always negative, the pattern is stable as long as both roots are negative, i.e., their product is positive. It is unstable when the product becomes negative, i.e., when

$$q_x^2 \leq 2(3\,\delta k^2 - r),$$

which obviously requires $|\delta k| \geq \sqrt{r/3}$; this defines the domain of the *Eckhaus instability*. The condition above also shows that the most unstable wavevector tends to zero when $|\delta k|$ tends to $\sqrt{r/3}$ from above. This long wavelength mode is an example of *hydrodynamic* mode linked to a broken continuous invariance, here the translational invariance. For a picture, return to Chapter 4.

2.3. Transverse Perturbations and Zigzag Instability

We now suppose $q_x = 0$. The two eigenmodes are uncoupled and we have $s_{(-)} = -2(r - \delta k^2) - q_y^2 \delta k - q_y^4 < 0$ for one of them. The other is amplified when

$$s_{(+)} = -q_y^2(q_y^2 + \delta k) > 0,$$

which implies $\delta k < 0$; this condition defines the domain of the *zigzag instability* and when δk tends to zero from below the wavevector q_y of the instability also tends to zero, while the growth rate varies as q_y^2. This second hydrodynamic mode is linked to the rotational invariance (no direction is favored by the primary instability mechanism in the xy plane).

Results of this elementary stability analysis are displayed in Fig. 2.

2.4. Stability of Waves

The starting point is now the complex Ginzburg–Landau equation quoted at the end of Chapter 8 (Equation (35)), which describes modulations to waves in a frame moving at the group velocity (but also uniform oscillatory instabilities with $k_c = 0$, as can be checked easily). The difference with the previous case only lies in the presence of complex coefficients, which is made apparent by performing changes of variables similar to those that lead to Equation (5). Here,

9. Applications of the Envelope Formalism

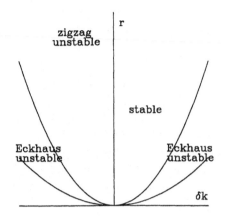

Fig. 2. The Eckhaus instability takes place when $\sqrt{r/3} < |\delta k| < \sqrt{r}$; zigzag perturbations grow when $\delta k < 0$.

with $b = b' + ib'' = \xi_0^2(1 + i\beta)$ and $g_0 = g_{\text{eff}}(1 + i\gamma)$, we obtain

$$\partial_t A = rA + (1 + i\beta)\partial_{x^2} A - (1 + i\gamma)|A|^2 A. \quad (9)$$

Phase winding solutions are obtained by inserting

$$A = \tilde{A}_0 \exp\left(i(\delta k\, x - \delta\omega\, t)\right)$$

in (9). Separation of real and imaginary parts yields

$$|\tilde{A}_0|^2 = r - \delta k^2,$$
$$\delta\omega = \beta\, \delta k^2 + \gamma|\tilde{A}_0|^2 = \gamma r + (\beta - \gamma)\, \delta k^2.$$

The stability of these solutions can be studied using a modified envelope equation analogous to (6). Here, we assume a modulated solution in the form: $A(x,t) = \tilde{A}(x,t)\exp\left(i(\delta k\, x - \delta\omega\, t)\right)$, which leads to

$$\partial_t \tilde{A} = (1 + i\gamma)\left[r - \delta k^2 - |\tilde{A}|^2\right]\tilde{A} + (1 + i\beta)\left[2i\,\delta k\partial_x + \partial_{x^2}\right]\tilde{A}.$$

Inserting $\tilde{A} = \tilde{A}_0 + a$ and linearizing with respect to a, we obtain

$$\partial_t a = -(1 + i\gamma)(r - \delta k^2)(a + \bar{a}) + (1 + i\beta)(2i\,\delta k\partial_x + \partial_{x^2})a.$$

Assuming further that $a = u + iv$, after separation of real and imaginary parts we get

$$(\partial_t + 2\beta\,\delta k\,\partial_x - \partial_{x^2})u = -2(r - \delta k^2)u - (2\,\delta k\,\partial_x + \beta\,\partial_{x^2})v,$$
$$(\partial_t + 2\beta\,\delta k\,\partial_x - \partial_{x^2})v = \bigl(-2\gamma(r - \delta k^2) + 2\,\delta k\,\partial_x + \beta\,\partial_{x^2}\bigr)u$$

(when $\beta = \gamma = 0$ the case of the Eckhaus instability is recovered). The general study is complicated by the fact that normal modes can no longer be taken in the form (7) since partial derivatives mix sines and cosines in an intricate way except when $\delta k = 0$.

In this latter case, taking $(u,v) = (U,V)\exp(st)\cos(q_x x)$, we obtain

$$(s + 2r + q_x^2)U - \beta q_x^2 V = 0,$$
$$(2\gamma r + \beta q_x^2)U + (s + q_x^2)V = 0,$$

which yields

$$s^2 + 2(r + q_x^2)s + q_x^2\bigl(2r(1+\beta\gamma) + q_x^2(1+\beta^2)\bigr) = 0. \qquad (10)$$

Since for $r > 0$ the sum of the roots is always negative, there will be an instability only if their product can become negative, i.e.,

$$2r(1+\beta\gamma) + q_x^2(1+\beta^2) < 0,$$

which will be possible only if (*Newell criterion*)

$$1 + \beta\gamma < 0$$

In the general case, we are obliged to assume $(u,v) = (U,V)\exp(st + iq_x x)$. In replacement of (10), a straightforward analysis yields:

$$s^2 + 2(r - \delta k^2 + q_x^2 + 2i\beta\,\delta k\,q_x)s$$
$$+ 2(r - \delta k^2)\bigl(q_x^2(1+\beta\gamma) + 2i\,\delta k\,q_x(\beta - \gamma)\bigr)$$
$$+ q_x^2(1+\beta^2)(q_x^2 - 4\,\delta k^2) = 0.$$

Since this equation has complex coefficients, though still formally simple, the problem of checking the sign of the real parts of the two roots becomes more complicated.

9. Applications of the Envelope Formalism   349

The instability of waves against long wavelength longitudinal modes is often called the *Benjamin–Feir instability* though this instability was originally derived for waves at the surface of a liquid, a case that corresponds to taking the "conservative limit" of the envelope equation: $r \to 0$ and $(\beta, \gamma) \to \infty$. Real terms become irrelevant in (9), which is then better read as a nonlinear Schrödinger equation: $i\,\partial_t A = \partial_{x^2} A + |A|^2 A$.

3. Lateral Boundary Effects

Up until now the envelope equations have been written for a laterally unbounded system. The presence of side boundaries that confine the system can modify the instability mechanism. For example, in convection, viscous friction at a lateral wall increases the damping, which inhibits the instability. By contrast, a thermal perturbation in the wall or simply a difference of thermal conductivity between the fluid and the wall, can build up a horizontal temperature gradient that may induce a fluid motion, thus encouraging the instability. In both cases, we expect lateral boundary effects to impose modulations and act as boundary conditions on the envelope. Technically, they are obtained by *matching* two expressions of the solutions. The first one, in terms of the envelope $A(x, y)$, i.e., $\frac{1}{2}\bigl(A\exp(ik_c x) + \text{c.c.}\bigr)$, is asymptotically exact far from the wall $(x \gg \mathcal{O}(r^{-1/2}))$. The other expression holding in a vicinity $\mathcal{O}(\lambda_c) \ll \mathcal{O}(r^{-1/2})$ is sought directly as a solution of the primitive problem.

The general form of boundary conditions on the envelope, nearly trivial at lowest order (see also Section 5 below), can be obtained phenomenologically using symmetry requirements, but can of course also be derived explicitly from the primitive equations.

3.1. Rolls Parallel to a Lateral Wall

In the direction of the pattern's wavevector, the envelope equation defines a well-posed second-order differential problem if we have one condition at each end of the interval corresponding to the width of the system. At lowest order we usually have

$$A(x = \pm \ell_x/2) = 0 \qquad (11)$$

or, for a semi-infinite system,

$$A(x = 0) = 0 \quad \text{and} \quad A(x \to \infty) = \sqrt{r}.$$

Writing $A(x) = |A(x)| \exp(i\phi(x))$, we can get the steady solution explicitly; the equation for ϕ reads

$$|A| \frac{d^2\phi}{dx^2} + 2 \frac{d|A|}{dx} \frac{d\phi}{dx} = \frac{1}{|A|} \frac{d(|A|^2 \, d\phi/dx)}{dx} = 0,$$

which yields

$$|A|^2 \frac{d\phi}{dx} = C,$$

where C is an integration constant. Since $|A|$ must vanish at $x = 0$, the divergence of $d\phi/dx$ can be avoided only if $C = 0$ and therefore $d\phi/dx = 0$. But since the phase is irrelevant, it can be chosen so that A is real. The envelope equation then simply reads

$$\frac{d^2 A}{dx^2} = -rA + A^3$$

and for a semi-infinite system we easily obtain

$$A(x) = \sqrt{r} \, \tanh(x\sqrt{r}/2).$$

The value and the variation of the coherence length $\xi = 2/\sqrt{r}$ ($\times \xi_0$ when restoring the longitudinal scales) are in good agreement with experimental results.

The condition $d\phi/dx = 0$ seems to mean that the phase is kept constant, i.e., $k = k_c$ (strict wavelength selection at k_c). In fact, this holds only at order $r^{3/2}$, the order at which the amplitude equation is valid. $C = 0$ actually means $C = \mathcal{O}(r^2)$, which implies $d\phi/dx = \mathcal{O}(r)$; the wavelength can vary linearly with r.

9. Applications of the Envelope Formalism   351

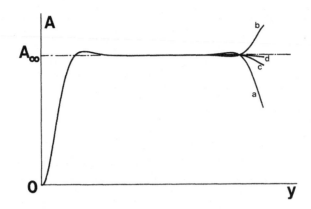

Fig. 3. Profile of the amplitude through a boundary layer perpendicular to the roll axis as obtained numerically by a "shooting method," i.e., adjustment of d^3A/dy^3: (a) 0.585401895, (b) 0.585401905, (c) 0.585401900, (d) 0.585401901,...

3.2. Rolls Perpendicular to a Lateral Wall

For rolls perpendicular to a lateral wall, we have a fourth-order differential equation so that we need two conditions at each end of the interval $\pm \ell_y$. As shown by Brown and Stewartson (1978), at lowest order we expect

$$A(\pm \ell_y/2) = 0 \quad \text{and} \quad \frac{dA}{dy}(\pm \ell_y/2) = 0 \qquad (12)$$

or, for a semi-infinite y-interval $[0, \infty)$,

$$A = \frac{dA}{dy} = 0 \quad \text{at } y = 0, \qquad A \to \sqrt{r} \quad \text{for } y \to \infty.$$

The solution of the corresponding envelope equation for $k = k_c$,

$$\frac{d^4 A}{dy^4} = rA - A^3,$$

is not known analytically but can be obtained numerically (Fig. 3). The equation above can be taken as an initial value problem with

$A = 0$ and $dA/dy = 0$. For a semi-infinite medium, the initial value of d^2A/dy^2 can be obtained from the value of the first integral:

$$\frac{dA}{dy}\frac{d^3A}{dy^3} - \frac{1}{2}\left(\frac{d^2A}{dy^2}\right)^2 - \frac{1}{2}rA^2 + \frac{1}{4}A^4 = C,$$

where the integration constant C can be evaluated at $y \to +\infty$ with $A \to r^{1/2}$ and all derivatives tending to zero. This gives $d^2A/dy^2 = r\sqrt{2}$ at $y = 0$. Since the only remaining parameter is the value of d^3A/dy^3 at $y = 0$, the solution can be obtained easily by a trial-and-error method. The numerical solution is allowed to diverge further and further from the origin as the value of d^3A/dy^3 is refined. We get $d^3A/dy^3 \simeq 0.5854019\ldots r^{5/4}$.

4. Structural Defects

In the absence of induction processes, dissipative structures in extended geometry are far from being ideal. The resulting pattern is made of patches of regular structures containing many defects. Using a terminology borrowed from the field of metallurgy, the regular patches are called "grains" and the "internal walls" separating two grains with different orientations are called *grain boundaries*. In addition, there may be *dislocations* and more complicated combined defects. The overall pattern is then called a *texture*.

Here we will study the simplest cases at lowest order: the steady grain boundary between two sets of rolls at right angle and the isolated dislocation.

4.1. Grain Boundaries

We consider the configuration depicted in Fig. 4a. To account for it, we clearly need two envelopes, one for the set of rolls with wavevector parallel to the x axis and one for the other set of rolls.

Let A_1 (A_2) be the envelope corresponding to the first (second)

9. Applications of the Envelope Formalism   353

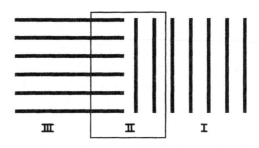

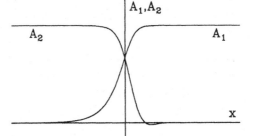

Fig. 4. Grain boundary between two systems of rolls at right angles. Top: sketch of the configuration. Bottom: numerical solution of Eq. (13).

set of rolls, so that

$$x \to +\infty \quad \Rightarrow \quad A_1 \to r^{1/2}, \quad A_2 \to 0,$$
$$x \to -\infty \quad \Rightarrow \quad A_2 \to r^{1/2}, \quad A_1 \to 0.$$

From the general expression of the set of equations governing a nonlinearly coupled set of envelopes (Section 2.4), we derive

$$\partial_t A_1 = rA_1 + \partial_{x^2} A_1 - (|A_1|^2 + \mu |A_2|^2) A_1, \qquad (13a)$$
$$\partial_t A_2 = rA_2 - \alpha \partial_{x^4} A_2 - (|A_2|^2 + \mu |A_1|^2) A_2, \qquad (13b)$$

where $1/\alpha = 4k_c^2 \xi_0^2$ and $\mu = g(\pi/2)/g_0$ (we make use of the assumed translational invariance in the y direction to impose $\partial_y \equiv 0$).

The very existence of a grain boundary between two roll systems requires the stability of the rolls themselves with respect to a square pattern. Otherwise, the region where the superposition is observed would not remain localized but, instead, invade the whole system. Recalling that the bifurcation is supercritical when $g_0 > 0$ and that when $|g(\alpha)| < g_0$ a bimodal structure is preferred (Chapter 8, Section 4.5), this yields $\mu > 1$. Here, we assume in addition no time dependence and no phase winding ($k_1 = k_2 = k_c$), so that A_1 and A_2 can be taken as real.

The aim of the theoretical analysis is to obtain an approximate functional description of the numerical solution depicted in Fig. 4b. From this figure, we can distinguish two "external" regions (I) and (III) where one of the sets of rolls dominates, the other being essentially negligible, and an "internal" region (II) where the two envelopes are of the same order of magnitude. The analytical approach involves a *matching* procedure of solutions obtained in the two external regions and a check for consistency in the internal region.

The solution at lowest order is obtained by noticing that the typical scale over which A_2 responds to an external forcing is $\mathcal{O}(r^{-1/4}) \ll \mathcal{O}(r^{-1/2})$, which is the typical scale of modulation for A_1. This implies that A_2 follows adiabatically the variations of A_1. From Equation (13b), assuming that $\partial_{x^4} A_2 \sim 0$, we obtain

$$A_2^2 = r - \mu A_1^2 \tag{14}$$

which has some meaning only for $A_1^2 < r/\mu$, i.e., in the region where rolls parallel to the y axis give way to the rolls at right angle. Inserting result (14) into (13a) we get the effective equation for A_1 in this region:

$$-(\mu - 1)r A_1 + \partial_{x^2} A_1 - (1 - \mu^2) A_1^3 = 0. \tag{15}$$

The solution that fulfills the boundary condition $A_1 \to 0$ when $x \to -\infty$ reads

$$A_1^{(\mathrm{III})}(x) = \sqrt{\frac{2r}{1+\mu}} \, \frac{1}{\cosh(\sqrt{r(\mu-1)}(x-x'))}, \tag{16}$$

where x' is a parameter left free for an adjustment of the position of the grain as seen from region (III). We observe that the penetration length of the rolls described by A_1 into the rolls described by A_2 is increased by a factor of $1/\sqrt{\mu-1}$ with respect to the usual coherence

9. Applications of the Envelope Formalism

length, in agreement with experimental results. This penetration length tends to infinity when $\mu \to 1$ from above, which is consistent with the fact that when $\mu < 1$ squares are preferred to rolls.

In region (I), rolls associated with amplitude A_1 dominate. Condition $A_1^2 < r/\mu$ is not fulfilled and, at lowest order, the trivial solution of (13b) must be chosen. Inserting $A_2 = 0$ into Equation (13a) for A_1 we get

$$rA_1 + \partial_{x^2} A_1 - A_1^3 = 0, \tag{17}$$

which yields the classical hyperbolic tangent

$$A_1^{(I)}(x) = \sqrt{r} \tanh\left(\sqrt{r/2}(x - x'')\right), \tag{18}$$

where x'' is another free parameter that fixes the absolute position of the grain boundary when coming from $+\infty$.

The point $x = x_*$, where $A_1^2 = r/\mu$, is a turning point where the different expressions for A_1 and its first derivative $\partial_x A_1$ must match (because (15) and (17) are second-order differential equations). This matching is achieved by varying $x'' - x'$. The solution at lowest order is then completely specified by expressions (14) and (16) (with adequate x') in region (III), and by (18) and $A_2 \equiv 0$ in region (I). This is the so called "external solution" (Fig. 5).

Things work well if this solution does not generate singularities that would be impossible to compensate for in the internal region (II) where A_2 varies rapidly and the fourth-order derivative is not negligible. In the vicinity of x_*, A_1 and $\partial_x A_1$ are continuous; A_2 is also continuous since it varies linearly with $x - x_*$ for $x < x_*$ and vanishes for $x > x_*$, but $\partial_x A_2$ is discontinuous and $\partial_{x^2} A_2$ diverges. In fact, replacing A_1 by the lowest order term of its Taylor expansion in Equation (13b), we get the equation for A_2 valid in region (II):

$$\alpha \partial_{x^4} A_2 = -\left(r^{3/2} c(x - x_*) + A_2^2\right) A_2, \tag{19}$$

where c is some numerical factor. Scale changes allow us to suppress all parameters from Equation (19). The amplitude scale is easily found to be $r^{3/5}$ and, r being small, since $3/5 > 1/2$, the corrections brought by (19) remain small. In the same way, lengths scale as

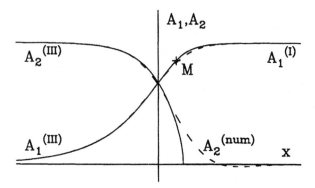

Fig. 5. The lowest order solution for a grain boundary as obtained from the matching of external solutions at M is robust to the introduction of corrections in the internal region as seen from the comparison with the numerical solution of (13), dotted line.

$r^{-3/10} \ll r^{-1/2}$ for r small, so that corrections remain confined in the internal region of width $\mathcal{O}(r^{-1/2})$, see Fig. 5. Everything being consistent, the lowest order solution is satisfactory and we need not look for an explicit solution to Equation (19). The complete solution can be found by a numerical simulation (see Fig. 4b).

4.2. Dislocations

Grain boundaries ensuring the connection between domains with different orientations take the form of rows of dislocations when the orientations of the grains become comparable (see Fig. 6a). In a similar way, two roll patterns with identical orientations but slightly different wavelengths are connected by dislocations (Fig. 6b) and the spacing between dislocations is easily seen to vary as the inverse of the difference between the wavelengths in the bulk of each domain, which leads to the concept of *isolated dislocation* studied in the following.

Let us consider a dislocation row of the type (Fig. 6b) with wavevectors in a $n/(n-1)$ ratio (n large). We can assume periodic boundary conditions in the x direction at a (large) distance ℓ_x such

9. Applications of the Envelope Formalism

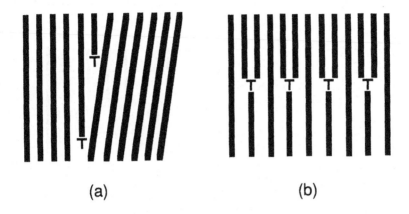

Fig. 6. Rows of dislocations: a) between two grains of similar orientations; b) between grains with wavevectors in a 3/2 ratio.

that the pattern is asymptotic to a set of rolls with $k_{(\mathrm{I})} = k_c + \delta k_{(\mathrm{I})}$ for $y \to +\infty$ (region (I)) and to a set with $k_{(\mathrm{II})} = k_c + \delta k_{(\mathrm{II})}$ for $y \to -\infty$ (region II)). Wavevector shifts $\delta k_{(\mathrm{I})}$ and $\delta k_{(\mathrm{II})}$ are related by the condition that $2n$ rolls (n wavelengths) in region (I) match with $2(n-1)$ rolls in region (II), i.e.,

$$\ell_x = \frac{2\pi n}{k_{(\mathrm{I})}} = \frac{2\pi(n-1)}{k_{(\mathrm{II})}}.$$

Accordingly, the asymptotic rates of phase winding will be different in the two regions: $\exp(i\,\delta k_{(\mathrm{I})}\,x)$ for $y \to -\infty$ and $\exp(i\,\delta k_{(\mathrm{II})}\,x)$ for $y \to +\infty$.

Following a circuit far from the location of the dislocation, we see that the phase of the envelope must experience a 2π discontinuity corresponding to the loss of a pair of rolls (by analogy with the metallurgical case, a *Burgers vector* can be defined characterizing the circuit, see Fig. 7). This discontinuity has a topological origin and cannot be removed. Its presence implies the existence of a singularity of the modulus of the envelope. Though this is not obvious from experimental results or numerical simulations of the Boussinesq equations, it is generally assumed that the phase jump takes place in

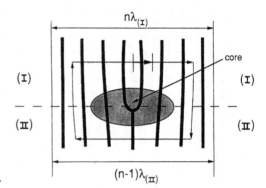

Fig. 7. Isolated dislocation.

a region where the modulus of the envelope decreases to zero. This singular region is called the *core* of the dislocation.

Within the framework of the envelope equation formalism, we do not have to worry about the way the singularity is resolved at the scale of the roll wavelength. The dimensions of the region perturbed by the presence of the core can be derived from the scalings implied by the envelope equation itself: $r^{-1/2}$ along x and $r^{-1/4}$ along y.

In general dislocations do not stay at rest. Their motion can be decomposed into two components: one parallel to the average wavevector of the structure is called the *gliding* motion, the other parallel the roll axis is the *climbing* motion.

The climbing motion is a wavelength selection process since when the dislocation moves along the roll axis, one of the regions grows at the expense of the other. The wavevector difference acts as a force on the dislocation. Owing to the symmetry of the texture considered, the gliding motion is expected to be damped; this would no longer be the case in the presence of a global curvature.

After a short transient, the climbing motion of an isolated dislocation becomes uniform. Therefore, we can assume that the dependence on y and t involves only the combination $y - Vt$, where V is the velocity of the dislocation. It is convenient to search for V by using the equation for the reduced envelope $\tilde{A}$ describing a structure with the average phase winding corresponding to $\delta k = \frac{1}{2}(\delta k_{(\mathrm{I})} + \delta k_{(\mathrm{II})})$.

9. Applications of the Envelope Formalism   359

Inserting $\tilde{A}(x, y - Vt)$ in Equation (6), we obtain

$$-V\,\partial_y A = (r - \delta k^2)A + (\partial_x - i\,\partial_{y^2})^2 A$$
$$+ 2i\,\delta k\,(\partial_x - i\,\partial_{y^2})A - |A|^2 A.$$

Dislocation motion is a slow process that relates to the dynamics of the phase so that we can admit without much discussion that the modulus of the envelope has relaxed to their local equilibrium value, i.e., that the first and the last term on the r.h.s. compensate exactly (the time required for this is of the order of r^{-1}, whereas the motion of the dislocation can be made arbitrarily slow). Therefore, we are left with

$$-V\,\partial_y \tilde{A} = (\partial_x - i\,\partial_{y^2})^2 \tilde{A} + 2i\,\delta k\,(\partial_x - i\,\partial_{y^2})\tilde{A}.$$

A simple dimensional argument then gives the dislocation velocity as a function of the average wavevector shift. Indeed the change of variables $x = \hat{x}/\delta k$, $y = \hat{y}/\sqrt{\delta k}$ yields

$$-V\,\delta k^{-3/2}\partial_{\hat{y}} A = (\partial_{\hat{x}} - i\,\partial_{\hat{y}^2})^2 A + 2i(\partial_{\hat{x}} - i\,\partial_{\hat{y}^2})A.$$

The solution of this equation, whatever it is, depends only on a boundary condition at large distance (circulation of the phase = 2π). Therefore, the combination $V\,\delta k^{-3/2}$ scales as a constant, which yields

$$V \sim \delta k^{3/2}.$$

Of course this is valid only if δk is positive, which is consistent with the fact that, according to the envelope formalism at lowest order, when δk is negative the system is unstable against the zigzag mode.

5. Pattern Selection at Lowest Order

Following Cross (1982), we develop here a theory of pattern selection based on the envelope formalism at first nontrivial order. This theory, which already accounts for many experimental facts, rests on the remark that at lowest order Equation (5) derives from a potential

$$\partial_t A = -\frac{\delta G(A, \bar{A})}{\delta \bar{A}}, \tag{20}$$

where symbol δ denotes the functional differentiation (see below).

Generalizing the approach developed in the previous chapter, Section 4.5, we define the potential G as the integral of a density $\mathcal{G}$:

$$G = \int_\mathcal{D} \mathcal{G}(A, \bar{A})\, dx\, dy, \tag{21}$$

with

$$\mathcal{G}(A, \bar{A}) = \frac{r^2}{2} - r|A|^2 + \frac{1}{2}|A|^4 + \left|(\partial_x - i\,\partial_{y^2})A\right|^2. \tag{22}$$

To prove this property, we compute the first order variation δG under the change $A \to A + \delta A$, obtaining

$$\begin{aligned}
\delta G =& G(A + \delta A, \bar{A} + \delta \bar{A}) - G(A, \bar{A}) \\
=& \int_\mathcal{D} \Big(-rA\,\delta\bar{A} + A^2 \bar{A}\,\delta\bar{A} \\
& + [(\partial_x - i\,\partial_{y^2})A][(\partial_x - i\,\partial_{y^2})\delta\bar{A}] + \text{c.c.} \Big)\, dx\, dy.
\end{aligned}$$

Considering all terms containing $\delta\bar{A}$, we see that some of them also involve $\partial_x\,\delta\bar{A}$ and $\partial_{y^2}\,\delta\bar{A}$, which are not independent variations but can be transformed by integrations by parts. Assuming a rectangular domain for simplicity, we have, for example,

$$\begin{aligned}
& \int_\mathcal{D} (\partial_x A\,\partial_x\,\delta\bar{A})\, dx\, dy \\
& = \int_{\ell_y} \Big([A\,\delta\bar{A}](\ell_x, y) - [A\,\delta\bar{A}](0, y)\Big)\, dy - \int_\mathcal{D} (\partial_{x^2} A)\,\delta\bar{A}\, dx\, dy
\end{aligned}$$

where boundary terms $[\cdots]$ do not contribute if $A \equiv 0$. In the same way, $\partial_{y^2}\,\delta\bar{A}$ can be transformed by two successive integrations by

9. Applications of the Envelope Formalism

parts and boundary terms disappear if in addition $\partial_y A \equiv 0$. Finally, we get

$$\delta G = \int_{\mathcal{D}} \Big(\big(-rA - (\partial_x - i\,\partial_{y^2})^2 A + A^2 \bar{A} \big) \delta \bar{A} + \text{c.c.} \Big) dx\,dy\,. \quad (23)$$

The meaning of the functional derivative of G with respect to its variables $(A, \bar{A})$ is enlightened by discretizing the integral:

$$\delta G \simeq \sum_{(i,j)\in\mathcal{D}} \Big(\big(-rA_{i,j} - (\partial_x - i\,\partial_{y^2})A\big|_{i,j} + A_{i,j}^2 \bar{A}_{i,j} \big) \delta \bar{A}_{i,j} + \text{c.c.} \Big) \Delta x\,\Delta y\,.$$

Considering each value of the function at a given node as a separate variable, the total variation is then understood as a sum of individual variations. Since we have taken care to avoid partial derivatives of $(\delta A, \delta \bar{A})$, the derivation with respect to local values $\delta A_{i,j}$ and $\delta \bar{A}_{i,j}$ keeps its usual meaning and then becomes trivial. As a result, the functional derivation comes to an extraction of the integrand relative to the relevant variations δA and $\delta \bar{A}$. When applied to (20) using (23), this procedure directly yields Equation (5) and implicitly involves boundary conditions (11) and (12). With less care, we could have simply assumed that the contribution corresponding to the boundary terms was negligible with respect to that of the bulk terms.

Owing to this variational property, the evolution then merely amounts to a monotonic relaxation toward a steady solution minimizing the potential. Indeed, we have

$$\frac{dG}{dt} = \int_{\mathcal{D}} \left(\frac{\delta G}{\delta \bar{A}} \partial_t \bar{A} + \text{c.c.} \right) dx\,dy = -2 \int_{\mathcal{D}} |\partial_t A|^2 dx\,dy \leq 0\,,$$

so that G, which is bounded from below on a finite domain, can only decrease toward a local minimum.

In the absence of boundary effects, the absolute minimum corresponds to a uniform solution with wavevector $k = k_c$. Any modulation leads to an increase of the potential. Expression (21) suggests that the variation of $G(A, \bar{A})$ will be proportional to the surface affected by the modulation and to the typical variation of the local

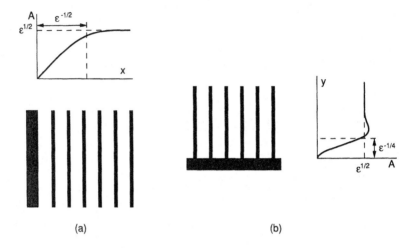

Fig. 8. The boundary layer for rolls perpendicular to the lateral wall (a) is narrower than that for rolls parallel to it (b); the corresponding potential increase is smaller, which makes this configuration more favorable.

potential density $\mathcal{G}$, which is easily seen to be of order r^2 since all terms in the integrand contribute to this order.

All sources of modulations studied before can be re-examined from the point of view of their contribution to the total potential.

(1) Close to a lateral boundary, for rolls parallel to a piece of wall of length ℓ_y, the boundary layer being of order $r^{-1/2}$, the potential will increase by about $r^2 \times r^{-1/2} \times \ell_y = r^{3/2}\,\ell_y$.

(2) If rolls are perpendicular to a piece of wall of length ℓ_x, the boundary layer being of width $r^{-1/4}$, the potential increase will be $r^2 \times r^{-1/4} \times \ell_x = r^{7/4}\,\ell_x$.

(3) The contribution of a grain boundary of length ℓ_{gb} will be $r^2 \times r^{-1/2} \times \ell_{gb} = r^{3/2}\,\ell_{gb}$ since the width of the grain boundary is of order $r^{-1/2}$; however, since the intensity of the instability does not fall completely to zero in the internal region, one expects a smaller increase than for rolls parallel to a lateral wall. Only a detailed calculation can give the correct prefactor.

9. Applications of the Envelope Formalism 363

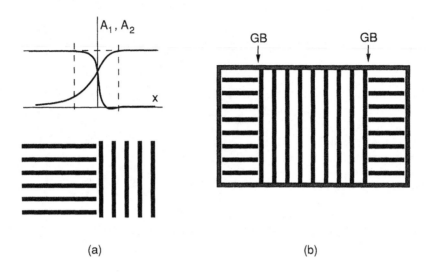

(a)   (b)

Fig. 9. A grain boundary contributes to the potential slightly less than the boundary layer for rolls parallel to a wall (a); a solution with two grain boundaries and rolls everywhere perpendicular to the lateral wall is well adapted to a rectangular geometry (b).

Since $7/4 > 6/4 = 3/2$, when r is small we have $r^{7/4} \ll r^{3/2}$; the potential increase for rolls perpendicular to the boundary is much smaller then for rolls parallel to it (Fig. 8). Therefore we expect patterns with rolls mostly perpendicular to the lateral walls.

A seemingly favorable pattern in a rectangular container is depicted in Fig. 9; rolls are everywhere perpendicular to the lateral walls and grain boundaries parallel to the short sides of the rectangle resolve the orientation change in the bulk. In fact, this very symmetric solution can be observed above threshold only if some induction process is used during the growth of the pattern. "Natural" textures are far more disordered, especially when the container has a circular geometry, since then geometrical constraints brought by the orthogonality condition cannot be resolved without large scale curvature and structural defects (see simulations presented in Chapter 8, Section 2).

In any sufficiently homogeneous region, the structure can be de-

Fig. 10. A director field can be defined in a texture with widely varying roll orientation except in the core region of structural defects.

scribed simply by specifying the length and direction of the local wavevector that plays the role of a *director* for the orientation field (Fig. 10), a description derived from that of liquid crystals (see Appendix 1). In this picture, the critical wavelength defines the microscopic scale and the coherence length, of order $r^{-1/2}$ or $r^{-1/4}$, plays the role of a macroscopic length, whereas the horizontal size of the container is expected to be even much larger. The director field cannot be defined in regions where several structures overlap (e.g., grain boundaries) or at places where the envelope becomes singular (e.g., at dislocation cores) but the surface of such regions remains small when compared with the total surface of the system. Moreover, variations of order $r^{1/2}$ of the length of the wavevector are prohibited by the large potential increase that they imply; this amounts to saying that the length of the wavevector is kept fixed and equal to k_c up to higher order corrections.

Using the terminology of nematic liquid crystals, we see that two kinds of deformations are possible: *splay*, corresponding to a curvature of the rolls at constant wavelength (Fig. 11a), and *bend*, implying large wavelength variations (Fig. 11b) forbidden by the previous considerations. In highly compressed (dilated) regions, the amplitude of the rolls is small, which makes it easy for the system to suppress (create) a pair of rolls. Therefore, the way to avoid costly bend deformation is by nucleating dislocations as suggested in Fig. 12a.

9. Applications of the Envelope Formalism

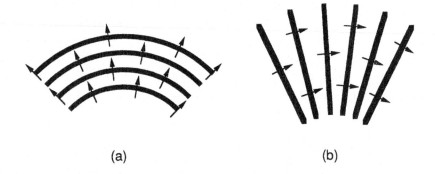

Fig. 11. Canonical deformations of the director field: a) splay; b) bend.

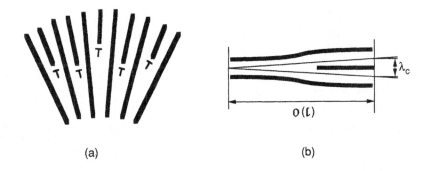

Fig. 12. a) A distribution of dislocations allows accommodation of a global bend; b) the angular change introduced by a single dislocation in a region of typical dimension ℓ scales as λ_c/ℓ.

The density of dislocations required to deal with a global bend of the pattern can be estimated from the circulation of the director along a contour $\mathcal{C}$ enclosing the region of interest (Fig. 12a). The contribution of a single dislocation to the circulation is an angle easily seen to scale as λ_c/ℓ where ℓ is the length of the contour (Fig. 12b). If $\Delta\theta$ is the angle measuring the total disorientation over the domain of interest (assumed to be of order unity, i.e., not scaling as a power of r), the typical number of dislocations is then given by $\Delta\theta/(\lambda_c/\ell)$, which therefore scales as ℓ.

The contribution of a single dislocation is r^2 times the surface of its core $r^{-1/2} \times r^{-1/4}$, that is to say $r^{5/4}$. Accordingly the contribution of a density of dislocations able to accommodate a change of orientation of order unity will be $r^{5/4} \times \ell$ where ℓ scales as the perimeter of the cell, to be compared, e.g., with the $r^{7/4} \times \ell$ contribution of the boundary layer associated with rolls perpendicular to a lateral wall.

At lowest order, the selection process should thus favor a pattern with rolls mainly perpendicular to the lateral boundaries, with curvature present since splay deformations cost nothing at lowest order and $r^{7/4} \ll r^{5/4}$ for r small. The prediction of the most probable texture at a given horizontal geometry is however difficult owing to the unavoidable presence of isolated defects such as dislocations and segments of grain boundaries. This result is in agreement with laboratory experiments in convection using large Prandtl number fluids already presented in Chapter 1 or numerical simulations on simple models such as those of Chapter 8. Unfortunately, it does not account for the nontrivial consequences of higher order effects responsible for wavelength selection and for feed-back couplings with secondary flows at the origin of weak turbulence in low Prandtl number convecting fluids.

6. Bibliographical Notes

A slightly different approach to the stability roll patterns based on envelope equations is developed by Newell and Whitehead, see BN **8** [7b]. The Eckhaus instability was first analyzed in:

[1] V. Eckhaus, *Studies in nonlinear stability theory*, Springer Tracts in natural Philosophy, Vol. 6 (Springer, 1965).

The stability of waves described by the complex envelope equation was reviewed by:

[2] A. C. Newell, "Envelope equations," Lectures in Applied Mathematics **15**, 157 (1974).

9. Applications of the Envelope Formalism

See also:

[3] J. T. Stuart and R. C. di Prima, "The Eckhaus and Benjamin-Feir resonance mechanisms," Proc. R. Soc. A **362**, 27 (1978).

A formal approach in terms of dynamical systems is developed by:

[4] P. Coullet and D. Repaux, "Models of pattern formation from a singularity theory point of view".

in the book edited by Tirapegui and Villaroel, BN **5** [5].

Boundary conditions on the envelope have been derived by:

[4] S. Brown and K. Stewartson, "On thermal convection in a large box," Stud. in Appl. Math. **57**, 187 (1977).

The relevance of the theoretical solution for convection experiments has been checked by J. E. Wesfreid, *et al*, BN **5** [1].

A first study of grain boundary using envelopes has been developed in:

[5] P. Manneville and Y. Pomeau, "A grain boundary in cellular structures near the onset of convection," Phil. Mag A, **48**, 607 (1983); (unfortunately, this reference contains many typographical errors left uncorrected).

Dislocation motion was analyzed in the same spirit by:

[6] E. D. Siggia and A. Zippelius, "Dynamics of defects in Rayleigh-Bénard convection," Phys. Rev. A **24** 1036 (1981).

and further studied by:

[7] Y. Pomeau, S. Zaleski, and P. Manneville, "Dislocation motion in cellular structures, Phys. Rev. A **27**, 2710 (1983).

The effects of drift flows have been considered by:

[8] E. Dubois-Violette, E. Guazzelli, and J. Prost, "Dislocation motion in layered structure," Phil. Mag. A **48**, 727 (1983).

The theory of structural defects was reviewed and checked against simulation of convection models (including drift flow) by:

[9] G. Tesauro and M. C. Cross, "Grain boundary in models of convective patterns," Phil. Mag. A **56**, 703 (1987),
"Climbing of dislocations in nonequilibrium patterns," Phys. Rev. A **34**, 1363 (1986).

Experiments on prepared rows of dislocations in convection have been performed by:

[10] J. A. Whitehead, "The propagation of dislocations in Rayleigh–Bénard flow and bimodal flow," J. Fluid Mech. **75**, 715 (1976).

The presentation adopted in Section 5 roughly follows that of:

[11] M. C. Cross, "Ingredients of a theory of convective textures close to onset," Phys. Rev. A **25**, 1065 (1982).

Chapter 10

Dynamics of Textures and Turbulence

The gross features of pattern selection discussed at the end of Chapter 9 have to be refined since the theory was based on an envelope formalism at lowest order valid close to the threshold only. Deeper in the nonlinear regime, universal modes related to continuous invariance properties play a dominant role in the range of control parameter where the system remains stable against specific secondary instability modes.

In Section 1, we present the formalism that leads to the equations governing the phase variables associated with these invariance properties in the case of steady roll patterns. We discuss further the nonlinear selection criteria (Section 2) which can be discussed in the same framework. Then we generalize the approach to more complicated cases and review briefly some results on the route to turbulence that develops beyond the threshold of phase instability up to *phase turbulence* (Section 3).

A different route can be followed when the instability is subcritical. This new scenario, called *spatio-temporal intermittency*, is characterized by a coexistence of turbulent and laminar regions in physical space and implies a new statistical approach to be described in Section 4.

Finally, we attempt to build a bridge between chaos as studied in weakly confined systems and more developed turbulence in the most conventional sense and to situate the problem of the *nature of turbulence* in the perspective of the rest of the book (Section 5).

1. Phase Diffusion in Steady Roll Patterns

Let us first summarize the results obtained in previous chapters. We were mostly interested in the vicinity of the primary instability threshold, a consistent description of which is given by envelope equations at lowest order. Neglecting the spatial dependence of the envelope, we could account for early nonlinear selection that "chooses" the locally stable pattern among all other possible wavevector superpositions. Then, we could deal with the local uniformization of the *texture* left after this first selection step by reintroducing the space dependence of the envelope. At this stage, it was suggested that the dynamics of modulus $|A|$ of the complex envelope could be separated from the dynamics of its phase (Chapter 9, Equations $(2a,b)$) on the basis of a comparison between characteristic times. Indeed, for modulations with sufficiently long wavelengths, the modulus $|A|$ was easily seen to be slaved to the gradient of the phase, i.e., the local value of the wavevector. In this section, we develop a formalism appropriate to this situation characteristic of the strongly nonlinear regime identified in the simulations presented in Chapter 8, Section 2.2.

1.1. Phase Diffusion Formalism

Let us examine the simplest case of a steady locally periodic structure solution as obtained, for example, from the Swift–Hohenberg model

$$\partial_t w = rw - \left(\nabla_\mathrm{h}^{\,2} + 1\right)^2 w - w^3, \qquad (1)$$

where $\nabla_\mathrm{h}^{\,2}$ is the two-dimensional ("horizontal") Laplacian. We assume that the system has condensed into a pattern of rolls everywhere roughly aligned along the y axis. The corresponding x-periodic reference solution $w_0(x)$ then fulfills

$$0 = rw_0 - \left(\partial_x^2 + 1\right)^2 w_0 - w_0^3. \qquad (2)$$

Periodic solutions with wavevectors $k_0 = 1 + \delta k$ inside the unstable band at given $r > 0$ (i.e., $|\delta k| \leq \frac{1}{2}\sqrt{r}$) can be obtained in terms of

10. Dynamics of Textures and Turbulence

their Fourier expansions

$$w_0(x) = W_1 \sin(k_0 x) + W_3 \sin(3k_0 x) + \ldots, \tag{3a}$$

where coefficients are obtained by identification:

$$W_1 = \left(\frac{4}{3}(r - 4\delta k^2)\right)^{1/2}, \quad W_3 = \frac{W_1^3}{256} \ldots \tag{3b}$$

(the cubic nonlinear term does not generate even harmonics of k_0, and odd harmonics become rapidly negligible for r not too large but finite, i.e., not infinitesimal; notice that this solution depends explicitly on the value of the wavevector, as it should generically).

Now, $w_0(x)$ being an exact solution, $w_0(x + \phi)$ is also an exact solution owing to translational invariance in the x direction. The phase ϕ is arbitrary at this stage and $w_0(x + \phi)$ can be expanded in powers of ϕ:

$$w_0(x + \phi) = w_0(x) + \phi \partial_x w_0(x) + \frac{\phi^2}{2} \partial_{x^2} w_0(x) + \ldots. \tag{4}$$

Inserting this solution in the primitive problem, expanding it in powers of ϕ, we obtain

$$\begin{aligned}\partial_t (w_0 + \phi \partial_x w_0 + \ldots) = 0 &= F(w_0 + \phi \partial_x w_0 + \ldots) \\ &= F(w_0) + \Lambda_0(\phi \partial_x w_0) + \ldots\end{aligned} \tag{5}$$

In this expression, F denotes the r.h.s. of Equation (1), which defines the primitive problem and $\Lambda_0 = \delta F/\delta w|_{w_0}$ is the differential operator derived from F by linearization around w_0:

$$\Lambda_0 = r - \left(\nabla_h^2 + 1\right)^2 - 3w_0^2.$$

As long as ϕ is uniform, i.e., independent of space, we have $\Lambda_0(\phi \partial_x w_0) = \phi \Lambda_0(\partial_x w_0)$. But, from the identity $F(w_0) \equiv 0$, we derive the identity $\partial_x(F(w_0)) = \Lambda_0(\partial_x w_0) \equiv 0$, so that (5) simply reads

$$\partial_x w_0 \, \partial_t \phi = \phi \Lambda_0(\partial_x w_0) \equiv 0,$$

that is to say $\partial_t \phi = 0$ —a uniform phase does not evolve. The mode associated with translational invariance, $\partial_x w_0$, is neutral and belongs to the kernel of Λ_0.

When ϕ is slowly varying in space, $w_0(x + \phi)$ is no longer an exact solution, but the error decreases as $\nabla_h \phi$ is made vanishingly small. Therefore, we look for the solution of the problem in the form

$$w(x, y, t) = w_0(x) + \phi \partial_x w_0 + w_1 + w_2 + \ldots,$$

where $w_1, w_2, \ldots$ are formally of the order of $\nabla_h \phi$, $\nabla_h \nabla_h \phi, \ldots$. On the r.h.s. of Equation (5), Λ_0 now acts on both terms of the product $\phi \partial_x w_0$, its x derivatives on both ϕ and $\partial_x w_0$ and its y derivatives on ϕ only. At first order in the phase gradient, we obtain

$$\partial_x w_0 \, \partial_t \phi + \partial_t w_1 = \Lambda_0 (\phi \partial_x w_0 + w_1).$$

Isolating the unknown correction w_1 on the l.h.s. and dropping the term $\partial_t w_1$ which is of higher order, we obtain

$$\Lambda_0 w_1 = \partial_t \phi \, \partial_x w_0 + \partial_x \phi \, g(x), \tag{6}$$

where $g(x) = 4(\partial_{x^2} + 1) \partial_{x^2} w_0$ comes from $\Lambda_0(\phi \partial_x w_0)$.

Since the kernel of Λ_0 is nontrivial (it contains $\partial_x w_0$), Equation (6) can be solved for w_1 only if its r.h.s. is orthogonal to the kernel of the adjoint operator Λ_0^+ (*Fredholm alternative*). As discussed in Appendix 2, Λ_0 is self-adjoint so that the r.h.s. has to be orthogonal to $\partial_x w_0$ itself. At this stage, $\partial_t \phi$ and $\partial_x \phi$ behave as constants with respect to the rapidly varying functions involved in the orthogonality condition. It is first easily checked that the contribution of $\langle \partial_x w_0 | g(x) \rangle$ vanishes automatically since $g(x)$ contains even derivatives of $w_0(x)$ only. On the other hand, since $\langle \partial_x w_0 | \partial_x w_0 \rangle \neq 0$, we obtain $\partial_t \phi = 0$, i.e., in fact $\partial_t \phi$ is of higher order than $\partial_x \phi$. Equation (6) can then be solved for $w_1(x)$ by isolating explicitly the slowly varying part proportional to $\partial_x \phi$. Inserting $w_1 = \partial_x \phi \, \tilde{w}_1(x)$ into (6), we get

$$\Lambda_0(\tilde{w}_1) = 4(\partial_{x^2} + 1) \partial_{x^2} w_0, \tag{7}$$

10. Dynamics of Textures and Turbulence

which is solved by identification.

At second order in $\nabla_h \phi$, terms coming from $\Lambda_0(\partial_x \phi \; \tilde{w}_1)$ must not be forgotten. We obtain

$$\Lambda_0 w_2 = \partial_t \phi \, \partial_x w_0 + \partial_{x^2} \phi \left[2(3\,\partial_{x^2} + 1) \, \partial_x w_0 + 4(\partial_{x^2} + 1) \, \partial_x \tilde{w}_1 \right] \\ + \partial_{y^2} \phi \left[2(\partial_{x^2} + 1) \, \partial_x w_0 \right] \tag{8}$$

The solvability condition for (8) is now completely nontrivial since all terms involve odd x derivatives of w_0. It takes the form of a constraint on the evolution of ϕ:

$$\partial_t \phi = D_\parallel \, \partial_{x^2} \phi + D_\perp \, \partial_{y^2} \phi \tag{9}$$

where two *diffusion coefficients* have been defined:

$$D_\parallel = -\frac{\langle \partial_x w_0 | 2(3\,\partial_{x^2} + 1) \, \partial_x w_0 + 4(\partial_{x^2} + 1) \, \partial_x \tilde{w}_1 \rangle}{\langle \partial_x w_0 | \partial_x w_0 \rangle}, \tag{10}$$

$$D_\perp = -\frac{\langle \partial_x w_0 | 2(\partial_{x^2} + 1) \, \partial_x w_0 \rangle}{\langle \partial_x w_0 | \partial_x w_0 \rangle}. \tag{11}$$

Here subscripts $\parallel$ and $\perp$ make an implicit reference to the wavevector of the underlying structure, therefore denoting diffusion in the direction of or perpendicular to the wavevector.

Using the expression of w_0 given in (3) and that of $\tilde{w}_1$ determined from (7), we obtain

$$D_\parallel = 4 \frac{r - 12\delta k^2}{r - 4\delta k^2},$$

$$D_\perp = 4 \left(\delta k - \frac{(r - 4\,\delta k^2)^2}{1024} \right).$$

In principle, once the solvability condition is fulfilled, the expansion can be continued but the calculations become extremely heavy. In practice, it is preferable to find higher order terms by symmetry considerations, therefore introducing phenomenological coefficients, see below.

1.2. Phase Instabilities

A negative diffusion coefficient is the signature of an instability. When $D_\parallel < 0$, we recover the *Eckhaus* instability, whereas $D_\perp < 0$ corresponds to the *zigzag* instability. These instability modes have already been obtained from the envelope formalism. The difference with the envelope calculation lies in the fact that it does not rely on a perturbation expansion in powers of the amplitude but on scalar products involving a nonlinear solution obtained at arbitrary distance from the primary instability threshold. This difference may not be obvious for the Eckhaus instability since $D_\parallel = 0$ yields the same result ($\delta k_{\rm E} = \pm(\frac{1}{2}\sqrt{r})/\sqrt{3}$) as that of the envelope formalism at lowest order. By contrast, for the zigzag instability we obtain $\delta k_{\rm ZZ} \simeq r^2/1024$, which would require an envelope equation consistent at order r^3, i.e., three orders in $r^{1/2}$ beyond the first nontrivial order $r^{3/2}$ that simply gives $\delta k_{\rm ZZ}$ independent of r. (It is not difficult to show that $D_\perp$ remains positive in the vicinity of the critical wavevector if the system is not rotationally invariant so that straight rolls remain stable against zigzags.)

To get an idea of the behavior close to the Eckhaus or zigzag instability threshold, we must supplement the phase diffusion equation with higher order terms. Since beyond threshold, $D_\parallel$ and $D_\perp$ do not vanish simultaneously, it is legitimate to separate the two cases. The phenomenological extension of (9) is easily derived by symmetry considerations (Kuramoto, 1984). All terms formally of order $(\partial_x^{m_x})(\partial_y)^{m_y}\phi^{m_\phi}$ preserving the following invariances:

- change $y \to -y$,
- simultaneous change $x \to -x$, $\phi \to -\phi$,

have to be included. In addition, relations between the phenomenological coefficients introduced with each of these terms can be derived from the fact that phase winding solutions with slightly different wavelengths or corresponding to tilted roll patterns should fulfill the so completed phase equation.

Let us examine first the case of the Eckhaus instability. By assumption, $D_\parallel$ is close to zero and negative, whereas $D_\perp > 0$ so

10. Dynamics of Textures and Turbulence

that zigzag fluctuations are damped. By inspection, we obtain

$$\partial_t \phi = D_\| \partial_{x^2}\phi - K_x \partial_{x^4}\phi + D_\perp \partial_{y^2}\phi + g\,\partial_x\phi\,\partial_{x^2}\phi. \tag{12}$$

Stabilization by the fourth-order derivative is effective if $K_x > 0$ (notice that y dependence is not essential and that the Eckhaus instability also takes place in strictly one-dimensional systems, in which case the y dependence does not appear at all). The coefficient g is obtained by inserting $\phi = (\delta k/k)x + \tilde\phi$ into (12) and identifying the equation for $\tilde\phi$ as the phase diffusion equation for a structure with underlying wavevector $k + \delta k$:

$$\partial_t \tilde\phi = (D_\|(k) + g\,\delta k/k)\,\partial_{x^2}\tilde\phi + D_\perp \partial_{y^2}\tilde\phi + g\,\partial_x\tilde\phi\,\partial_{x^2}\tilde\phi + \ldots,$$

so that

$$D_\|(k+\delta k) = D_\|(k) + \delta k\,\frac{dD_\|}{dk} \equiv D_\|(k) + \delta k/k\,g$$

$$\Rightarrow \quad g = k\,\frac{dD_\|}{dk}.$$

Higher order terms are excluded by a scaling argument assuming $|D_\||$ small and comparing terms retained in the equation, which yields $\partial_x \sim \sqrt{|D_\||}$, $\partial_y \sim |D_\||$, $\phi \sim \sqrt{|D_\||}$, and $\partial_t \sim |D_\||^2$, showing that the equation as its stands is consistent at order $|D_\||^{5/2}$.

The case of the zigzag instability is treated in the same way. We obtain

$$\begin{aligned}\partial_t \phi =& D_\|\partial_{x^2}\phi + D_\perp \partial_{y^2}\phi - K_y\partial_{y^4}\phi \\ &+ (g_1\partial_x\phi + g_2(\partial_y\phi)^2)\partial_{y^2}\phi.\end{aligned} \tag{13}$$

A similar scaling argument justifies the presence of the two nonlinear terms. Coefficients g_1 and g_2 in (13) are obtained by expressing that a solution with $\phi = (\delta k_x/k)x + (\delta k_y/k)y + \tilde\phi$ corresponds to a slightly tilted pattern with wavevector $\sqrt{(k+\delta k_x)^2 + (\delta k_y)^2}$. This gives

$$g_1 = 2g_2 = k\,\frac{dD_\perp}{dk}.$$

Nonlinearities introduced above in the phase diffusion equations are in some sense trivial: they account for the existence of a continuous branch of nearly marginal modes indexed by a wavevector that can vary both in length and in direction. It then becomes interesting to develop a formalism dropping the reference to an ideal underlying pattern and thus including this property from the start.

1.3. Rotationally Invariant Formulation of Phase Dynamics

The essential advantage of the calculation developed in Section 1.1 is to make the relevance of the neutral mode associated with translation completely explicit through assumption (4). However, progress may seem minor since the only achievement is a new discussion of the stability of a well-ordered pattern characterized by an underlying wavevector aligned along a given direction (x axis). In fact, a new approach to the dynamics of textures has been introduced, and to get a less limited perspective we have just to reconsider the notion of phase from a more general standpoint. To this aim, we notice first that a complete description of the considered roll pattern is usually understood in terms of level lines of $w(x, y, t)$ or more generally $V(x, y, z, t)$, i.e., in terms of iso-phase lines of the solution written as

$$V(x, y, z, t) = V_0(u(x, y, t); z), \qquad (14)$$

where $V_0(u, z)$ is the solution with period $\lambda_0 = 2\pi/k_0$ and $u(x, y, t)$ its generalized phase describing the "horizontal" dependence. The dynamics of the pattern is then reduced to that of iso-phase lines.

Up to now we have taken

$$u(x, y, t) = x + \phi(x, y, t), \qquad (15)$$

and, since the calculation rests on a power expansion, it is tacitly assumed that ϕ remains small. This forbids the consideration of large scale changes of the the *orientation* and *wavelength* of the rolls that can be expected in natural textures. Indeed, passing from an ideal pattern with wavevector $\mathbf{k} = k_0\,\hat{\mathbf{x}}$ to a neighboring pattern

10. Dynamics of Textures and Turbulence 377

with wavevector $\mathbf{k} = (k_0 + \delta k_x)\hat{\mathbf{x}} + \delta k_y\, \hat{\mathbf{y}}$ yields a "secular" phase variation $\phi(x,y) = k_0^{-1}(\delta k_x x + \delta k_y y)$ that is unbounded at large distances. Such ideal patterns are obviously better specified by their generalized phase $u = k_0^{-1}\mathbf{k} \cdot \mathbf{x}_\mathrm{h}$, where $\mathbf{k}$ is a constant vector of length k, $\lambda = 2\pi/k$ the corresponding wavelength, and $\mathbf{x}_\mathrm{h}$ denotes the position in the horizontal plane. Accordingly, we can specify a nonideal pattern by its wavevector field $\mathbf{k}(x,y,t)$ instead of $\phi(x,y,t)$, defining the local wavevector as

$$\mathbf{k} = k_0 \nabla_\mathrm{h} u\,. \tag{16}$$

This definition of $\mathbf{k}$, slowly varying in time and space, is possible everywhere (14) holds, i.e., everywhere the periodic structure can be identified without ambiguity. Since we consider the asymptotic limit of long times, well after the end of the early nonlinear selection stage, $\mathbf{k}$ is defined almost everywhere except at the core of point defects (e.g., dislocations) and in limited strips where several patterns overlap (grain boundaries).

The main assumptions used in the phase diffusion calculation, the existence of a globally well-aligned pattern with everywhere small position fluctuations, are now replaced by the condition that the gradients of $\mathbf{k}$ remain small. A description of the phase dynamics has now to be built using $\mathbf{k}$ and its gradient but making no reference to a specific direction in the horizontal plane, i.e., in terms of rotationally invariant quantities only. At order zero in $\nabla_\mathrm{h} \mathbf{k}$, everything can depend on the length k of the wavevector but not on its orientation. At first order we have simply $\nabla_\mathrm{h} \cdot \mathbf{k}$ (we may note that, being a gradient field by construction, $\mathbf{k}$ is irrotational: $\partial_{xy} u = \partial_x k_y = \partial_y k_x$).

We define the *director field* $\hat{\mathbf{n}}$ through

$$\mathbf{k} = k\,\hat{\mathbf{n}}\,,$$

and get

$$\nabla_\mathrm{h}\cdot\mathbf{k} = \nabla_\mathrm{h}\cdot(k\,\hat{\mathbf{n}}) = k\,\nabla_\mathrm{h}\cdot(\hat{\mathbf{n}}) + \hat{\mathbf{n}}\cdot\nabla_\mathrm{h} k\,,$$

i.e., a splitting into a contribution involving $\nabla_\mathrm{h}\cdot\hat{\mathbf{n}}$, which relates to the local curvature of the pattern (see Chapter 9, Fig. 10) and a term describing the local dilatation (normal derivative of the length of **k**).

The dynamics of the pattern is understood in terms of the motion of its iso-phase lines, i.e., contours defined by conditions of the form $u(x, y, t) = $ cst. The normal velocity V_n of such contours can thus be defined from

$$\partial_t u + V_n \, \hat{\mathbf{n}} \cdot \nabla u = 0, \tag{17}$$

where V_n is a function of rotationally invariant quantities only. The form of the result at lowest order is easily guessed to read

$$V_n = -D_\| \, \hat{\mathbf{n}} \cdot \nabla_h(k/k_0) - D_\perp \nabla_h \cdot \hat{\mathbf{n}}, \tag{18}$$

which generalizes (9). Coefficients $D_\|$ and $D_\perp$ introduced here are checked to be the same as those obtained in the previous calculation by considering a pattern nearly aligned along the x axis and with underlying wavevector k_0. Taking $u = x + \phi$, we get $\partial_x u = 1 + \partial_x \phi = k_x/k_0$, $\partial_y u = \partial_y \phi = k_y/k_0$ and, at lowest order, $(k/k_0) \simeq 1 + \partial_x \phi$ so that $\hat{\mathbf{n}} \cdot \nabla(k/k_0) \simeq \partial_{x^2} \phi$. In the same way, up to quadratic corrections we have $n_x = 1$, and $n_y = \partial_y \phi$ so that $\nabla_h \cdot \hat{\mathbf{n}} = \partial_{y^2} \phi$. Finally, $\partial_t u = \partial_t \phi$ and $\hat{\mathbf{n}} \cdot \nabla u \simeq 1$ complete the identification.

The calculation that yields the expression of the coefficients is closely similar to that used in Section 1.1 and need not be presented here. The game consists in fitting the given modulated pattern to an expression of the form "$V_0(u) + $ corrections," where the corrections are taken as products of the rotationally invariant slowly varying quantities $(\nabla_h \cdot \hat{\mathbf{n}}, k/k_0, \ldots)$ by unknown periodic functions of u with period $\lambda_0 = 2\pi/k_0$. The sought evolution equation is then obtained from solvability conditions ensuring the existence of these rapidly varying functions (for a detailed presentation and calculations including higher order derivatives, see Ohta, 1985).

In the long-time limit we expect a global texture made of large grains, the sizes of which scale with the size of system itself. The inverse aspect ratio is then the most natural expansion parameter and a slightly different expansion scheme yields the generalized phase dynamics equation (Cross and Newell, 1984) in the form

$$\tau(k) \, \partial_t \tilde{u} + \nabla_h \cdot \big(\mathbf{k} B(k) \big) = 0, \tag{19}$$

10. Dynamics of Textures and Turbulence   379

where $\tilde{u} = ku$, and $\tau(k)$ and $B(k)$ are two functions to be determined. Expanding Equation (19), we obtain

$$\tau(k)\,\partial_t \tilde{u} + kB(k)\nabla_\mathrm{h}\cdot\mathbf{n} + \frac{d\bigl(kB(k)\bigr)}{dk}\,\mathbf{n}\cdot\nabla k = 0,$$

which becomes equivalent to (17)+(18). Inserting further $\tilde{u} = k(x + \phi(x,y,t))$ (nearly aligned pattern) in this equation, we obtain

$$\tau\,\partial_t \phi + \frac{d\bigl(kB(k)\bigr)}{dk}\,\partial_{x^2}\phi + B(k)\,\partial_{y^2}\phi,$$

which forces the identification

$$D_\| = -\frac{1}{\tau}\frac{d\bigl(kB(k)\bigr)}{dk},\qquad D_\perp = -\frac{1}{\tau}B(k).$$

In fact, (17)+(18) can be written in the form (19), provided $D_\|$ and $D_\perp$ are known functions of k since B then derives from

$$\frac{dB}{B} = \frac{dk}{k}\left(\frac{D_\|}{D_\perp} - 1\right),$$

obtained by elimination of τ which can be determined afterwards.

2. Nonlinear Selection and Weak Turbulence

2.1. Nonlinear Wavelength Selection Criteria

Several selection processes are closely related to the calculations developed above. Let us first consider model (1) which derives from the potential density $\mathcal{G}(w) = \frac{1}{2}(1-r)w^2 + \frac{1}{2}(\nabla_\mathrm{h}^2 w)^2 - (\nabla_\mathrm{h} w)^2 + \frac{1}{4}w^4$. During the evolution, the potential $G = \int_\mathcal{D} \mathcal{G}\,dx\,dy$ decreases and since it is bounded from below, it reaches a local minimum. The condition that it reaches its absolute minimum may be viewed as a first and somewhat obvious *nonlinear selection criterion*. For a one-dimensional pattern, we have to find the wavevector k_opt minimizing

$G_\lambda = (1/\lambda) \int_0^\lambda \mathcal{G}\, dx$, the potential *per* wavelength $\lambda = 2\pi/k$ and *per* unit length in the y direction. The application to solution (3) with $k_0 = 1 + \delta k$ readily gives

$$\begin{aligned} 4G_\lambda = &(1-r)(W_1^2 + W_3^2 + \ldots) \\ &+ \left(\frac{3}{8}(W_1^4 + W_3^4 + \ldots) + \frac{1}{2}W_1^2 W_3(W_1 + 3W_3) + \ldots\right) \\ &- 2(1+\delta k)^2(W_1^2 + 9W_3^2 + \ldots) \\ &+ (1+\delta k)^4(W_1^2 + 81 W_3^2 + \ldots). \end{aligned}$$

The values of coeffcients $W_1, W_3, \ldots$ are obtained from the extremum conditions

$$\frac{\partial G_\lambda}{\partial W_{(n)}} = 0, \qquad n = 1, 3, \ldots, \tag{20}$$

at given δk. To get the optimum wavevector we must compute the total derivative $dG_\lambda/d\,\delta k = 0$, which is reduced to

$$\frac{\partial G_\lambda}{\partial \delta k} = 0$$

when (20) is taken into account. After simplification, this yields

$$(1 + \delta k_{\mathrm{opt}})^2 = \frac{W_1^2 + 9W_3^2}{W_1^2 + 81 W_3^2}$$

or, at finite but small distance to threshold, using (3)

$$\delta k_{\mathrm{opt}} = -\frac{r^2}{1024}.$$

It can be seen immediately that this optimum precisely corresponds to the condition $D_\perp = 0$, i.e., $\delta k = \delta k_{\mathrm{ZZ}}$. A formal proof of this fact can be obtained by computing the first variation δG of the potential under the dilatation $x \to (1+\eta)x$. It is easily seen that $\delta G \propto \eta \int 2\left[(\partial_{x^2} w)^2 - (\partial_x w)^2\right] dx$, which can be transformed into the numerator of $D_\perp$ by means of an integration by parts ($\partial_{y^2}\phi$ plays exactly the same role in the phase diffusion formalism as η in the

10. Dynamics of Textures and Turbulence

present calculation). Since, by definition, the potential should be stationary at the optimum, $\delta G = 0$ implies $\delta k_{\text{opt}} = \delta k_{\text{ZZ}}$.

The idea that states *marginally stable* against transverse phase modulations are optimal is acceptable since an infinitesimal curvature that does not change the wavelength at lowest order does not relax precisely when $\delta k = \delta k_{\text{ZZ}}$. By contrast, when $\delta k \neq \delta k_{\text{ZZ}}$ the perturbation evolves. As already noted (Chapter 4, Section 3.2), when $\delta k < \delta k_{\text{ZZ}}$ the perturbation increases in a way that tends to decrease the wavelength, i.e., to increase the wavevector so as to get closer to the zigzag instability threshold. On the other hand, rolls with $\delta k > \delta k_{\text{ZZ}}$ presenting some finite curvature tend to straighten, i.e., to decrease the wavevector again getting closer to the marginal condition at given r.

This suggests to us to examine the existence condition for stationary axisymmetric patterns. In cylindrical coordinates (ρ, α) the Swift–Hohenberg model reads

$$\partial_t w = rw - (\partial_{\rho^2} + \frac{1}{\rho}\partial_\rho + \frac{1}{\rho^2}\partial_{\alpha^2} + 1)^2 w - w^3.$$

Stationary axisymmetric solutions are given by

$$0 = rw - (\partial_{\rho^2} + \frac{1}{\rho}\partial_\rho + 1)^2 w - w^3. \tag{21}$$

Expanding the differential operator

$$(\partial_{\rho^2} + \frac{1}{\rho}\partial_\rho + 1)^2$$
$$\equiv (\partial_{\rho^2} + 1)^2 + \frac{2}{\rho}(\partial_{\rho^2} + 1)\partial_\rho - \frac{1}{\rho^2}\partial_{\rho^2} + \frac{1}{\rho^3}\partial_\rho, \tag{22}$$

we guess that, sufficiently far from the center, terms in $1/\rho^n$ become negligible so that the analytic solution can be searched as an expansion in powers of $1/\rho$

$$w(\rho) = w_0(\rho) + \frac{1}{\rho}w_2(\rho) + \ldots. \tag{23}$$

The subscript "2" for the first correction is chosen to stress the parallel with the previous calculation since the curvature of the rolls is given by the second transverse derivative of the phase variable.

Inserting (23) into (21), at lowest order we recover Equation (2) now written for the radial coordinate ρ. At order $1/\rho$, we obtain

$$\Lambda_0 w_2 = -\frac{2}{\rho}(\partial_{\rho^2} + 1)\partial_\rho w_0 \,.$$

As before, a solvability condition bearing on the rapidly varying part of the r.h.s. is obtained:

$$0 = -\langle \partial_\rho w_0 | 2(\partial_{\rho^2} + 1)\partial_\rho w_0 \rangle \,,$$

which is again easily seen to be equivalent to the condition $D_\perp = 0$, as expected.

The dynamics of the phase is also clearly involved in the selection processes involving the motion of dislocations. As already seen in Chapter 9, Section 4.2, a region containing a dislocation connects two domains with parallel orientation but slightly different wavelengths. Here we consider the simplest case of a dislocation climbing parallel to the roll axis in a system deriving from a potential. The dislocation moves to decrease the total potential, which obviously tends to decrease the surface of the domain with the least optimal wavevector. The dislocation remains at rest if the average wavevector shift δk characterizing the pattern is equal to δk_{opt} since there is no potential gain in moving the dislocation. This provides us with a new selection criterion.

Among other processes, selection by grain boundary motion is also of interest. For two systems of rolls at right angles as in Chapter 9, Section 4.1, in a potential system, the grain boundary can stay at rest only if both domains have the same wavelength and, among all possibilities, that with both wavevector shifts equal to δk_{opt} is the most stable.

Unfortunately, systems driven far from equilibrium are not expected to derive from a potential. Accordingly, criteria that were

10. Dynamics of Textures and Turbulence

seen to collapse into a single condition in the variational case should become different. At this stage, it is not yet clear whether selection associated with curvature does or does not coincide with the marginal stability condition against zigzags but the conditions bearing on dislocations and grain boundary motions yield separate solvability conditions generically.

Combining these results with those of the preliminary discussion of pattern selection at the end of Chapter 9, we see that, since usually lateral boundary conditions force the rolls to arrive perpendicularly to the wall, criteria related to curvature and defect motion should play a major role. On the other hand, nontrivial wavelength selection derived from amplitude equations at high order for rolls parallel to the boundary should play a modest role except in special well-controlled situations (e.g., axisymmetric rolls). In any case, since each criterion tends to impose its own preferred wavelength in the bulk, the lack of compatibility between them in the nonvariational case can be at the origin of a residual unsteadiness in the long-time limit. However, since this *spatio-temporal chaos* results from a *geometric frustration* and evolves basically according to a diffusive phase dynamics, it remains very weak. Taking into account *drift flows* will drastically change this picture.

2.2. Drift Flows and the Transition to Weak Turbulence

Up to now, the physical nature of the pattern forming system was not essential, only the fact that there was a single instability mode well described by a scalar field depending on "horizontal" coordinates, whereas the "vertical" dependence could be eliminated. It may be a little surprising to notice that this can apply to fluid systems only when the velocity vector field is entirely slaved to the scalar (e.g., temperature) field involved in the instability mechanism. As discussed in Chapter 3, this happens when viscous characteristic time is infinitely short when compared with all other relevant times. For convection in a single component fluid, this corresponds to $\tau_\theta/\tau_v = \nu/\kappa = P \to \infty$. At finite Prandtl numbers, the situation is more complicated than that described by plain phase diffusion mainly because the horizontal velocity field contains a slowly varying part that becomes a relevant degree of freedom able to shift the pattern.

This relevance is best understood in the case of stress-free horizontal boundary conditions since a uniform flow independent of z in the (x, y) plane is neutral owing to Galilean invariance of the problem in that plane. A slightly nonuniform flow is therefore strongly coupled to large scale envelope modulations accounting for roll curvature that induce slowly varying terms in the vertical vorticity equation. The phase formalism developed in Section 1 can be extended to include the contribution of this flow responsible for Busse oscillatory, instability but it may be useful to consider a generalized formalism that places the translational and Galilean invariances on an equal footing (Fauve et al., 1987).

With no-slip boundary conditions, a uniform flow with parabolic profile is damped in the absence of an externally applied pressure gradient but a trace of this coupling remains. In fact, a proper treatment of phase fluctuations requires the introduction of drift flows, which renders the gradient expansion singular.

Things become rapidly very technical so that we will only quote points of interest for pattern selection in Rayleigh–Bénard convection with realistic boundary conditions (i.e., no-slip). The first interesting result relates to transverse phase diffusion, which can still be solved analytically in spite of the presence of drift flows, ending in an expression of the form

$$D_\perp = \frac{\xi_0^2}{\tau_0} \left[\frac{k - k_c}{k_c} + \frac{d(P)}{r_2(P)} \frac{R - R_c}{R_c} \right],$$

with ξ_0^2/τ_0 and $r_2(P)$ as determined in Chapters 3 (Section 2.3) and 8 (Section 3.3) respectively, and $d(P) = 0.166 + 23.04/P + 6.196/P^2$. The curvature is seen to induce a divergence-free flow whose vertical profile is roughly parabolic as seen in Fig. 1a. It is parallel to the wavevector of the undistorted rolls (x direction) and directed so as to straighten the rolls. It brings the major contribution to $d(P)$ for P moderate to small ($23.13/P + 6.43/P^2$), and its stabilizing effect

10. Dynamics of Textures and Turbulence

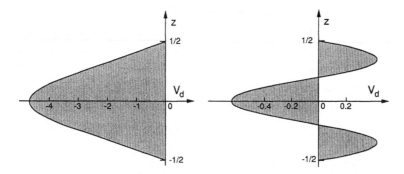

Fig. 1. Vertical profiles of drift flows induced by curvature at lowest order. For transverse phase diffusion (left), the drift flow closely resemble a Poiseuille flow. By contrast, for axisymmetric selection (right), the drift flow averages to zero over the depth of the layer.

fully explains that the zigzag instability becomes irrelevant at low Prandtl numbers.

The companion problem of axisymmetric selection can be solved by the same technique, which leads to a strikingly different result. Since the x direction in the phase diffusion problem here becomes the ρ direction, the secondary flow that appears is now purely radial. As such, it must vanish on average over the height since otherwise a net inward flux would take place. Fortunately, there is enough freedom in the calculation to block this net flux by an outward pressure gradient. As a result, the vertical velocity profile depicted in Fig. 1b is obtained and $d(P)$ is replaced by a similar expression: $d'(P) = 0.166 + 1.42/P - 1.22/P^2$. The P-independent term is unchanged as expected, whereas the residual secondary flow is also seen to contribute dominantly to the selection criterion when P is small $(1.51/P - 0.98/P^2)$.

The case of arbitrary modulations is much more complicated since they induce a large scale flow whose average over the thickness must be divergence-free to ensure the continuity of matter. As just seen, this generally imposes the building of internal pressure gradients able to cancel the divergent component of the net flow so as to make it purely rotational (vertical vorticity). This phenomenon is at the very root of the singularity of the gradient expansion in convecting fluids. To go a little further, it may be found interesting to work with generalized models of convection including drift flows.

On general grounds, the horizontal flow $\mathbf{v}_h = (v_x, v_y)$ can be split into an irrotational component and a divergence-free part by writing

$$v_x = \partial_x \psi_{\text{irr}} + \partial_y \psi_{\text{rot}},$$
$$v_y = \partial_y \psi_{\text{irr}} - \partial_x \psi_{\text{rot}}.$$

The irrotational component is obtained from the continuity equation:

$$\nabla_h \cdot \mathbf{v}_h + \partial_z v_z = 0 \quad \Rightarrow \quad \nabla_h^2 \psi_{\text{irr}} = -\partial_z v_z.$$

which can be solved from the knowledge of v_z.

The general equation for the vertical vorticity is obtained by taking the z component of the curl of the Navier–Stokes equations

$$(\partial_t + P\nabla^2) \nabla_h \psi_{\text{rot}} = \partial_x (\mathbf{v} \cdot \nabla v_y) - \partial_y (\mathbf{v} \cdot \nabla v_x), \qquad (24)$$

where ∇ and $\mathbf{v}$ are fully three-dimensional. Notice that the source term for the vertical vorticity is formally quadratic in $\mathbf{v}$ so that it vanishes at first order in $\mathbf{v}$.

In the present context, we are looking for a reduced expression involving only the projection of the velocity components on the most unstable mode periodic with spatial period $\sim \lambda_c = 2\pi/k_c$ and their lowest order combinations including the induced slowly varying contributions.

Assuming that w in model (1) represents the amplitude of vertical velocity, i.e., $v_z(x, y, z, t) = w(x, y, t) f(z)$, we can derive the horizontal field from the continuity equation. At first order in $\mathbf{v}$, we obtain

$$\nabla_h^2 \psi_{\text{irr}} = -k_c^2 \psi_{\text{irr}} = -(df/dz) w$$

($k_c = 1$ in (1)). Forgetting the vertical space dependence, we obtain

$$\mathbf{v}_h = (c/k_c^2) \nabla_h w, \qquad (25)$$

10. Dynamics of Textures and Turbulence

where c is a numerical constant of order unity accounting for the projection of the vertical dependence (π in the stress-free case with $f(z) = \sin(\pi z)$). Inserting $\mathbf{v}_h$ given by (24) and w into the vertical vorticity equation (24) averaged over the height, we obtain after some rearrangements:

$$(\partial_t + P(d - \nabla_h^2))\nabla_h^2 \psi_{\text{rot}}^{\text{slow}} = g(\partial_y w\, \partial_x \nabla_h^2 w - \partial_x w\, \partial_y \nabla_h^2 w) \quad (26)$$

where g and d are two constants ($d = 0$ in the stress-free case). Velocity components deriving from $\psi_{\text{rot}}^{\text{slow}}$ are given by

$$v_x^{\text{slow}} = \partial_y \psi_{\text{rot}}^{\text{slow}} \quad \text{and} \quad v_y^{\text{slow}} = -\partial_x \psi_{\text{rot}}^{\text{slow}}.$$

In turn, this velocity field is able to translate the pattern so that $\mathbf{v}_h^{\text{slow}} \cdot \nabla_h w$ has to be added to $\partial_t w$ to describe the time evolution of the field w at a formally cubic level consistent with the nonlinear term usually kept.

A theoretical approach to the role of the drift flow can be developed within the framework of envelope equations, and in a next step, using the Cross–Newell equations for phase dynamics (Equation (19) completed by a term accounting for the convection of the phase and a supplementary equation deriving from (26) to govern the drift). This approach helps in interpreting the nature of the *skew-varicose instability* as an Eckhaus instability modified by drift flow effects which turn out to be destabilizing in this case in contrast with the case of the zigzags. It can also be used to explain the off-centering instability of axisymmetric patterns observed in experiments and the structure of the drift flow in more general circumstances.

In practice, it has been found interesting to resort to numerical simulations to get a better understanding of the transition to turbulence in large aspect ratio systems at moderate to low Prandtl numbers. Swift–Hohenberg type of models have been used with possibly different nonlinear terms but including the drift flow contribution generated by (26). Simulations show that, for sufficiently small Prandtl number, the time behavior is much wilder that for models without drift which, as seen in Chapter 8, is basically relaxational.

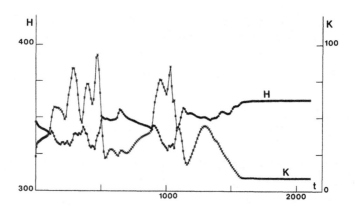

Fig. 2. Evolution of the "heat flux" H and large-scale-flow kinetic energy K during a turbulent transient observed in numerical simulations on a modified Swift–Hohenberg model including drift flows ($P = 1.6$, $c = 0$ (stress-free model), rectangular geometry with aspect-ratios $\Gamma_x \simeq 16$ and $\Gamma_y \simeq 11.5$. Notice that the behavior of H is strikingly different from that of w_2, the corresponding quantity in Model (4–5) in Chapter 8, Fig. 4.

Figure 2 displays the evolution of the heat flux H and the kinetic energy K contained in the drift flow during a "turbulent transient." An intermittent alternation of calm periods and more agitated sequences is observed. Accidents on the curves correspond to the nucleation of dislocations as illustrated in Fig. 3 and calm periods to a slow reorganization of the pattern.

The transient reported here is very long when compared with transients in relaxational systems with similar aspect ratios. Sustained weak turbulence has not been observed in the case presented but can well appear for different geometries or different values of the control parameters (r and P). Taking the drift flow into account seems essential to understanding the details of the transition to turbulence in convection at low Prandtl number and especially the cyclic nucleation of dislocations observed as a preliminary step on the route to turbulence in cylindrical geometry (see Chapter 4, Section 4.2).

At a qualitative level, we can say that pattern selection in con-

10. Dynamics of Textures and Turbulence

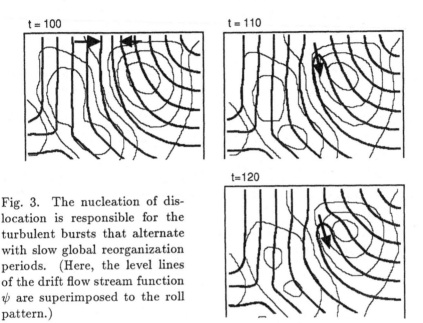

Fig. 3. The nucleation of dislocation is responsible for the turbulent bursts that alternate with slow global reorganization periods. (Here, the level lines of the drift flow stream function ψ are superimposed to the roll pattern.)

vecting fluids is controlled not only by *geometrical frustration* associated with the horizontal shape of system but also by *dynamical frustration* since the drift flow generated by curvature must be compatible with the global texture retained. A weakly turbulent spatio-temporal chaos may result from a lack of compatibility between drift flow and curvature that would be more pronounced than that induced by a wavelength mismatch. In addition, the intermittent nature of this turbulence could explain the power law decrease of the noise power spectra measured experimentally.

3. Transition to Turbulence in Oscillating Patterns

The transition to turbulence in convection close to the threshold may seem progressive. This is essentially due to the fact that, as seen in Chapter 9, Section 5, modulations to an ideal structure are governed at lowest order by an envelope equation that derives from a potential. Of course, this property is not preserved at higher orders but the nonvariational contributions introduced, even the large scale flow, come and correct a basically time-independent regime.

However, this is but one of the possible codimension-one cases of bifurcations toward a dissipative structure. Otherwise we can have:

- uniform and stationary instabilities ($k_c = \omega_c = 0$) breaking only an internal symmetry; this rather infrequent and relatively trivial case will not be considered further;
- uniform and time periodic modes with $k_c = 0$ but $\omega_c \neq 0$, breaking time translational invariance (e.g., oscillatory chemical instabilities);
- time-periodic space-periodic modes with both $k_c \neq 0$ and $\omega_c \neq 0$, leading to the formation of waves (e.g., convection in binary mixtures).

In contrast with steady rolls, oscillating systems are generically described by envelope equations with complex coefficients that cannot derive from a potential, even at lowest order. In the simplest (one-dimensional) case, a single wave is present, yielding a complex Ginzburg–Landau equation in a frame moving at the group velocity (Chapter 8, Equation (35)):

$$\partial_t A = rA + (1 + i\alpha)\, \partial_{x^2} A - (1 + i\beta)|A|A\,, \qquad (27)$$

and we already know that uniform solutions can be stabilized only if Newell's criterion is fulfilled. A gradient expansion can be developed for the temporal phase of A in (27), which leads to the Kuramoto–Sivashinsky equation (Kuramoto and Tsuzuki, 1976; Sivashinsky, 1977)

$$\partial_t \theta = D\, \partial_{x^2}\theta - K\, \partial_{x^4}\theta - g(\partial_x \theta)^2\,, \qquad (28)$$

where the diffusion coefficient $D = 1 + \alpha\beta$ is negative when Newell's criterion is not fulfilled so that a modulational (Benjamin–Feir) instability is expected.

10. Dynamics of Textures and Turbulence

Equation (28) and its variant for $\phi = \partial_x \theta$ obtained by derivation with respect to x, which reads after rescaling

$$\partial_t \phi + \phi \partial_x \phi = -\partial_{x^2}\phi - \partial_{x^4}\phi \tag{29}$$

have been much studied mainly because of the existence of spontaneously chaotic solutions as soon as a sufficient number of Fourier modes are unstable (growth rate $s = k^2 - k^4$, with $k = 2n\pi/\ell$, where ℓ is the length of the system, periodic boundary conditions assumed).

In addition, Equation (29) bears a strong resemblance with the Burgers equation, the one-dimensional reduction of Navier–Stokes equations, and has some relevance for hydrodynamic problems where a long wavelength instability leads to negative viscosity effects. The Burgers equation can be changed into a heat diffusion equation by a nonlinear Hopf–Cole transformation and as such can have only steady asymptotic states. By contrast, the Kuramoto-Sivashinsky equation is particularly well suited for studying the growth of spatio-temporal disorder, i.e., the passage from temporal chaos typical of a low-dimensional dissipative system to *turbulent regimes* specific to partial differential equations. These chaotic regimes, usually called *phase turbulence*, have been mostly studied in the framework of dissipative dynamical systems theory (see Bibliographic Notes). Results worth noting here relate to the existence of long-lived turbulent transients and to a linear growth of the "dimension of chaos" with the number of unstable modes.

In two space dimensions, Equation (27) has singular solutions in the form of spirals. In the stable case, these solutions persist indefinitely owing to their character of topological defect. In the unstable case, they nucleate spontaneously as finite time singularities of the evolution of the temporal phase. The spatio-temporally chaotic regime that sets in has been called *topological turbulence* (Coullet et al., 1988).

Systems in which propagative waves develop can be described by an equation of the form (27) only when one can change to a frame in translation at the group velocity of the waves in a given direction. If the medium is space-reversal invariant (e.g., in binary fluid convection) a competition between left and right propagating waves takes place, which has to be described in the laboratory frame. Two coupled complex envelopes are then necessary with interesting consequences on nonlinear pattern selection when the full horizontal space dependence is included. For example, specific defects called "zipper-modes" correspond to grain boundaries between left and right propagating waves "running side by side." A particularly rich dynamical behavior develops.

Finally, the phase formalism can be adapted to the phenomenological description of situations where more than one mode is marginal. The case of Busse oscillations in Rayleigh–Bénard convection with stress-free boundary conditions, with underlying translational and Galilean invariance, has already been quoted. Examples can be found in secondary instabilities of the Taylor vortices in cylindrical Couette flow. For example, two phase variables are necessary to account for slow modulations of the wavy mode that presents itself as undulating Taylor vortices. The first variable gives the azimuthal position of the undulation and the second one the mean position of the vortices along the cylinder axis. Symmetry considerations and scaling arguments are essential in deriving the corresponding evolution equations (see Bibliographic Notes).

From a general point of view this generalized phase-dynamics approach seems well-adapted to study the growth of spatio-temporal chaos in extended geometry.

4. Spatio-temporal Intermittency

4.1. Introduction

Up until now, we have considered mainly systems displaying a progressive transition to turbulence. The main steps were: (1) a supercritical primary instability saturating gently above threshold; (2) possible secondary instabilities associated with a phase diffusion coefficient going continuously through zero; (3) a transition to temporal chaos for the phase variable best understood as weak spatio-temporal

10. Dynamics of Textures and Turbulence

chaos owing to the spatial meaning of the phase. This "progressivity" can be given a quantitative value through the measurement of the gradients of the modulations supposed to remain "small" (validity of the approach in terms of envelope and phase). Of course this is only schematic since the presence of large-scale secondary flows and structural defects comes and slightly obscures this picture. Anyway, the resulting global scenario can be termed "supercritical" in an enlarged sense.

In fact, the transition to turbulence can be much "wilder" when the first step is subcritical instead of being supercritical (in the ordinary sense). The key fact is then the possible coexistence of different regimes for a same set of the control parameters. In the field of dynamical systems with a small number of degrees of freedom, this situation has already been encountered, which leads generically to *crises* and *intermittency* (Chapter 6) contrasting with the subharmonic route and the original Ruelle–Takens scenario for which everything occurs rather continuously in phase space. When spatial modulations are allowed, the coexistence between different regimes can happen in physically different regions of space separated by walls. The presence of strong spatial modulations is permitted by the absence of long-range coherence characterizing the vicinity of a linear instability threshold, so that the "distance" between the basic state and the bifurcated state can become very large when measured naively.

For the transition to turbulence, the most intersting situation corresponds to one where the competition takes place between a laminar basic state and a turbulent regime. In the simplest case, the laminar regime is linearly stable, i.e., stable against infinitesimal (unlocalized) fluctuations, but becomes unstable to finite amplitude localized perturbations. In contrast with the supercritical case, the existence of the transition therefore relies on the presence of residual perturbations or spatial inhomogeneities in the initial conditions.

As a concrete experimental example of this situation, let us quote the Couette flow between two concentric cylinders rotating in the same direction. This flow configuration is thought to remain linearly stable at all shear rates when the outer cylinder rotates in the same direction as and faster than the inner one (stable stratification of angular momentum, see Chapter 3) but it is known to become turbulent at sufficiently large shears. The aspect of the turbulent regime is quite special; it takes the form of spiral bands "filled" with small scale turbulence separated by whole domains of laminar flow ("barber pole" regime).

4.2. A Case Study: The Modified Swift-Hohenberg Model

Here we illustrate this transition to turbulence with results of numerical simulations on the modified Swift-Hohenberg model used in Chapter 8, Section 4.2. Up to a rescaling of variables and parameters, this model, which reads

$$\partial_t w = rw - (\partial_{x^2} + 1)^2 w - w\partial_x w, \qquad (30)$$

is equivalent to a damped Kuramoto-Sivashinsky equation written in the form

$$\partial_t \phi = -\eta\phi - \partial_{x^2}\phi - \partial_{x^4}\phi - \phi\partial_x\phi.$$

From the relation between the control parameters r and η, $\eta = \frac{1}{4}(1-r)$, it is easily understood that the conduction regime, $w \equiv 0$ for $r < 0$, corresponds to $\eta > \frac{1}{4}$ and that the pure Kuramoto-Sivashinsky equation $\eta = 0$ is recovered when $r = 1$. In this limit, a single control parameter remains: the length ℓ of the interval at which boundary conditions are imposed; and we know that for ℓ large enough, most solutions are turbulent. By contrast, for r positive but small enough, i.e., for η slightly smaller that $1/4$, stationary cellular solutions are selected. Therefore, we have two parameters to "unfold" the transition to turbulence, parameter r and the *aspect ratio* of the system. Recalling that the critical wavelength in model (30) is $\lambda_c = 2\pi$ and that, for r small and ℓ large, the width of a cell is expected to be of the order of $\frac{1}{2}\lambda_c = \pi$, we can define the aspect ratio of the system as the number of cells: $\Gamma = \ell/\pi$.

Figure 4 displays a qualitative picture of this unfolding in terms of a phase diagram in the (r, Γ) plane. Three regions can be distinguished according to the value of Γ. When Γ is small, viz. ~ 5,

10. Dynamics of Textures and Turbulence 395

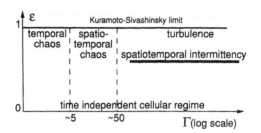

Fig. 4. Schematic bifurcation diagram for the modified Swift–Hohenberg model.

the transition to turbulence is typical of low-dimensional dynamical systems as discussed in Chapter 5, Section 1.1. Increasing Γ, viz. ~ 50, we enter a region where the transition involves truly spatio-temporal processes. The situation is complicated owing to still sizable end effects that can stabilize nicely regular periodic or quasi-periodic regimes. By contrast, for Γ very large the transition to turbulence occurs systematically *via* spatio-temporal intermittency.

Before going into a presentation of the transition process, let us begin with a description of *spatio-temporal intermittency* as a regime of turbulence observed in model (30) sufficiently close to the Kuramoto-Sivashinsky limit. Figure 5 displays a space–time representation of the dynamics and explains how a black-and-white reduced description can be derived from the remark that, at a given time t, the solution $w(x)$ can be decomposed into a series of intervals where the local structure is close to a perfect cellular structure, separated by strongly distorted portions. Places where the solution is regular (distorted) are painted in black (white) and it appears clearly that, as times goes on, the widths of the two kinds of regions fluctuate. When observed at a given point, this turbulent regime has an intermittent signature with *laminar* intermissions as long as the considered point belongs to a regular portion of the solution and *turbulent* bursts when a distorted region passes through it. An *intermittency factor* can be defined as the ratio of the time spent in turbulent bursts to the total duration of the experiment. However, it is clear that this *intermittency* has a *spatio-temporal* origin (in contrast with strictly temporal intermittency studied in Chapter 6).

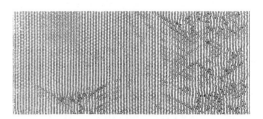

Fig. 5. Top: evolution of the solution as detected from the position of successive extremas of $w(x,t)$. Bottom: Black-and-white compression of the information contained in the full solution.

Let us now turn to the transition between this turbulent regime and the stationary state that takes place for r small (Fig. 6). The global state of the system at given r is best specified by the *turbulent fraction*, i.e., the ratio of the cumulated widths of turbulent domains to the total length (finer characterizations can be obtained from the distributions of widths of the turbulent and laminar domains or from standard correlation functions). When r is large, the turbulent fraction is also large and the typical size of the laminar domains is small. When r is lowered, the turbulent fraction decreases and the size of the laminar domains increases. Further decreasing r, we reach a point where some laminar domains acquire macroscopic sizes, i.e., of the order of the length of the system, and for even lower r invade the whole system, at the expense of the domain left spatio-temporally intermittent. For ℓ sufficiently large the decay of spatio-temporal intermittency takes place at a reasonably well-defined ℓ-independent threshold r_{sti}.

The situation is different when r_{sti} is approached from below. Two cases can be distinguished according to the nature of the initial condition, either regular or distorted.

Regular initial conditions obtained by preparing the system close to the convection threshold (r positive but small) and then increas-

10. Dynamics of Textures and Turbulence 397

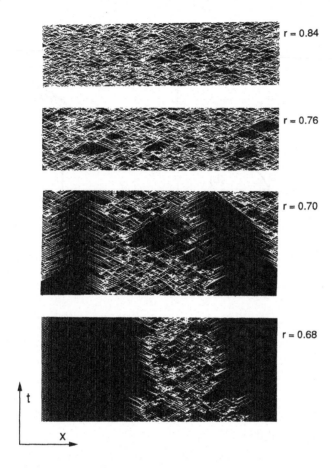

Fig. 6. Transition to turbulence *via* spatio-temporal intermittency in model (30). From top to bottom: well above threshold, $r = 0.84$; closer to threshold, $r = 0.76$; "at" threshold, $r = 0.70$; slightly below threshold, $r = 0.68$.

ing slowly the control parameter are seen to remain stable against small perturbations well above r_{sti}. The stationary regular pattern is in fact *metastable* up to the threshold r_{osc} of an inverse Hopf bifurcation that can be observed only if the length ℓ of the system is not too large since confinement effects contribute to the stabilization of the bifurcated regime. By contrast, when ℓ is large, the spatio-temporally intermittent regime is obtained immediately.

398   DISSIPATIVE STRUCTURES AND WEAK TURBULENCE

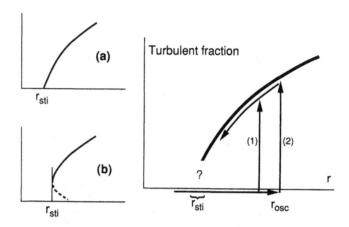

Fig. 7. Qualitative sketch of the variation of the turbulent fraction as a function of the control parameter at given ℓ expected for a continuous (a) or slightly discontinuous (b) transition.

Using disordered initial conditions, below r_{sti}, we observe spatio-temporal intermittency as a transient turbulent regime only, the final state being stationary and perfectly regular. The duration of the transient increases as r_{sti} is approached from below and beyond this value, the spontaneous decay towards the reference laminar regime is never observed (at least as far as one can tell from finite cost numerical simulations).

The situation is summarized in Fig. 7 where the variation of the turbulent fraction is tentatively sketched as a function of the control parameter. Numerical results are compatible with either a continuous (Fig. 7a) or a slightly discontinuous (Fig. 7b) variation of the turbulent fraction at r_{sti}.

4.3. Towards a Theory

Simulations presented above suggest that the turbulent regime result from a *contamination* of the laminar state by the chaotic bursts. This fits well a previous conjecture by Y. Pomeau (1986) who proposed to cast the description of such transitions in terms of *directed percolation*.

10. Dynamics of Textures and Turbulence   399

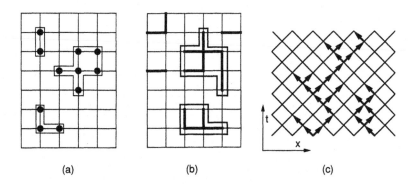

Fig. 8. Percolation: a) site percolation; b) bond percolation; c) directed bond percolation.

Ordinary percolation is a stochastic process defined on a lattice. For *site percolation*, the presence of active sites at the lattice nodes is drawn at random with a given probability p and one asks for the existence of an infinite cluster of sites seen to exist for p greater than some critical probability p_c depending on the topology of the lattice (see Fig. 8a). If *bond percolation* is considered, the presence of links between nodes is the random variable and the existence of a continuous path going to infinity is tested (Fig. 8b). For *directed bond percolation*, we subject the determination of the percolating path to the additional condition that links are one-way, all oriented along the same diagonal of the lattice (Fig. 8c).

Directed percolation is of particular interest to us since it enables us to describe the propagation in space (direction x) of a given information by a probabilistic contamination rule as time proceeds, hence the application to flow through porous media, epidemics, forest fires, etc. Taking the first example cited, we can describe the medium as a network of pores that can be open or closed with a given probability p and monitor the penetration of a fluid by detecting whether pores at some level are wet or dry. A crucial point to note is that a site at the intersection of pores that have not been reached by the fluid cannot become wet. The dry state is said to be *absorbing*. By contrast, the other state is often called *active*. As a result, when $p > p_c \simeq 0.6445$ the "fluid" percolates through the medium to infinity whereas, below p_c, only finite clusters of wet sites exist.

The analogy between the process just described and the transition to turbulence *via* spatio-temporal intermittency is striking. Moreover, the fact that turbulent bursts are seen to propagate through laminar domains but can collapse into a laminar state forces the identification "laminar $\to$ absorbing" and "turbulent $\to$ active." However, if we are to consider turbulence as the result of deterministic processes, we may be worried by this direct reduction to a probabilistic cellular automaton as directed percolation presents itself.

A cellular automaton is an *assembly* of subsystems j, each in one of a *finite set* of *discrete states* $\{X_\alpha \mid \alpha = 0, 1, \ldots, k-1\}$, evolving according to a *rule* whose output depends on the states of the individuals belonging to a fixed *neighborhood* of each subsystem (see Bibiographic Notes). Of course, rules can be deterministic or probabilistic but in the present context we would certainly prefer a reduction to a deterministic cellular automaton rather than to a probabilistic one. In fact, several intermediate levels of modeling are possible between the time-space continuous level of partial differential equations and the fully discrete level of cellular automata. In this respect, *coupled map lattices* are particularly interesting systems. Their local phase space is no longer discrete but continuous and the evolution remains deterministic. They can be written as

$$X_j^{n+1} = \sum_{j' \in \mathcal{V}_j} W_{jj'} f\left(X_{j'}^n\right),$$

where j is the space index, n the (discrete) time index, $\mathcal{V}_j$ the relevant neighborhood of site j, and $W_{jj'}$ the coupling coefficient between site j and $j' \in \mathcal{V}_j$. In one space dimension, with a diffusive coupling on a three-site neighborhood, we have

$$X_j^{n+1} = f\left(X_j^n\right) + g\left(f\left(X_{j-1}^n\right) - 2f\left(X_j^n\right) + f\left(X_{j+1}^n\right)\right).$$

Such systems are able to mimic certain aspects of nonlinear pattern formation with a great economy of numerical resources and among

10. Dynamics of Textures and Turbulence

other processes, the transition to turbulence *via* spatio-temporal intermittency can be obtained by an adequate choice of the local map f.

In order to describe this regime in a realistic way, we must first assume a splitting of the local phase space into two different regions: one corresponding to the laminar regime, the other to a chaotic dynamics. The absorbing character of the laminar state is achieved by demanding a transient chaotic state, e.g., at the vicinity of a crisis point. As seen in Fig. 9, a sufficiently large coupling is able to convert the *locally transient* chaotic dynamics into a *globally sustained* turbulent state strongly reminiscent of what is obtained with the partial differential equation. This similarity is not surprising since the continuous-time continuous-space system is kept far from a linear instability threshold so that space-time coherence is reduced, which makes legitimate a decomposition of the whole system into finite length sub-units interacting at discrete times. On the other hand, the connection to probabilistic cellular automata can be understood from the interplay between the deterministic divergence of trajectories in the local phase space and the coupling that ensures the transfer of the so-randomized information "laminar/turbulent."

The importance of Pomeau's remark comes from the fact that with the concept of directed percolation, the statistical physics of critical phenomena enters the field of turbulence in a new way. In this context, the first questions relate to the existence of an order parameter and to the order of the transition. First-order transitions are characterized by a jump of the order parameter at the transition point whereas it varies continuously for second-order transitions.

Second-order transitions are characterized by sets of critical exponents, such as exponent β controlling the growth of the order parameter as a function of the distance to the critical point. Here the fraction f_t of active (turbulent) sites plays the role of the order parameter, so that, it the transition is second-order, we may expect $f_t = (p - p_c)^\beta$. Moreover, *universality classes* can be defined, i.e., broad classes of systems that behave quantitatively in the same way close to the critical point, i.e., with the same set of *critical exponents*.

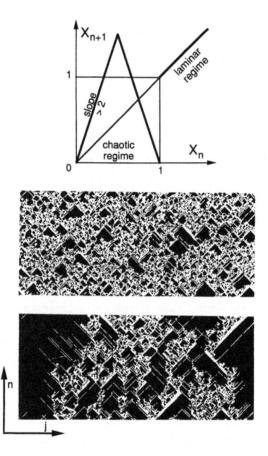

Fig. 9. Spatio-temporal intermittency observed in a one-dimensional coupled map lattice with variable coupling. Top: local map. Middle and bottom: simulations well above and close to the spatio-temporal intermittency threshold.

Directed percolation is believed to be the prototype of a large class of probabilistic cellular automata with a single absorbing state. In this respect, a quantitative study of spatio-temporal intermittency shows that it does not belong to this universality class and a rich manifold of situations holds. Depending on details of the problem considered, the transition to turbulence can be second-order with variable critical exponents different from those of directed percolation or even first-order.

All the statistical features of the transition to turbulence *via* spatio-temporal intermittency are not fully understood at present but this scenario appears to be specific to large aspect-ratio systems. Moreover, it seems generic for systems where the transition is not supercritical in any enlarged sense. This is confirmed by the fact that it can be observed not only in specially designed models (coupled map lattices) but also in typical partial differential equations (e.g., model (30)) and in laboratory experiments (e.g., in convection). At a conceptual level, in addition to building a new bridge between statistical physics and turbulence, this scenario helps us to reconcile local short-term determinism with spatio-temporal chaos for weakly confined systems in much the same way as scenarios based on strange attractors do for temporal chaos in low-dimensional dynamical systems.

5. Hydrodynamics and Turbulence

5.1. Introduction

In this concluding section, we enlarge our field of study to the problems posed by *open flows*, i.e., flows with a global transport of matter from an intake region (*upstream*) to some outlet region (*downstream*). This situation, realized for example in pipe or channel flows, shear layers, or boundary layers, may be of much greater importance than most of the situations considered up to now, at least from a practical point of view.

In closed systems driven out of equilibrium, nonlinearities can of course have an intrinsically local origin (e.g., a chemical reaction) but we have seen that, in most cases involving fluids, the primary role was played by the *advection term* $\mathbf{v} \cdot \nabla(\cdots)$ at both the linear and the nonlinear stage. At least at lowest order, this role may seem to remain somewhat trivial since advection appears involved in instability mechanisms and couplings in a straightforward way. As a result, a universal picture is obtained (envelope equations).

In fact, at higher orders, things become more intricate when *drift flows* cannot be neglected. These large-scale secondary flows able to transport small-scale cellular structures are the first manifestations of complex nonlinear behavior in hydrodynamics.

Purely hydrodynamic shear flows, i.e., flows without other relevant fields than the velocity field itself, can be characterized by a dimensionless parameter called the Reynolds number R. This number can be viewed as the product of the damping time for velocity fluctuations on a given scale $\delta\ell$, $\tau_v = \delta\ell^2/\nu$, where ν is the kinematic velocity, by the average shear rate on that scale $\delta V/\delta\ell$; therefore,

$$R = \frac{\delta V \delta\ell}{\nu}.$$

Applied constraints usually fix the scale of δV and $\delta\ell$ to "external" values V_0 and ℓ_0, which defines an external Reynolds number $R_0 = V_0 \ell_0/\nu$. After rescaling, the Navier–Stokes equation reads

$$\partial_t \mathbf{v} + \mathbf{v} \cdot \nabla \mathbf{v} = -\nabla p + \frac{1}{R_0} \nabla^2 \mathbf{v} \tag{31}$$

(to which the incompressibility condition $\nabla \cdot \mathbf{v} = 0$ must be added). The competition between advection and diffusion appears clearly in (31). When $R_0 \to 0$, i.e., for small velocity, small scale, or large viscosity, viscous diffusion controls the flow (Stokes limit) while when $R_0 \to \infty$, inertia effects dominate through the advection term (inviscid or Euler limit). In this last case, purely mechanical instabilities can occur, e.g., the Kelvin–Helmholtz instability, the discussion of which goes beyond our present scope.

Taking for granted that a primary hydrodynamical instability has taken place, we can analyze the role of the advection term along two major directions. We can either look at the actual process of *transition to turbulence* in more or less concrete situations, or else consider *developed turbulence*. In the first case, advection keeps its primitive meaning and one must look at the effect of the coupling between intrinsic instability modes (usually waves) and externally imposed transport. This leads to the distinction between *absolute*

10. Dynamics of Textures and Turbulence

and *convective* instabilities briefly sketched in Section 5.2 below. In the second case, the $\mathbf{v}\cdot\nabla\mathbf{v}$ term is better viewed as an internal eddy generator at the origin of a cascading transfer of energy between large scales where shear is imposed and small scales where dissipation takes place. This point of view is taken in Section 5.3 and leads to the discussion of the number of degrees of freedom effectively excited in developed turbulence.

5.2. Advection and Absolute/Convective Instabilities

When a macroscopic flow downstream is imposed from the outside, as in conventional situations of fluid dynamics, the system is not space-reversal invariant so that a single envelope is sufficient to describe waves resulting from a primary instability mechanism. If the system were space-translation invariant, a Galilean change of frame would help us to write the problem as a complex Ginzburg–Landau equation (27), but this is rarely realistic. Indeed, external perturbations at rest in the laboratory frame most often come and break this Galilean invariance. Consider, for example, the stability of the boundary layer that develops along a plate in a parallel flow. Under certain conditions, a subtle instability mechanism leads to the formation of "Tollmien–Schlichting waves." Not only are downstream propagating waves inequivalent to upstream propagating waves but the very existence of the edge of the plate breaks the global translational invariance and the boundary layer is seen to thicken all along the plate. This obliges us to write the envelope equation in the laboratory frame, i.e., as

$$\partial_t A + V\,\partial_x A = rA + (1+i\alpha)\,\partial_{x^2} A - (1+i\beta)|A|^2 A\,, \qquad (32)$$

where parameter V accounts for the downstream transport and, as usual, r controls the intensity of the instability mechanism (here a conveniently rescaled Reynolds number).

In this context, the attention has been drawn recently on the important distinction between *absolute* and *convective* instabilities, i.e., between instabilities that develop in spite of the downstream

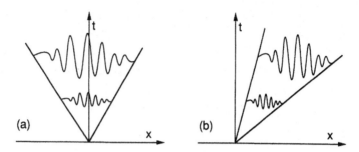

Fig. 10. Absolute (a) and convective (b) instabilities.

transport, therefore invading the whole available domain, and those that grow while being carried downstream but are just seen to pass when observed from a given fixed point in the laboratory frame (see Fig. 10). This distinction rests on a new reading of the linear dispersion relation of the unstable modes adapted to systems with a nonvanishing global flux of matter.

The linear stability problem can be written formally as

$$L(\partial_x, \partial_t; r)v(x,t) = S(x,t), \qquad (33)$$

where r is the control parameter and S a source term. Fourier transforming equation (33), we obtain

$$L(ik, -i\omega; r)v(k,\omega) = S(k,\omega).$$

The instability threshold is reached when the response to S diverges, i.e., when

$$L(ik, -i\omega, r) = 0. \qquad (34)$$

This relation can be solved in the time domain, i.e., in ω, as done up until now, which yields $\omega = \omega_n(k,r)$, with k real for spatially periodic perturbations in the form $\exp(ikx)$. In this approach, we have $i\omega \equiv s$, our earlier notation for the complex growth rate.

Solving (34) in the space domain, i.e., in k, yields $k = k_n(\omega, r)$, with ω real for a temporally periodic excitation of the form $\exp(-i\omega t)$. ik is now a complex spatial growth rate without equivalent in our previous approaches.

10. Dynamics of Textures and Turbulence

In fact, instead of making reference to one or the other of these harmonic analyses, it is preferable to consider the linear response to a general excitation obtained by integration from the Green function $G(x,t)$, which is the response to a perturbation δ localized in both time and space: $S(x,t) = \delta(x)\delta(t)$.

The system is said to be *absolutely unstable* if $G(x,t) \to \infty$ when $t \to \infty$ at given x (the instability develops everywhere), and *convectively unstable* if $G(x,t) \to 0$ under the same conditions (x fixed and $t \to \infty$), in spite of the fact that $G(x,t) \to \infty$ when $x = vt$ in a certain range of velocity v. Equivalently, we can look at the stable/unstable character of the complex wavevector k_0 at which the group velocity vanishes, i.e., $\partial\omega/\partial k(k_0) = 0$. When the imaginary part of $\omega(k_0)$ is positive, the instability is absolute since the perturbation can grow on the spot. On the contrary, when it is negative, the system is at most convectively unstable. In the general context of this book, the term "convective" is somewhat unfortunate since according to this terminology Rayleigh–Bénard convection is an absolute instability. The distinction is nevertheless essential for open flows since it allows us to account for the sensitivity to residual turbulence and perturbations in the intake region.

As an example, let us return to Equation (32). The dispersion relation reads

$$\omega = (Vk + \alpha k^2) + i(r - k^2)$$

and the trivial solution $A \equiv 0$ is unstable provided $r > k^2$. The group velocity of the unstable modes is given by

$$\frac{\partial \omega}{\partial k} = V + 2(\alpha - i)k,$$

which goes through zero for

$$k_0 = \frac{V}{2(i - \alpha)} = \frac{V}{2}\frac{\alpha + i}{1 + \alpha^2}$$

so that

$$\omega(k_0, r) = ir - \frac{V^2}{4(\alpha - 1)} = -\frac{V^2 \alpha}{4(\alpha^2 + 1)} + i\left(r - \frac{V^2}{4(\alpha^2 + 1)}\right).$$

Therefore, the instability is absolute if $r > V^2/4(\alpha^2+1)$ and convective if $0 < r < V^2/4(\alpha^2+1)$, i.e., when V is large enough, which is in agreement with the intuition (of course the trivial solution remains stable as long as $r < 0$).

In the nonlinear regime, when the instability is convective, localized perturbations are transported downstream while growing. They can saturate before leaving the observation region. In that case, a selective amplification of the residual turbulence takes place, which leads to the growth of *intermittently perturbed coherent structures*. More complex cases can occur in which the character of the instability, either convective or absolute, can change with the position in space. As far as applications to hydrodynamics are concerned, this approach appears to be promising.

5.3. Homogeneous, Isotropic, Developed Turbulence

Let us assume that the flow under consideration is unstable under external shear of typical magnitude V_0/ℓ_0. Then a simple argument shows that the term $\mathbf{v} \cdot \nabla \mathbf{v}$ generates eddies with typical size $\ell_0/2$ or, in Fourier space, $k_0 \to 2k_0$. This happens when the Reynolds number $R = R_0 = U_0\ell_0/\nu$ is large enough, i.e., when the viscous diffusion time is large so that advection is the dominant factor.

A given eddy of size $\ell = 2\pi/k$ can be characterized by its velocity V_ℓ, its turn-over time $\tau_\ell = \ell/V_\ell$, its kinetic energy $K_\ell = \frac{1}{2}V_\ell^2$, and its viscous dissipation $D_\ell = \nu(V_\ell/\ell)^2$, yielding the definition of a local Reynolds number $R_\ell = \ell V_\ell/\nu$, which can be read as the ratio of the kinetic energy to the dissipation. When R_ℓ is large, dissipation is negligible, which is the case for the largest eddies by assumption.

As discussed previously, the nonlinear coupling is expected to generates daughter eddies at twice the wavevector of their mother. Some kinetic energy is therefore transferred from scale ℓ_0 to scale $\ell_0/2$. But if R_0 is large enough, the local Reynolds number for scale $\ell_0/2$ is still large and the process can repeat itself. A whole energy cascade takes place toward smaller and smaller scales, down to a point where dissipation becomes dominant (Fig. 11).

The problem is then to estimate the velocity at scale ℓ and to

10. Dynamics of Textures and Turbulence

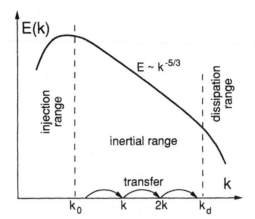

Fig. 11. Qualitative picture of the energy transfer among scales from the injection range to the dissipation range (inertial cascade).

derive from it the spectrum of the velocity fluctuations in the *inertial range*, between the *injection* scale $\sim \ell_0$ and the *dissipation* scale ℓ_d where the energy transfer is expected to be replaced by viscous diffusion. The solution can be obtained easily provided that we assume a steady rate of energy transfer independent of the scale for a homogeneous and isotropic fully-developed turbulent flow. The argument due to Kolmogorov (1941, hence the frequently used notation K41) is mainly dimensional.

The basic assumption is that the amount of energy transferred to scale $\ell/2$ is a fraction of the amount of energy present at scale ℓ, that this fraction is independent of the scale ℓ itself, but that the transfer takes a time of the order of the turn-over time at that scale.

Calling ϵ the rate of transfer by unit mass through the cascade, dimensionally we have

$$\epsilon \sim V_\ell^2/\tau_\ell = V_\ell^3 k_\ell$$

and according to our assumption, at steady state, this rate is equal to both the injection rate at scale ℓ_0 and the dissipation rate at the smallest scales where gradients are large. Therefore, we have

$$\epsilon \sim V_0^3 k_0$$

and

$$V_\ell = \left(\epsilon \frac{\ell}{2\pi}\right)^{1/3} = V_0 \left(\frac{\ell}{\ell_0}\right)^{1/3}$$

The corresponding Reynolds number

$$R_\ell = R_0 \left(\ell/\ell_0\right)^{4/3} = R_0 \left(k_0/k_\ell\right)^{4/3}$$

decreases as ℓ decreases (or equivalently as k_ℓ increases), i.e., as the cascade proceeds. The viscous dissipation is no longer negligible when $R_\ell \sim 1$, which defines the dissipation scale ℓ_d, also called *Kolmogorov scale* ($k_d = 2\pi/\ell_d$):

$$R_d \sim 1 \sim R_0 \left(k_0/k_d\right)^{4/3} \Rightarrow k_d \sim k_0 R_0^{3/4} \Rightarrow k_d \sim \left(\epsilon/\nu^3\right)^{1/4}. \quad (35)$$

The energy spectrum of velocity fluctuations $E(k)$ is defined from

$$K_{\text{total}} = \int_0^\infty E(k)\, dk. \quad (36)$$

Splitting the reciprocal space into concentric shells $k_0 2^{p-1} < k < k_0 2^p$, we can write

$$K_{\text{total}} = \sum_p K_p = \sum_p E(k_p) k_p,$$

where the factor k_p in the last expression stands to ensure dimensional consistency as required from (36) where $E(k)$ is defined as an energy density by unit wavevector bandwidth.

Shell by shell, we have

$$K_p \sim \frac{1}{2} V_p^2 \sim V_0^2 (k_0/k_p)^{2/3} \sim E(k_p) k_p.$$

After dropping the indices, the last equivalence yields the celebrated $k^{-5/3}$ spectrum:

$$E(k) \sim \epsilon^{2/3} k^{-5/3}.$$

Interpreting Kolmogorov embedded eddies in much the same way as juxtaposed convective cells in weakly confined systems, we can have a first estimate of the number of degrees of freedom excited in developed turbulence. Since structures of size $\ell \ll \ell_d \sim 1/k_d$ are ironed out by viscous dissipation and slaved to larger scales, we have

10. Dynamics of Textures and Turbulence

only to count the number of presumably independent structures of size $\sim \ell_d$ in a domain of volume $\sim \ell_0^3$. This gives

$$N_* \sim \left(\frac{\ell_0}{\ell_d}\right)^3 \sim \left(\frac{k_d}{k_0}\right)^3 \sim k_0^3 \, R^{9/4}.$$

But nonlinear interactions are expected to reduce this number in much the same way as in weakly confined systems. Furthermore, the assumption of a constant energy transfer rate all along the cascade implicitly contains the idea that the energy transferred was equally shared by all the daughter eddies at half scale. Slight deviations from the K41 spectrum and more pronounced deviations for higher order quantities suggest that the energy transfer is sporadic and that only a fraction of smaller scale eddies are "active," the others remaining "quiescent." This process is called *intermittency in the inertial range*. Geometrical and stochastic models of such a transfer can be built, according to which the effective number of degrees of freedom is greatly reduced.

Intermittency in the inertial range can be attributed to the persistence of coherent structures. Actually, such structures have already been observed in shear flows and one can wonder whether methods developed for dynamical systems with a small number of degrees of freedom would not be appropriate to get a picture of their dynamics. On the other hand, the study of weakly confined systems, especially phase turbulence and spatio-temporal intermittency, could provide insight into the nonlinear behavior of highly excited small scale structures. At present, a theory combining these different approaches harmoniously has not been built yet but we may well be on the path of reconciling the viewpoint of Ruelle and Takens with that of Landau on the *nature of turbulence*.

6. Bibliographical Notes

Recent references to the topics treated in this chapter can be found in the proceedings of several workshops on pattern formation, notably:

[1] J. E. Wesfreid and S. Zaleski, eds., *Cellular Structures in Instabilities*, Lecture Notes in Physics **210** (Springer-Verlag, 1984).

[2] A. R. Bishop, G. Grüner, and B. Nicolaenko, eds., *Spatio-temporal coherence and chaos in physical systems* (North-Holland, 1986).

[3] J. E. Wesfreid, et al., eds., *Propagation in systems far from equilibrium*, Springer Series in Synergetics, Vol. 41 (Springer-Verlag, 1988).

[4] P. Coullet and P. Huerre, eds., *New trends in nonlinear dynamics and pattern forming phenomena: the geometry of nonequilibrium*, (Plenum Press, 1990).

General aspects of nonlinear dynamics in large aspect ratio systems are presented by:

[5a] A. C. Newell, "The dynamics of patterns: a survey," in [3] p. 122.

[5b] S. Fauve, "Large scale instabilities of cellular flows", in BN 5 [5], p. 63.

whereas a review of the phase dynamics approach is given by:

[6] H. R. Brand, "Phase dynamics — a review and a perspective," in [3] p. 206.

Though the introduction of phase variables is more ancient (see e.g., Newell and Whitehead, BN 8 [8]), the way to obtain phase equations at finite distance from onset by means of a gradient expansion was first discussed in:

[7] Y. Pomeau and P. Manneville, "Stability and Fluctuations of a spatially periodic flow," J. Physique-Lettres **40**, L-609 (1979).

The phenomenological nonlinear extension of the phase diffusion equation has been reviewed by:

[8] Y. Kuramoto, "Phase dynamics of weakly unstable periodic structures," Prog. Theor. Phys. **71**, 1182 (1984).

who also considers the case of waves. The inclusion of rotational invariance is treated by:

[9] T. Ohta, "Euclidian invariant formulation of phase dynamics. 1) non-propagating periodic patterns," Prog. Theor. Phys. **73**,1377 (1985).

[10] M. C. Cross and A. C. Newell, "Convection patterns in large aspect ratio systems," Physica **10D**, 299 (1984).

In [10], drift flow effects are taken into account at a phenomenological

10. Dynamics of Textures and Turbulence

level and consequences on phase instabilities and nonlinear selection are discussed.

The existence of a selection criterion derived from the phase formalism for curved roll pattern was shown in:

[11] Y. Pomeau and P. Manneville, "Wavelength selection in axisymmetric cellular structures" J.Physique **42**, 1067 (1981).

In fact, the presence of drift flows modifies the criterion as pointed by:

[12] M. C. Cross: "Phase dynamics of convective rolls," Phys. Rev. A **27**, 490 (1983).

[13] P. Manneville and J. M. Piquemal, "Zigzag instability and axisymmetric rolls in Rayleigh-Bénard convection: the effects of curvature," Phys. Rev. A **28**, 1174 (1983).

(the values of the coefficients quoted in the text are extracted from [13] which gives a full derivation of the coherence length and relaxation time at threshold, Chapter 4, Section 2.3 and the nonlinear solution for rolls, Chapter 8, Section 3.3; similarities and difference between no-slip and stress-free boundary conditions relative to these selection criteria are also discussed in detail). The explicit calculation of the wavelength selected by curvature has been extended in the far-from-threshold regime by:

[14] J. C. Buell and I. Catton, "Wavenumber selection in large-amplitude axisymmetric convection," Phys. Fluids **29**, 23 (1986).

A unified treatment of the interplay between the drift flow and the phase of a roll pattern (already apparent in [13]) can be found in:

[15] S. Fauve, E. W. Bolton, and M. E. Brachet, "Nonlinear oscillatory convection: a quantitative phase dynamics approach," Physica **29D**, 202 (1987).

Some selection criteria have not been explicitly mentioned in the text, e.g., the selection by lateral boundary studied by Cross *et al.*, BN **8** [9], or the selection by a slow space variation of control parameter analyzed by, e.g.,

[16] L. Kramer, E. Ben Jacob, H. Brand, and M. C. Cross, "Wavelength selection in systems far from equilibrium," Phys. Rev. Lett. **49**, 1891 (1982).

[17] Y. Pomeau and S. Zaleski, "Pattern selection in a slowly varying environment," J. Physique-Lettres **44**, L-135 (1983).

Experimental tests of individual nonlinear selection criteria have been performed, e.g.,

[18] V. Croquette and A. Pocheau, "Wave number selection in Rayleigh-Bénard convective structures," in [1] above.

[19] D. Cannell, M. A. Domingez-Lerma, and G. Ahlers, "Experiments on wave number selection in rotating Couette–Taylor flow," Phys. Rev. Lett. **50**, 1365 (1983).

The program contained in the Cross–Newell theory has been carried out in convection experiments at moderate Prandtl number by:

[20] M. S. Heutmaker and J. P. Gollub, "Wave-vector field of convective flow patterns," Phys. Rev. A **35**, 242 (1987).

An idea of the complexity of the nonlinear dynamics involving drift flows is best obtained from numerical simulations of models such as those quoted in Chapter 8 (e.g., BN 8 [4]). A more detailed presentation of results in Fig. 2–3 can be found in:

[21] P. Manneville, "Towards an understanding of weak turbulence close to the convection threshold in large aspect ratio systems," J. Physique-Lettres **44**, L-903 (1983).

More recent simulations in circular geometry have been presented by:

[22] H. S. Greenside, M. C. Cross, and W. M. Coughran, Jr., "Mean flows and the onset of chaos in large-cell convection," Phys. Rev. Lett. **60**, 2269 (1988).

whereas, on the theoretical side, solutions of the Cross-Newell phase equations with drift flow in circular geometry can be found in:

[23] A. Pocheau, "Phase dynamics attractors in an extended cylindrical convective layer," J. Physique **50**, 2059 (1989).

helping us to understand dynamical frustration effects at the origin of weak turbulence in low Prandtl number, large aspect-ratio, convecting systems.

The Kuramoto-Sivashinsky equation has been derived independently by:

[24] Y. Kuramoto and T. Tsuzuki, "Persistent propagation of concentration waves in dissipative media far from thermal equilibrium," Prog. Theor. Phys. **55**, 356 (1976).

[25] G. I. Sivashinsky, "Nonlinear analysis of hydrodynamic instability in laminar flames. Part I. Derivation of basic equations," Acta Astronautica **4**, 1177 (1977).

10. Dynamics of Textures and Turbulence

The transition to chaos in the Kuramoto-Sivashinsky equation has been studied in the spirit of dissipative dynamical systems especially by:

[26] J. M. Hyman, B. Nicolaenko, and S. Zaleski, "Order and complexity in the Kuramoto-Sivashinsky model of weakly turbulent interfaces," in [2] p. 265. For a review with references, see:

[27] P. Manneville, "The Kuramoto-Sivashinsky equation: a progress report," in [3], p. 265.

The general description of oscillating systems using envelope equations is given in:

[28] Y. Kuramoto, *Chemical oscillations, waves, and turbulence* (Springer-Verlag, 1984).

Two complex envelopes such as those introduced by Coullet, Fauve, and Tirapegui, BN **8** [13], are required for the description of quasi one-dimensional waves in convecting binary fluids described by, e.g.,

[29] C. M. Surko, P. Kolodner, A. Passner, and R. W. Walden, "Finite-amplitude traveling-wave convection in binary fluid mixtures," in [2] p. 220.

In moderately confined geometry, localized states can appear, see:

[30] R. Heinrichs, G. Ahlers, D. S. Cannell, "Traveling waves and spatial variation in the convection of a binary mixture," Phys. Rev. A **35**, 2761 (1987) 2761.

[31] E. Moses and V. Steinberg, "Flow patterns and nonlinear behavior of traveling waves in a convective binary mixture," Phys. Rev. A **34**, 693 (1986).

These states can be understood within the envelope formalism, as shown by:

[32] M. C. Cross, "Traveling and standing waves in binary fluid convection in finite geometries," Phys. Rev. Lett. **57**, 2935 (1986).

In more extended geometry, turbulence occurring in strongly nonlinear regimes are associated with the presence of structural defects and their nucleation/motion/annihilation, see:

[33] P. Coullet, C. Elphick, L. Gil, and J. Lega, "Topological defects of wave patterns," Phys. Rev. Lett. **59**, 884 (1987) 884.

[34] P. Coullet and J. Lega, "Defect-mediated turbulence in wave patterns," Europhys. Lett. **7**, 511 (1988).

The occurrence of *barber pole turbulence* in Couette flow is mentioned in the very last pages of Volume II of Feynman's Lecture Notes in Physics. This regime, and the transition that leads to it, is remakably described by:

[35] D. Coles: "Transitions in circular Couette flow," J. Fluid Mech. **21**, 285 (1965).

who points out the hysteretic character of the instability. For recent observations, consult:

[36] C. D. Andereck, S. S. Liu, and H. L. Swinney, "Flow regimes in a circular Couette system with independently rotating cylinders," J. Fluid Mech. **164**, 155 (1986).

and for a theoretical interpretation:

[37] J. J. Hegseth, C. D. Andereck, F. Hayot, and Y. Pomeau "Spiral turbulence and phase dynamics," Phys. Rev. Lett. **62**, 257 (1989).

In a seminal paper,

[38] Y. Pomeau: "Front motion, metastability, and subcritical bifurcations in hydrodynamics," in [2], p. 3.

proposed to interpret the transition to turbulence by subcritical instabilities in extended systems as a contamination process akin to directed percolation, critical properties of which are reviewed by:

[39] W. Kinzel, "Directed percolation," in: *Percolation structures and processes*, G.Deutcher et al., eds., Annals of the Israel Phys.Soc., Vol.5, p. 425 (1983).

This fully stochastic system is the simplest example of probabilistic cellular automata but deterministic cellular automata are also of interest since, though utterly simple, they are already capable of displaying complex space-time behavior, see, e.g.,:

[40] S. Wolfram, ed., *Theory and Applications of Cellular Automata* (World Scientific, 1986).

Results on the transition to turbulence *via* spatio-temporal intermittency in the damped Kuramoto-Sivashinsky equation have been presented in:

[41] H. Chaté and P. Manneville, "Transition to turbulence via spatio-temporal intermittency," Phys. Rev. Lett. **58**, 112 (1987).

However, the term *spatio-temporal intermittency* has been introduced earlier by

10. Dynamics of Textures and Turbulence

[42] K. Kaneko, "Spatio-temporal intermittency in coupled map lattices," Prog. Theor. Phys. **7**, 1033 (1985).

to describe the turbulent regime reached by an array of identical maps coupled by diffusion, all just below some temporal intermittency threshold.

The connection with directed percolation suggested by Pomeau has been explored in details by:

[43] H. Chaté, P. Manneville, "Spatio-temporal intermittency in coupled map lattices," Physica D **32**, 402 (1988).

For a review with a broader scope, see:

[44] H. Chaté and P. Manneville, "Transition to turbulence *via* spatio-temporal intermittency: modeling and critical properties," in [4].

Experimental evidence of spatio-temporal intermittency in the sense of Pomeau has been given by:

[45] S. Ciliberto, P. Bigazzi, "Spatio-temporal intermittency in Rayleigh-Bénard convection," Phys. Rev. Lett. **60**, 286 (1988).

[46] F. Daviaud, M. Dubois, and P. Bergé, "Spatio-temporal intermittency in quasi one-dimensional Rayleigh-Bénard convection," Europhys. Lett. **9**, 441 (1989).

For an introduction to the distinction between absolute and convective instabilities see:

[47] P. Huerre, "Spatio-temporal instabilities in closed and open flows,". in BN **5** [5], p.141, or "On the absolute/convective nature of primary and secondary instabilities," in [3].

A general review of developed turbulence, Kolmogorov's theory, and intermittency in the inertial range can be found in Chapter 6 of:

[48] M. Lesieur, *Turbulence in fluids* (Martinus Nijhoff, 1987).

Simple models of intermittent cascades in the inertial range have been proposed by:

[49] B. Mandelbrot, "Intermittent turbulence in self-similar cascades: divergence of high moments and dimension of the carrier," J. Fluid Mech. **62**, 331 (1974).

[50] U. Frisch, P. L. Sulem, and M. Nelkin, "A simple model of intermittent fully developed turbulence," J. Fluid Mech. **87**, 719 (1978).

The subsequent reduction of the number of degrees of freedom has been analyzed by:

[51] R.H. Kraichnan, "Intermittency and attractor size in isotropic turbulence," Phys. Fluids **28**, 10 (1985).

Coherent structures appearing in turbulent shear flows are described in:

[52] C.-H. Ho and P. Huerre, "Perturbed free shear layers," Annual Review of Fluid Mechanics **16**, 365 (1984).

Finally, recent discussions of interesting issues relative to the *nature of turbulence* are provided by:

[53] A. C. Newell, "Chaos and turbulence: is there a connection?," in: *Perspective in nonlinear dynamics*, Shlesinger et al., eds. (World Scientific, Singapore, 1986), p. 38.

[54] U. Frisch, "Fully developed turbulence: where do we stand?" in: *Dynamical systems: a renewal of mechanism*, S. Diner, D. Fargue, and G. Lochak, eds. (World-Scientific, Singapore, 1986), p. 13.

Appendix 1

Macroscopic Dynamics

1. General Fluid Systems

1.1. Densities and General Balance Equations

The description of *continuous media* is basically an extension to weakly nonuniform systems of the theory of macroscopic systems at thermodynamic equilibrium. The state of such systems can be characterized by a set of extensive variables $\{X\}$. The amount of X at a given time t contained in a domain $\mathcal{D}$ limited by some surface $\mathcal{S}$ can therefore be given as an integral

$$X(t) = \int_{\mathcal{D}} \rho_X(x, y, z; t)\, dx\, dy\, dz, \qquad (1)$$

where ρ_X is the corresponding density *per* unit volume. This implies the concept of *material point*, a volume element $dV = dx\, dy\, dz$ around point $M(x, y, z)$, infinitesimal from a mathematical viewpoint but large at the microscopic scale so that $dX = \rho_X\, dV$ is already a macroscopic variable and the thermodynamic relations valid for an infinite system remain applicable. The corollary is that thermal fluctuations are essentially negligible and that ρ_X evolves according to deterministic laws.

The amount of X in domain $\mathcal{D}$ can vary by creation/destruction at distributed sources/sinks and by addition/subtraction through the surface $\mathcal{S}$. With Σ_X denoting the source/sink density (positive for sources, negative for sinks, identically vanishing for a conserved quantity), $\mathbf{J}_X$ being the flux defined as the amount of X which leaves

$\mathcal{D}$ by unit surface and unit time, and $\hat{\mathbf{u}}$ being the unit vector normal to $\mathcal{S}$ pointing outwards, the general balance equation reads

$$\frac{d}{dt}\int_\mathcal{D} \rho_X \, dV + \int_\mathcal{S} \mathbf{J}_X \cdot \hat{\mathbf{u}} \, dS = \int_\mathcal{D} \Sigma_X \, dV.$$

This equation is turned into differential form by means of the *flux-divergence theorem*:

$$\partial_t \rho_X + \nabla \cdot \mathbf{J}_X = \Sigma_X. \tag{2}$$

1.2. Fluid Particles and the Continuity of Matter

For the description of flowing media, the key notion is that of *fluid particle*, a material point to which a velocity can be attributed. The velocity field is then defined as

$$\mathbf{v} = (v_x, v_y, v_z) = \left(\frac{dx_\mathrm{M}}{dt}, \frac{dy_\mathrm{M}}{dt}, \frac{dz_\mathrm{M}}{dt}\right), \tag{3}$$

where x_M, y_M, and z_M are the three coordinates of the fluid particle followed along its trajectory; these *Lagrangian coordinates* are functions of time and initial positions, i.e., $x_\mathrm{M} = x_\mathrm{M}(t; x_0, y_0, z_0), \ldots$. The physical properties of the corresponding material point remain governed by the thermodynamic laws applicable to the uniform and homogeneous system.

Another description called *Eulerian* is obtained by staying in the *laboratory frame*. The relation between the Lagrangian and Eulerian pictures involves the notion of *material derivative*, i.e., the total derivative with respect to time

$$\begin{aligned}\frac{d(\cdot)}{dt} &= \partial_t(\cdot) + \frac{dx_\mathrm{M}}{dt}\partial_x(\cdot) + \frac{dy_\mathrm{M}}{dt}\partial_y(\cdot) + \frac{dz_\mathrm{M}}{dt}\partial_z(\cdot)\\ &= \big[\partial_t + \mathbf{v}\cdot\nabla\big](\cdot).\end{aligned} \tag{4}$$

At this stage, the flux term $\mathbf{J}_X$ of the general balance equation (2) can be split into a *convective* part associated with the passive

A1. Macroscopic Dynamics

transport of the extensive quantity, $\mathbf{J}_{X,\text{conv}} = \rho_X \mathbf{v}$, and a *diffusive* part present even in the absence of global motion:

$$\mathbf{J}_X = \rho_X \mathbf{v} + \mathbf{J}_{X,\text{diff}}. \qquad (5)$$

Let us consider first the balance equation for mass which is a conserved quantity, $\Sigma_m \equiv 0$, transported only by convection, $\mathbf{J}_m \equiv \rho \mathbf{v}$ (by definition $\rho_m \equiv \rho$). From (2) we obtain simply

$$\partial_t \rho + \nabla \cdot (\rho \mathbf{v}) = 0,$$

which, using (4), also reads

$$\frac{d\rho}{dt} + \rho \nabla \cdot \mathbf{v} = 0. \qquad (6)$$

Equation (6) is called the *continuity equation* of matter. This relation is always valid, even in multicomponent fluids for which fractional mass densities ρ_i are defined with $\sum_i \rho_i = \rho$.

The content of the general balance equation is enlightened by passing from the density *per* unit volume ρ_X to the density *per* unit mass or *specific density* $\tilde{\rho}_X = \rho_X/\rho$. Then, taking into account Definition (5) for the diffusive flux and Equation (6), (2) can be rewritten as:

$$\rho \frac{d\tilde{\rho}_X}{dt} + \nabla \cdot \mathbf{J}_{X,\text{diff}} = \Sigma_X, \qquad (7)$$

which clearly suppresses all reference to the extrinsic variations of the density due to the motion of the fluid particle itself. In the following, we will always write balance equations for fluid particles (i.e., as (7)). The convective part of the fluxes being removed automatically in such a formulation, the subscript "diff" becomes useless and will be suppressed.

1.3. Fundamental Equation of Dynamics

The equation of motion of a continuous medium can be derived in a straightforward way from Newton's law of dynamics applied to a fluid particle. Like the total mass, the total linear momentum **p** of a fluid particle is entirely carried by the matter from which it is made, so that the corresponding diffusive flux $\mathbf{J_p}$ vanishes identically. The immediate translation of $m\,d\mathbf{v}/dt = \mathbf{F}_{\text{ext}}$ valid for an isolated particle of mass m submitted to an external force $\mathbf{F}_{\text{ext}}$ will be $\rho\,d\mathbf{v}/dt = \mathbf{f}_{\text{ext}}$, where $\mathbf{f}_{\text{ext}}$ is now a force *per* unit volume. However, some care is required in the evaluation of this force since the fluid particle is surrounded by the rest of the medium so that it is submitted to surface forces (stresses) from its neighborhood in addition to possible body forces: $\mathbf{f}_{\text{ext}} = \mathbf{f}_{\text{surf}} + \mathbf{f}_{\text{bulk}}$.

Stresses are defined in terms of a second order-tensor σ. By definition, $\sigma_{\alpha\beta}$ is the component along direction α of the force *per* unit surface applied by the exterior world on a surface element normal to direction β. Normal and tangential components correspond to $\alpha = \beta$ and $\alpha \neq \beta$, respectively, e.g., σ_{zz} and σ_{xy}. When applied to a vector equation, the flux-divergence theorem yields a contribution of the form $\sum_\beta \partial_\beta \sigma_{\alpha\beta}$ to its α component. The fundamental equation of dynamics therefore reads

$$\rho \frac{dv_\alpha}{dt} = \sum_\beta \partial_\beta \sigma_{\alpha\beta} + f_\alpha, \tag{8}$$

where f_α is the α component of the applied body force by unit volume, denoted **f** in the following.

From the definition of the stress tensor, it is easily checked that the α component of the torque exerted on the fluid particle by its environment is given by

$$\Gamma_\alpha = \sum_{\beta\gamma} \epsilon_{\alpha\beta\gamma} \sigma_{\beta\gamma},$$

where $\epsilon_{\alpha\beta\gamma}$ is the completely antisymmetric third-order tensor. For example $\Gamma_x = \sigma_{yz} - \sigma_{zy}$. But external torques cannot be applied to the bulk of an isotropic fluid. Since the internal stress torque cannot be compensated for by an external bulk torque, the stress tensor must be symmetric to fulfill the angular momentum balance equation, i.e., $\sigma_{\alpha\beta} = \sigma_{\beta\alpha}$. This is in contrast with the case of anisotropic fluids and especially nematic liquid crystals to be examined later.

2. Thermohydrodynamics

Here we consider first the simplest possible case of a single component, isotropic fluid, i.e., a medium whose thermodynamic properties *at equilibrium* entirely derive from a state function relating the entropy S to the internal energy U and the volume V at given mass, the temperature T and pressure p being the conjugate intensive parameters. Then we extend the description to passive mixtures adding a concentration variable and a conjugate chemical potential and add a note on boundary conditions at a fluid interface. The case of nematic liquid crystals is presented in the next section.

2.1. Energy Continuity and First Principle

The total energy of a fluid particle contains three contributions, a kinetic part associated with its velocity, a potential part linked to its position in an external force field deriving from a potential $\tilde{\mathbf{f}} = -\nabla\tilde{\phi}$, and an internal part

$$\rho\tilde{e} = \rho\frac{1}{2}\mathbf{v}^2 + \rho\tilde{\phi} + \rho\tilde{u}$$

($\tilde{e}$, $\tilde{\phi}$, and $\tilde{u}$ are densities *per* unit mass).

The total energy is conserved. From Equation (7) we obtain

$$\rho\frac{d\tilde{e}}{dt} + \nabla\cdot\mathbf{J}_E = 0. \tag{9}$$

Notice that both the kinetic energy and the potential energy are attached to the fluid particle and therefore cannot diffuse, which is not the case of the internal energy; $\mathbf{J}_E$ is therefore reduced to $\mathbf{J}_U$.

The variation of the kinetic energy density $\frac{1}{2}\mathbf{v}^2$ is derived from the fundamental equation of mechanics by taking the scalar product with $\mathbf{v}$ itself:

$$\rho\frac{d}{dt}\left(\frac{1}{2}\mathbf{v}^2\right) = \rho\sum_\alpha v_\alpha\frac{dv_\alpha}{dt} = \sum_\alpha\sum_\beta v_\alpha\,\partial_\beta\sigma_{\alpha\beta} + \sum_\alpha v_\alpha f_\alpha. \tag{10}$$

The last term in (10) represents the work of external forces, which is just the opposite of the potential energy variation:

$$\mathbf{v}\cdot\mathbf{f} = \mathbf{v}\cdot(\rho\tilde{\mathbf{f}}) = \rho(-\mathbf{v}\cdot\nabla\tilde{\phi}) = -\rho\frac{d\tilde{\phi}}{dt}$$

($d\tilde{\phi}/dt = \partial_t\tilde{\phi} + \mathbf{v}\cdot\nabla\tilde{\phi}$, but $\tilde{\phi}$ is supposed independent of t so that $\partial_t\tilde{\phi} \equiv 0$). The continuity equation for the sum $\frac{1}{2}\mathbf{v}^2 + \tilde{\phi}$ is therefore easily rewritten as

$$\rho\frac{d}{dt}\left(\frac{1}{2}\mathbf{v}^2 + \tilde{\phi}\right) = -\sum_{\alpha\beta}\sigma_{\alpha\beta}\partial_\beta v_\alpha + \sum_\beta \partial_\beta\left(\sum_\alpha v_\alpha \sigma_{\alpha\beta}\right). \quad (11)$$

Subtracting (11) from (9) yields the continuity equation for the internal energy:

$$\rho\frac{d\tilde{u}}{dt} + \nabla\cdot\mathbf{J}_U = +\sum_{\alpha\beta}\sigma_{\alpha\beta}\partial_\beta v_\alpha - \sum_\beta \partial_\beta\left(\sum_\alpha v_\alpha \sigma_{\alpha\beta}\right). \quad (12)$$

Some more work is necessary to turn this expression of the *first principle* of thermodynamics into a more transparent form. It is first easily realized that $\sum_\alpha v_\alpha \sigma_{\alpha\beta}$, which appears in the last term at the r.h.s. of (12) is the β component of the flux of mechanical work *received* by the fluid particle during its motion, and that the last term is the divergence of this flux which we denote $-\mathbf{J}_W$ according to the sign convention chosen at the beginning. We can then define the *heat flux*, the flux of heat *given* to the environment, as $\mathbf{J}_Q = \mathbf{J}_U - \mathbf{J}_W$.

To go still further, we isolate the contribution of the thermodynamic pressure p to the stress tensor:

$$\sigma_{\alpha\beta} = -p\,\delta_{\alpha\beta} + \sigma'_{\alpha\beta}, \quad (13)$$

so that

$$\sum_{\alpha\beta}\sigma_{\alpha\beta}\partial_\beta v_\alpha = -p\sum_\alpha \partial_\alpha v_\alpha + \sum_{\alpha\beta}\sigma'_{\alpha\beta}\partial_\beta v_\alpha$$

$$= -p\nabla\cdot\mathbf{v} + \sum_{\alpha\beta}\sigma'_{\alpha\beta}\partial_\beta v_\alpha$$

$$= -\rho\left(p\frac{d\varpi}{dt}\right) + \sum_{\alpha\beta}\sigma'_{\alpha\beta}\partial_\beta v_\alpha,$$

A1. Macroscopic Dynamics

where the last expression derives from the use of the continuity equation (6) written for $\varpi = 1/\rho$:

$$\rho\, d\varpi/dt = \nabla\cdot\mathbf{v}. \tag{6'}$$

Inserting these intermediate results in (12), we obtain

$$\rho\frac{d\tilde{u}}{dt} = -\nabla\cdot\mathbf{J}_Q - p\frac{d\varpi}{dt} + \sum_{\alpha\beta}\sigma'_{\alpha\beta}\partial_\beta v_\alpha. \tag{14}$$

Finally $-\nabla\cdot\mathbf{J}_Q$, which accounts for the increase of the fluid particle's heat content, may be written equally well as $\rho\, d\tilde{q}/dt$. Equation (14) can therefore be recast in the form

$$\frac{d\tilde{u}}{dt} = \frac{d\tilde{q}}{dt} - p\frac{d\varpi}{dt} + \frac{1}{\rho}\sum_{\alpha\beta}\sigma'_{\alpha\beta}\partial_\beta v_\alpha. \tag{15}$$

Except for the third term on the r.h.s., which is new and accounts for the generation of internal energy due to internal forces (viscous friction), Equation (15) directly derives from the usual formulation of the first principle: $dU = dQ + dW = dQ - p\, dV$ (ϖ being the specific volume, the second term on the r.h.s. obviously corresponds to the compression work).

2.2. Entropy Source and Second Principle

From (7), the balance equation for the entropy reads

$$\rho\frac{d\tilde{s}}{dt} + \nabla\cdot\mathbf{J}_s = \Sigma_s, \tag{16}$$

where the entropy source term $\Sigma_s \geq 0$ is positive for general *irreversible* processes tending to restore equilibrium on a large scale and vanishes only for *reversible* processes.

For an infinitesimal quasi-static process, the *second principle* simply reads

$$dQ = T\, dS, \tag{17}$$

which, using the first principle for a simple fluid, also reads

$$dS = \frac{1}{T}(dU + p\,dV).$$

For the fluid particle, the local equilibrium assumption directly leads to

$$T\frac{d\tilde{s}}{dt} = \frac{d\tilde{u}}{dt} + p\frac{d\tilde{\omega}}{dt}.$$

Taking into account the internal energy balance (14), we get

$$\rho\frac{d\tilde{s}}{dt} + \frac{1}{T}\nabla\cdot\mathbf{J}_Q = \frac{1}{T}\sum_{\alpha\beta}\sigma'_{\alpha\beta}\partial_\beta v_\alpha$$

or, using the identity $\nabla\cdot(\mathbf{J}_Q/T) = (1/T)\nabla\cdot\mathbf{J}_Q + \mathbf{J}_Q\cdot\nabla(1/T)$,

$$\rho\frac{d\tilde{s}}{dt} + \nabla\cdot\left(\frac{\mathbf{J}_Q}{T}\right) = \mathbf{J}_Q\cdot\nabla\left(\frac{1}{T}\right) + \frac{1}{T}\sum_{\alpha\beta}\sigma'_{\alpha\beta}\partial_\beta v_\alpha. \qquad (18)$$

By comparison with (16), this leads to the identification

$$\mathbf{J}_s = \frac{\mathbf{J}_Q}{T}, \qquad \Sigma_s = \frac{1}{T}\sum_{\alpha\beta}\sigma'_{\alpha\beta}\partial_\beta v_\alpha + \mathbf{J}_Q\cdot\nabla\left(\frac{1}{T}\right). \qquad (18')$$

The definition of the entropy flux may be considered as natural since (17) results from the integration of the diffusive flux through the 'walls' of the fluid particle. Note also that only the anisotropic part of the stress tensor appears in this expression and that the second term is the entropy source associated with thermal diffusion.

2.3. Constitutive Equations

The set of equations (6, 8, 12) is not closed. Relations between the gradients of intensive parameters (e.g., the temperature T), the so-called *thermodynamic forces*, and the fluxes of extensive quantities (e.g., the heat flux $\mathbf{J}_Q$) remain to be determined. If gradients are

A1. Macroscopic Dynamics

weak at the molecular scale, a *linear response* assumption can be made. The fluid is said to be *Newtonian*. Accordingly, we expect

$$\mathbf{J}_{Y_i} = \sum_j L_{ij}\, \nabla X_j,$$

where Y_i is any extensive quantity and X_j any intensive parameter. However, the matrix L_{ij} is further constrained by several conditions. First, the entropy source must be positive definite so that the equilibrium state corresponds to a maximum of the entropy reached for $\nabla X_j \equiv 0$. Second, the symmetries of the medium forbid the coupling of quantities with different tensorial characters (*Curie's principle*).

For a simple fluid, this implies that the heat flux (a polar vector) is coupled only to the temperature gradient and that the viscous dissipation associated with the deformation tensor has to be treated separately. Therefore, we obtain first the *Fourier law* written as

$$\mathbf{J}_Q = -\chi \nabla T \qquad (19)$$

which defines the thermal conductivity χ.

The intensive parameter conjugate to the stress tensor $\sigma_{\alpha\beta}$ is the rate-of-deformation tensor $\partial_\beta v_\alpha$, which has to be split into irreducible components before writing down the *Stokes law*. This decomposition reads

$$\partial_\beta v_\alpha = \frac{1}{3}\, \delta_{\beta\alpha}\, \nabla\cdot\mathbf{v} + d_{\alpha\beta} + \frac{1}{2}(\partial_\beta v_\alpha - \partial_\alpha v_\beta).$$

The first term describes isotropic compression, the second term

$$d_{\alpha\beta} = \frac{1}{2}\left(\partial_\alpha v_\beta + \partial_\beta v_\alpha - \frac{2}{3}\, \delta_{\beta\alpha}\, \nabla\cdot\mathbf{v}\right)$$

is a traceless symmetric tensor corresponding to *pure deformations* and the last, antisymmetric, term accounts for *pure rotations* (vorticity). Each irreducible component is then coupled to the corresponding term in the stress tensor. Owing to isotropy, the antisymmetrical part of the stress tensor vanishes identically, so that we obtain

$$\sigma'_{\alpha\beta} = \eta_v\, \nabla\cdot\mathbf{v} + 2\eta\, d_{\alpha\beta}. \qquad (20)$$

The coefficient η of the second term is the *shear viscosity*. The first term accounts for the irreversible response to isotropic compression (*bulk viscosity* coefficient), the reversible part being included in the definition of the thermodynamic pressure p.

Inserting (19) and (20) into the expression of the entropy source yields

$$T\Sigma_s = \frac{\chi}{T}(\nabla T)^2 + \eta_v(\nabla\cdot\mathbf{v})^2 + 2\eta\sum_{\alpha\beta} d_{\alpha\beta}^2.$$

The positivity of the entropy production implies that χ, η, η_v are all positive.

2.4. Summary of the Equations for a Simple Fluid

In addition to Equation (6) accounting for the continuity of matter, inserting (19) and (20) into Equations (8) and (13), we obtain

- equation of motion:

$$\rho\frac{d\mathbf{v}}{dt} = -\nabla p + \eta\nabla^2\mathbf{v} + \left(\frac{1}{3}+\eta_v\right)\nabla(\nabla\cdot\mathbf{v});$$

- continuity of internal energy:

$$\rho\frac{d\tilde{u}}{dt} = \chi\nabla^2 T - p\,\nabla\cdot\mathbf{v} + 2\eta\sum_{\alpha\beta} d_{\alpha\beta}d_{\alpha\beta} + \eta_v(\nabla\cdot\mathbf{v})^2.$$

To these equations we must add the thermodynamic state equations relating the local values of intensive parameters and the density of extensive quantities:

$$p = p(\varpi, T), \qquad \tilde{u} = \tilde{u}(\varpi, T)$$

(with $\varpi = 1/\rho$), e.g., for an ideal gas

$$p\varpi = RT, \qquad \tilde{u} = C_v T.$$

An important limit corresponds to the case of incompressible fluids for which the variations of ρ are negligible. The continuity and motion equations then simply read

$$\nabla\cdot\mathbf{v} = 0 \tag{21}$$

$$\rho\frac{d\mathbf{v}}{dt} = -\nabla p + \eta\nabla^2\mathbf{v} \tag{22}$$

A1. Macroscopic Dynamics

but in (22) the pressure p has lost all thermodynamic significance and must rather be understood as a Lagrange multiplier associated with condition (21). Equations (21) and (22) are called the *Navier–Stokes* equations, The inviscid limit "$\eta \to 0$" yields the so-called *Euler equations*.

2.5. Passive Mixtures

We consider now a multi-component fluid in the absence of chemical reaction so that the amount of each component is a conserved quantity. The partial densities of each species being ρ_i, we define the densities *per* unit mass as $\tilde{\rho}_i = \rho_i/\rho$ with $\rho = \sum_i \rho_i$. The corresponding conservation equations then read

$$\rho \frac{d\tilde{\rho}_i}{dt} + \nabla \cdot \mathbf{J}_i = 0,$$

with $\sum_i \tilde{\rho}_i = 1$ and $\sum_i \mathbf{J}_i = 0$. Nothing new is brought to the global motion equation and to the internal energy equation if external forces act in the same way on all species. The entropy equation is derived in the same way as for the single component fluid, except that we have to insert $dQ = T\,dS$ in $dU = dQ - p\,dV + \sum_i \mu_i N_i$, where μ_i is the chemical potential associated with the mass fraction N_i of species i, which, once translated to the fluid particle, yields

$$T \frac{d\tilde{s}}{dt} = \frac{d\tilde{u}}{dt} + p\frac{d\tilde{\varpi}}{dt} - \sum_i \mu_i \frac{d\tilde{\rho}_i}{dt}.$$

Taking into account the energy equation, after manipulations analogous to those leading to (18), we obtain by identification

$$\mathbf{J}_s = \frac{1}{T}\left(\mathbf{J}_Q - \sum_i \mu_i \mathbf{J})_i\right),$$

$$\Sigma_s = \mathbf{J}_Q \cdot \nabla\left(\frac{1}{T}\right) + \frac{1}{T}\sum_{\alpha\beta} \sigma'_{\alpha\beta}\,\partial_\beta v_\alpha - \sum_j \mathbf{J}_i \cdot \nabla\left(\frac{\mu_i}{T}\right).$$

Notice that in an n component mixture, owing to the relation between the fluxes quoted above, only $n-1$ are independent. In a binary mixture, only one diffusion field is relevant. For a dilute system, neglecting all other processes, a gradient of chemical potential, i.e., a concentration gradient since $\mu \sim RT \log(\tilde{\rho})$, implies a flux of solute: $\mathbf{J} = D\nabla\tilde{\rho}$ (*Fick law*). But if temperature variations are allowed, cross effects are expected (*thermo-diffusion*) since the heat flux and the solute flux on the one hand, the temperature gradient and the solute concentration gradient on the other, all have the same tensorial character (polar vectors).

2.6. Fluid Interfaces

A solid medium responds elastically to imposed displacements at its boundary. By contrast, at an interface between two fluid medias, say (a) and (b), motion in medium (a) below induces motion in medium (b) above. Boundary conditions to be applied at the interface must express the continuity of material properties. They can be obtained by taking the limit of an infinitely thin layer around the interface and moving with it. First of all, the absence of overlap or empty space between the two media (continuity of matter) implies the continuity of the velocity normal to the interface:

$$\mathbf{v}^{(a)} \cdot \widehat{\mathbf{u}} = \mathbf{v}^{(b)} \cdot \widehat{\mathbf{u}},$$

where, by convention, $\widehat{\mathbf{u}}$ denotes the unit vector normal to the interface and oriented from (a) to (b). The continuity of linear momentum yields, in the same way,

$$\sum_{\beta}\left(\sigma^{(a)}_{\alpha\beta}\widehat{u}_{\beta}\right) = \sum_{\beta}\left(\sigma^{(b)}_{\alpha\beta}\widehat{u}_{\beta}\right).$$

At a curved interface with a uniform surface tension A, we have

$$p^{(b)} - p^{(a)} = A\left(\frac{1}{R_1} + \frac{1}{R_2}\right),$$

where A is the surface tension, and R_1 and R_2 are the principal radii of curvature, positive when the concavity is directed towards medium (b). Recalling that the viscous stress is defined by $\sigma_{\alpha\beta} = \sigma'_{\alpha\beta} - p\delta_{\alpha\beta}$, we obtain

$$\sum_\beta \left(\sigma'_{\alpha\beta}{}^{(b)} - \sigma'_{\alpha\beta}{}^{(a)}\right) = \left(p^{(b)} - p^{(a)} - A\left(\frac{1}{R_1} + \frac{1}{R_2}\right)\right)\widehat{u}_\alpha.$$

When the surface tension is allowed to vary but the interface remains plane, the tangential stress is discontinuous and the size of the discontinuity is given by the gradient of the surface tension:

$$\sum_\beta \left(\sigma_{\alpha\beta}^{(b)} - \sigma_{\alpha\beta}^{(a)}\right) + \partial_\alpha A = 0.$$

3. Nematic Liquid Crystals

3.1. The Nematic Phase

Liquid crystals are organic materials displaying intermediate phases between the usual isotropic liquid state and the solid state (*mesomorphic*). These phases are more or less ordered and, accordingly, more or less fluid. Here, we consider only the simplest case of uniaxial *nematics* usually made of relatively rigid elongated molecules characterized by long-range orientational order only (no position ordering of the molecules as in *smectic* phases).

In nematics, the orientational order is quadrupolar with an order parameter $Q_{\alpha\beta} \propto (n_\alpha n_\beta - \frac{1}{3}\delta_{\alpha\beta})$, where the n_αs are the components of a unit vector $\mathbf{n}$, the *director*, characterizing the direction of local anisotropy (Fig. 1a). Only the orientation is relevant, not the direction, so that all properties must be invariant under the change $\mathbf{n} \to -\mathbf{n}$. Moreover, the response of vector quantities to applied vector fields is given by second-order tensor susceptibilities with two distinct eigenvalues relative to eigendirections parallel and perpendicular to $\mathbf{n}$, respectively. Considering for example the case of the magnetization $\mathbf{M}$ induced by a magnetic field $\mathbf{H}$, writing $\mathbf{H} = \mathbf{H}_\| + \mathbf{H}_\perp$ and $\mathbf{H}_\| = (\mathbf{H} \cdot \mathbf{n})\mathbf{n}$, we obtain

$$\mathbf{M} = \chi_\perp \mathbf{H} + \chi_a (\mathbf{H} \cdot \mathbf{n})\mathbf{n}, \tag{23}$$

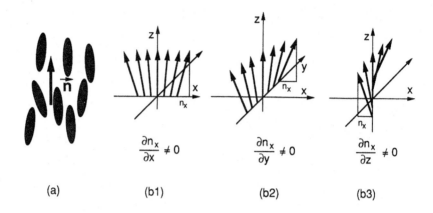

Fig. 1. a) Director field describing the nematic order. b) The three basic distortions: b1) splay; b2) twist; b3) bend.

where $\chi_\parallel$ and $\chi_\perp$ are the two eigenvalues of the magnetic susceptibility ($\chi_{\alpha\beta}$) and $\chi_a = \chi_\parallel - \chi_\perp$ is called the anisotropic part.

3.2. Elasticity

Besides anisotropy, an important difference with simple fluids comes from the presence of the internal degree of freedom associated with orientation. At thermodynamic equilibrium, **n** is uniform in space and any disorientation costs some energy. At first order, the expression of a slight distortion of the director field **n** parallel to $\hat{z}$ at the origin is given by

$$dn_z = 0, \qquad dn_\alpha = \sum_{\beta=x,y,z} \partial_\beta n_\alpha \, d\alpha \quad \text{for } \alpha = x, y.$$

The three basic distortions are illustrated in Fig. 1:

- *splay*: $\partial_x n_x \neq 0$ or/and $\partial_y n_y \neq 0$, Fig. 1b1;
- *twist*: $\partial_x n_y \neq 0$ or/and $\partial_y n_x \neq 0$, Fig. 1b2;
- *bend*: $\partial_z n_x \neq 0$ or/and $\partial_z n_y \neq 0$, Fig. 1b3.

A1. Macroscopic Dynamics

The distortion energy density reads (Franck, 1958)

$$\begin{aligned}\mathcal{E}_d &= \frac{1}{2}\Big\{K_1\big((\partial_x n_x)^2 + (\partial_y n_y)^2\big) + K_2\big((\partial_x n_y)^2 + (\partial_y n_x)^2\big) \\ &\quad + K_3\big((\partial_z n_x)^2 + (\partial_z n_y)^2\big)\Big\} \\ &= \frac{1}{2}\Big\{K_1(\nabla\cdot(\mathbf{n}))^2 + K_2(\mathbf{n}\cdot(\nabla\times\mathbf{n}))^2 + K_3(\mathbf{n}\times(\nabla\times\mathbf{n}))^2\Big\},\end{aligned}$$

with the supplementary condition $|\mathbf{n}| = 1$.

Usually, one has $K_2 < K_1 \sim K_3$, e.g. for MBBA[†] at room temperature $K_1 \simeq 6.10^{-7}$, $K_2 \simeq 3.10^{-7}$, $K_3 \simeq 7.10^{-7}$ in CGS units.

3.3. Statics

The total distortion energy is obtained by integration over the domain $\mathcal{D}$: $E_d = \int_\mathcal{D} \mathcal{E}_d\, dx\, dy\, dz$. At equilibrium, this total energy should remain stationary with respect to fluctuations of the orientation. The variation of elastic energy associated with an infinitesimal deformation $\mathbf{n} \to \mathbf{n} + \delta\mathbf{n}$ reads

$$\delta E_d = \int_\mathcal{D} \delta\mathcal{E}_d\, dx\, dy\, dz = \int_\mathcal{D} \big\{-\mathbf{h}\cdot\delta\mathbf{n}\, dx\, dy\, dz\big\}$$

$$\text{with}\quad h_\alpha = -\frac{\partial\mathcal{E}_d}{\partial n_\alpha} + \sum_\beta \partial_\beta\left(\frac{\partial\mathcal{E}_d}{\partial(\partial_\beta n_\alpha)}\right).$$

A deformation $\delta\mathbf{n}$, which preserves the length of $\mathbf{n}$, is locally an infinitesimal rotation $\delta\theta$ around some axis $\hat{\mathbf{u}}$ so that $\delta\mathbf{n} = \delta\theta\,\hat{\mathbf{u}}\times\mathbf{n}$. Using $\mathbf{h}\cdot(\hat{\mathbf{u}}\times\mathbf{n}) = \hat{\mathbf{u}}\cdot(\mathbf{n}\times\mathbf{h})$, we obtain

$$\delta E_d = -\int_\mathcal{D} \delta\theta\,\mathbf{h}\cdot(\hat{\mathbf{u}}\times\mathbf{n})\, dx\, dy\, dz = -\int_\mathcal{D} \delta\theta\,\hat{\mathbf{u}}\cdot(\mathbf{n}\times\mathbf{h})\, dx\, dy\, dz,$$

where $\mathbf{n}\times\mathbf{h} = \boldsymbol{\Gamma}_{el}$ is the *elastic torque*. At equilibrium, the variation of elastic energy must vanish, which implies

$$\boldsymbol{\Gamma}_{el} \equiv 0 \quad\Rightarrow\quad \mathbf{n}\times\mathbf{h} = 0 \quad \text{i.e., } \mathbf{h}\propto\mathbf{n}\ \text{ everywhere.}$$

[†] N-(p-methoxybenzylidene)-p-n-butylaniline!

Other contributions to the equilibrium condition are introduced by the coupling with external fields. For example, the magnetization **M** induced by the field **H** according to (23) adds to the energy a magnetic term $\mathcal{E}_m$ from which a magnetic torque Γ_m derives:

$$\mathcal{E}_m = -\frac{1}{2}\chi_a(\mathbf{H}\cdot\mathbf{n})^2 \quad \text{and} \quad \Gamma_m = \chi_a(\mathbf{n}\cdot\mathbf{H})(\mathbf{n}\times\mathbf{H}).$$

In the same way, an electric field couples with the director through the electric polarizability, so that

$$\mathcal{E}_e = -\frac{1}{2}\epsilon_a(\mathbf{E}\cdot\mathbf{n})^2 \quad \text{and} \quad \Gamma_e = \epsilon_a(\mathbf{n}\cdot\mathbf{E})(\mathbf{n}\times\mathbf{E}).$$

The equilibrium condition in the bulk then reads

$$\Gamma_{\text{stat}} = 0 = \Gamma_{el} + \Gamma_m + \Gamma_e.$$

Molecular anchoring at an interface also influences the orientational ordering in the bulk. The control of surface orientation is achieved by mechanical treatment (rubbing), chemical treatment (deposition of surfactants), or more sophisticated processes (oblique evaporation). Generally, a specific direction $\mathbf{n}_s$ is favored by the anchoring mechanism. The anchoring energy density (by surface unit) $\mathcal{E}_s$ is minimum when the director stands along this direction, e.g., $\mathcal{E}_s \propto (\mathbf{n}|_s - \mathbf{n}_0)^2$, where $\mathbf{n}|_s$ is the orientation in the nematic "at" the wall and $\mathbf{n}_0$ is the favored orientation. The quantity to be minimized at equilibrium is the sum of all contributions to the total energy: bulk (elastic, magnetic, dielectric) + surface ($\int_S \mathcal{E}_s \, dS$). The anchoring is said to be *strong* when deviations from the favored orientation are practically excluded owing to a large energy surface. In that case the anchoring acts as a rigid boundary condition on the director field. Otherwise the minimization of the total energy "bulk + surface" cannot be avoided.

The most often encountered situations are displayed in Fig. 2:
- *planar* anchoring: $\mathbf{n}_0$ parallel to the wall, Fig. 2a;
- *homeotropic* anchoring: $\mathbf{n}_0$ perpendicular to the wall, Fig. 2b;

A1. Macroscopic Dynamics

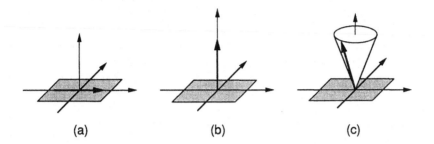

Fig. 2. Usual anchoring conditions: a) planar; b) homeotropic; c) conical.

- *oblique* anchoring (Fig. 2c) is often *weak*, especially when it is *conical*.

3.4. Dynamics

The complexity of nematodynamics comes essentially from the nontrivial coupling of the orientation field with the strain field. In contrast with the case of isotropic liquids, bulk torque can be applied to nematics which also transmit torques associated with boundary conditions on the director. The continuity of the angular momentum density must then be written down explicitly, which leads to supplementary equations of the form

$$(\Gamma_{\text{stat}})_\alpha + \sum_{\beta\gamma} \epsilon_{\alpha\beta\gamma}\, \sigma_{\beta\gamma} = 0,$$

where Γ_{stat} represents the contributions introduced above.

Here we summarize the Ericksen–Leslie–Parodi formulation for an incompressible nematic ($\nabla \cdot \mathbf{v} = 0$) in the limit of weak local curvature (which is sufficient in usual conditions). The equation of motion reads

$$\rho \frac{dv_\alpha}{dt} = -\partial_\alpha p + \partial_\beta \sigma'_{\alpha\beta}.$$

On general grounds, the stress tensor is the sum of an elastic part and a viscous part. Hydrostatics is concerned with the equilibrium between the pressure term and the elastic contribution of the stress tensor (reversible part). In the presence of flow, if the local curvature remains weak, the elastic part is of higher order and can be neglected. In the bulk of the nematic, two kinds of distortion imply viscous dissipation: usual deformations (displacements) $d_{\alpha\beta} = \frac{1}{2}(\partial_\alpha v_\beta + \partial_\beta v_\alpha)$, and rotations of the director with respect to the fluid

$$\mathbf{N} = \frac{d\mathbf{n}}{dt} - \mathbf{\Omega} \times \mathbf{n},$$

where $\mathbf{\Omega} = \frac{1}{2} \nabla \times (\mathbf{v})$ is the rotation rate of the fluid.

Assuming linear relations between deformations and stresses yields
$$\begin{aligned}\sigma'_{\alpha\beta} = &\alpha_4 d_{\alpha\beta} + \alpha_1(n_\gamma d_{\gamma\delta} n_\delta)n_\alpha n_\beta \\ &+ \alpha_5 n_\alpha(n_\gamma d_{\gamma\beta}) + \alpha_6(n_\gamma d_{\gamma\alpha})n_\beta \\ &+ \alpha_2 n_\alpha N_\beta + \alpha_3 N_\alpha n_\beta\end{aligned}$$

(Leslie stress tensor). From the six viscosity coefficients introduced above, only five are independent owing to Onsager reciprocity relations (Parodi)

$$\alpha_2 + \alpha_3 = \alpha_6 - \alpha_5.$$

Coefficient α_4 is independent of the orientation (isotropic component); α_1 is associated with a symmetric stretching and does not contribute to the viscous torque.

Coefficients α_5 and α_6 contribute to the stress associated with the irrotational component of the flow, $d_{\alpha\beta}$. Assuming $\partial_x v_y = \partial_y v_x = s$ (s is the shear rate), with $\mathbf{n}$ along the y axis, we obtain

$$\sigma_{xy} = \alpha_6 s \quad \text{and} \quad \sigma_{yx} = \alpha_5 s,$$

which implies the viscous torque component

$$\Gamma_{v,z} = \sigma_{xy} - \sigma_{yx} = (\alpha_6 - \alpha_5)s = \gamma_2 s$$

by definition of γ_2.

Coefficients α_2 and α_3 contribute to the stress associated with the rotational component of the flow $\mathbf{\Omega}$. Assuming $\partial_x v_y + \partial_y v_x = 0$

A1. Macroscopic Dynamics

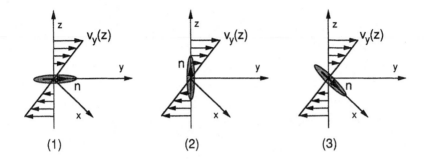

Fig. 3. The three basic Miesowicz geometries of a nematic under shear.

so that $\Omega_z = \frac{1}{2}(\partial_x v_y - \partial_y v_x) = \partial_x v_y = s$ and therefore $N_x = \Omega_z = s$, we have

$$\sigma_{xy} = \alpha_3 s \quad \text{and} \quad \sigma_{yx} = \alpha_2 s$$

and, accordingly

$$\Gamma_z = \sigma_{xy} - \sigma_{yx} = (\alpha_3 - \alpha_2)s = \gamma_1 s$$

by definition of γ_1, which can be shown to be positive owing to one of the thermodynamic inequalities ensuring the positivity of the entropy source.

3.5. Viscometry

Real shear-flow experiments involve a superposition of contributions leading to the definition of effective shear viscosities depending on the geometry, i.e., on the orientation of the director with respect to the shear (*Miesowicz geometries*, fig 3):

- Case 1, **n** along the y axis: $\eta_1 = \frac{1}{2}(\alpha_4 + \alpha_6 + \alpha_3)$, $\Gamma_1 = -\alpha_3 s$;
- Case 2, **n** along the z axis: $\eta_2 = \frac{1}{2}(\alpha_4 + \alpha_5 - \alpha_2)$, $\Gamma_2 = \alpha_2 s$;
- Case 3, **n** along the x axis: $\eta_3 = \frac{1}{2}\alpha_4$, $\Gamma_3 = 0$.

Intuitively, we can guess that $\alpha_2 < 0$, η_2 large, η_1 small and η_3 intermediate. For MBBA at room temperature, one gets (in Poise, the CGS unit): $\eta_1 = 0.238$, $\eta_2 = 1.035$, $\eta_3 = 0.416$, $\alpha_1 = 0.065$,

$\alpha_2 = -0.785$, and $\alpha_3 = -0.012$. For comparison, the viscosity of water is ~ 0.01 and that of glycerol ~ 8.5.

In spite of the linear relations between stresses and strains, nematics appear to be non-Newtonian because the effective viscosity depends on the orientation, the equilibrium of which results from a complex interplay between elastic, magnetic,..., and viscous contributions to the torque equation.

4. Bibliographical Notes

The presentation adopted in Sections 1 to 3 is extensively developed in Chapters 2 to 6 of:

[1] S. R. de Groot, P. Mazur, *Non-equilibrium thermodynamics* (North-Holland, 1962).

For an introduction to surface tension effects, see Chapter 7 of:

[2] L. D. Landau, E. M. Lifshitz, *Fluid mechanics* (Pergamon Press, 1959).

The continuous theory of nematic liquid crystals is reviewed in:

[3] P. G. de Gennes, *The physics of liquid crystals* (Oxford University Press, 1974)

Appendix 2

Differential Calculus

1. Initial Value Problems

1.1. Introduction

Let us consider an initial value problem for some first-order differential system

$$\frac{dX}{dt} = F(X), \qquad X(t_0 = 0) = X_0. \tag{1}$$

The solution can be written as $X(t) = U(t,0)X_0$, where $U(t,0)$ is the evolution operator. The formal expression of U is obtained by integrating (1) between 0 and t:

$$X(t) = X_0 + \int_0^t F(X(t')) \, dt',$$

replacing $X(t')$ by: $X(t') = X_0 + \int_0^{t'} F(X(t'')) \, dt''$, then $X(t'')$ by $X_0 + \int_0^{t''} F(X(t''')) \, dt'''$, and so on.

This recursive process defines U as a *time-ordered* multiple integral which helps us to prove in a constructive way the existence and uniqueness of solutions by Picard's method of successive approximations. It is also useful in perturbative approaches when F can be split into two parts, a dominant one integrable by hand and a "small" perturbation which is not.

The simplest case obviously corresponds to a linear dynamical system with constant coefficients,

$$\frac{dX}{dt} = MX. \tag{2}$$

Owing to the fact that $\int_0^t MX_0\, dt' = tMX_0$, the solution can be obtained explicitly from the general expression, which yields

$$X(t) = \left(I + \sum_{k=1}^{\infty} \frac{t^k}{k!} M^k\right) X_0 = \exp(tM)\, X_0\,, \qquad (3)$$

where I denotes the identity operator and the sum of the series is by definition the exponential of the constant linear operator M. The evaluation of this exponential is more easily performed after the structure of M has been resolved. Before doing this, we will briefly fix notations and summarize useful results about vectors and matrices.

1.2. Linear Maps, Matrices, and Change of Bases

Elements of a d-dimensional vector space V over a field K (here R or C, the fields of real or complex numbers) will be denoted X. A basis is a set $\{E_i\}$ of d linearly independent vectors such that

$$X = \sum_i x_i E_i\,. \qquad (4)$$

A linear operator M can be *represented* by its action on basis vectors:

$$ME_j = \sum_i m_{ij} E_i\,,$$

which defines a $(d \times d)$ matrix $[M]$ (column vectors are the coordinates of the images of the basis vectors). Applying M to an arbitrary vector yields

$$y_j = (MX)_j = \sum_i m_{ji} x_i\,. \qquad (5)$$

The representation of a linear operator changes with the basis. The new basis vectors can be defined in terms of the old basis vectors as

$$E'_j = \sum_i t_{ij} E_i\,. \qquad (6)$$

A2. Differential Calculus

The matrix [T] with elements t_{ij} is the representation of an *invertible* linear operator T; its column vectors are the coordinates of the new basis vectors in the old basis. It is easily found by substitution that the coordinates $\{x_i\}$ and $\{x'_i\}$ of a given vector X in basis $\{E_i\}$ and $\{E'_i\}$ are related by

$$x_i = \sum_j t_{ij} x'_j. \tag{7}$$

The matrix [T] can be inverted to yield

$$x'_j = \sum_k u_{jk} x_k, \tag{8}$$

with $\sum_j t_{ij} u_{jk} = \delta_{ik} = \sum_j u_{ij} t_{jk}$ by definition of the inverse $[U] = [T]^{-1}$. We recall that, if d is not too large, the inverse matrix is easily determined by hand as the transpose of the matrix of cofactors divided by the determinant, the cofactor A_{ij} of element a_{ij} being $(-1)^{i+j} \times$ the determinant of the $(d-1) \times (d-1)$ matrix obtained by suppressing the ith row and the jth column.

From (8), (5), and (7), we get

$$y'_l = \sum_j u_{lj} \sum_j m_{ji} \sum_i t_{ik} x'_k = \sum_j m'_{jk} x'_k,$$

so that the representation of operator M in the new basis is given by $m'_{ij} = \sum_k \sum_l u_{ik} m_{kl} t_{lj}$, that is to say, in matrix form,

$$[M'] = [T]^{-1}[M][T]. \tag{9}$$

The matrices [M'] and [M] are said to be *similar*. It will be useful to note that

$$[M']^k = [T]^{-1}[M]^k[T]. \tag{10}$$

1.3. Eigenvalues and Invariant Subspaces

Under the action of a linear operator M, some linear subspaces are left invariant. Trivial invariant subspaces are the null space spanned by the null vector, the total space V, the image of V under M, and the kernel, i.e., the subset mapped by M on the null vector. If the kernel of M is nontrivial, i.e., not reduced to the null vector, M is not invertible.

Nontrivial invariant subspaces are those associated with the eigenvalues of M defined by $MX = sX$, i.e., $M - sI$ has a nontrivial kernel. Eigenvalues are the roots of the *characteristic polynomial* given by $\det(M - sI) = 0$. The *fundamental theorem of algebra* states that this degree-d polynomial has d roots in C, possibly degenerate, i.e., we can write

$$a_0 + a_1 s + \ldots + a_{d-1} s^{d-1} + s^d = 0 = \prod_k (s - s_k)^{d_k},$$

with $\sum_k d_k = d$ (d_k is the multiplicity of eigenvalue $s = s_k$).

The recourse to the *complex extension* of a real vector space is necessary for complex eigenvalues of a real operator. This is best illustrated by the case of a two-dimensional rotation matrix that has two simple but complex conjugate eigenvalues $s_\pm = \pm i\omega$, roots of $s^2 + \omega^2 = 0$, and no real eigenvectors.

When all the eigenvalues are simple, operator M is diagonalizable. Invariant subspaces are all one-dimensional, each being spanned by the corresponding eigenvector, possibly in the complex extension of the vector space. The case of multiple eigenvalues is more delicate. The systematic search for invariant subspaces is performed in several steps sketched in Fig. 1. First the matrix is cast into upper triangular form with eigenvalues on the diagonal and then into a diagonal arrangement of irreducible triangular blocks (quasi-diagonal form). The final step consists in reducing each triangular block into its canonical *Jordan form*. The canonical Jordan form of the restriction of a linear operator M to the irreducible sub-space attached to a d_k-degenerate eigenvalue s_k reads

$$M_{s_k} = s_k I_{d_k} + N_{d_k},$$

where I_{d_k} is the d_k-dimensional identity matrix and N_{d_k} the d_k-dimensional Jordan matrix with zeroes everywhere except along the

A2. Differential Calculus

Fig. 1. Steps of the reduction of a matrix to its canonical Jordan form.

first diagonal above the main diagonal; see below for an example with $d_k = 3$.

1.4. Application to Linear Initial Value Problems

As already stated, the solution of $dX/dt = MX$ reads $X(t) = \exp(tM) X_0$, where "exp" is the shorthand notation introduced in (3) for the series of successive powers of M. Owing to (10), this is most easily evaluated in the basis where M is represented by a matrix in Jordan normal form. Let us consider an eigenvalue s_k and the restriction of M to its d_k-dimensional eigenspace, and drop unnecessary subscripts k and d_k. Since I and N obviously commute ($IN = NI$), we have $\exp(tM_s) = \exp(t(sI + N)) = \exp(tsI)\exp(tN)$ and, directly $\exp(tsI) = \exp(ts) I$. The computation of the second exponential is not much more complicated from the definition. Considering, for example, the case $d_k = 3$, we have

$$N = \begin{bmatrix} 0 & 1 & 0 \\ 0 & 0 & 1 \\ 0 & 0 & 0 \end{bmatrix}, \quad N^2 = \begin{bmatrix} 0 & 0 & 1 \\ 0 & 0 & 0 \\ 0 & 0 & 0 \end{bmatrix}, \quad N^3 = \begin{bmatrix} 0 & 0 & 0 \\ 0 & 0 & 0 \\ 0 & 0 & 0 \end{bmatrix} = [0]$$

(more generally $N^{d_k} \equiv 0$; the operator is said to be *nilpotent*). Therefore, we have

$$\exp(tN) = I + tN + \frac{1}{2}t^2 N^2 = \begin{bmatrix} 1 & t & \frac{1}{2}t^2 \\ 0 & 1 & t \\ 0 & 0 & 1 \end{bmatrix},$$

which finally yields

$$\exp(tL_s) = \begin{bmatrix} \exp(st) & t\exp(st) & \frac{1}{2}t^2\exp(st) \\ 0 & \exp(st) & t\exp(st) \\ 0 & 0 & \exp(st) \end{bmatrix}.$$

The origin of terms of the form $(t^k/k!)\exp(st)$, called *secular terms*, is best visualized on the following simple two-dimensional example: $dX/dt = Y$, $dY/dt = -X - 2Y$, preferably written as

$$\frac{d^2X}{dt^2} + 2\frac{dX}{dt} + X = \left(\frac{d}{dt} + 1\right)^2 X = 0.$$

This equation has $s = -1$ as double eigenvalue. Its general solution reads $X(t) = (\alpha + \beta t)\exp(-t)$. The term βt appears as a slow drift away from the exponential trend. Next, considering the nearly degenerate case

$$\frac{d^2X}{dt^2} + (2-\epsilon)\frac{dX}{dt} + (1-\epsilon)X = \left(\frac{d}{dt} + 1\right)\left(\frac{d}{dt} + 1 - \epsilon\right)X = 0$$

with eigenvalues $s_1 = -1$, $s_2 = -1 + \epsilon$, we obtain

$$X(t) = \alpha_1 \exp(-t) + \alpha_2 \exp(-(1-\epsilon)t)$$
$$= \alpha_1\left(1 + \frac{\alpha_2}{\alpha_1}\exp(\epsilon t)\right)\exp(-t),$$

which shows that the solution of the degenerate case is recovered by expanding $\exp(\epsilon t)$ at lowest order, i.e., as $1 + \epsilon t$ (valid as long as one consider times much smaller than $1/\epsilon$, that is to say, indefinitely when ϵ tends to 0).

2. Boundary Value Problems

2.1. Scalar Products and Adjoint Problems

Up until now, we have not made use of usual metric properties attached to vector spaces through the definition of *scalar products*, the correlative notion of *orthogonal projection* replacing that of *affine projection* implicit in (4).

A Hermitian structure can be given to a vector space on $\mathbb{C}$ by means of the canonical scalar product

$$\langle Y|X\rangle = \sum_i \bar{y}_i x_i \tag{11}$$

A2. Differential Calculus

(for a vector space on R this simply reads $\langle Y|X\rangle = \sum_i y_i x_i$, from which the usual Euclidian structure is obtained).

The adjoint $M^\dagger$ of an operator M is defined *via* the relation

$$\langle Y\,|\,M^\dagger X\rangle = \langle MY\,|\,X\rangle = \overline{\langle X\,|\,MY\rangle}, \qquad (12)$$

where the last equality derives directly from (11). Adjoint operators are represented by adjoint matrices such that $m^\dagger_{ij} = \bar{m}_{ji}$. An operator is said to be *Hermitian* or *self-adjoint* if it is identical to its adjoint: $M^\dagger \equiv M$. The corresponding matrix fulfills $m_{ij} = \bar{m}_{ji}$ (conjugate transposed matrix). Eigenvalues of a self-adjoint operator are real and eigenvectors associated with distinct eigenvalues are orthogonal.

Vectors forming an orthonormal basis fulfill $\langle E_i\,|\,E_j\rangle = \delta_{ij}$. The elements of the matrix [T] expressing the change between two orthonormal bases $\{E_i\}$ and $\{E'_i\}$ are given by $t_{ij} = \langle E_i\,|\,E'_j\rangle$, whereas for the inverse matrix we have $u_{ij} = \langle E'_i\,|\,E_j\rangle$. From (11), we get $u_{ij} = \bar{t}_{ji}$. A matrix such as [T] whose inverse is identical to its transposed conjugate matrix is called *unitary* (in the real case, matrices associated with the transformations describing orthonormal bases changes are called *orthogonal*; the inverse matrices are simply equal to the transposed matrices).

Adjointness properties of interest for the calculation of bifurcated solutions by perturbation are involved in the solution of inhomogeneous linear problems of the form

$$MX = F. \qquad (13a)$$

The solution is unique and given by $X = M^{-1}F$ for all F as long as M is invertible, i.e., as long as the kernel of M is trivial ($MX = 0 \Rightarrow X = 0$). In one dimension ($d = 1$), the situation is trivial since we have simply $mx = f$ which has a unique solution when $m \neq 0$. When $m = 0$, the problem has no solution if $f \neq 0$ and the indeterminacy is complete if $f = 0$. Things are less trivial in higher dimensions. As a result, either the solution of (13a) is unique, i.e., M is invertible and the homogeneous problem

$$MX = 0 \qquad (13b)$$

has only a trivial kernel, or the adjoint homogeneous problem

$$M^\dagger \tilde{X} = 0 \qquad (14)$$

has a nontrivial solution $\tilde{X} \neq 0$ (*Fredholm alternative*). In the latter case, the inhomogeneous problem (13a) has a solution if and only if the r.h.s. F is orthogonal to the kernel of $M^\dagger$ (*Fredholm theorem*). With $\tilde{X}$ being in the kernel of $M^\dagger$, this reads

$$\langle \tilde{V} \,|\, F \rangle = 0. \qquad (15)$$

Let us illustrate this in the simplest nontrivial case, $d = 2$. The homogeneous problem (13b) can be written as:

$$\begin{aligned} ax_1 + bx_2 &= 0, \\ cx_1 + dx_2 &= 0, \end{aligned} \qquad (16)$$

and has a nontrivial solution if the determinant of the (2×2) matrix vanishes, here:

$$ad - bc = 0.$$

(In a perturbation expansion around the threshold, Chapter 8, this corresponds to the threshold condition.) The solution is then a one parameter family, spanned by $X = (b, -a)$, the kernel of M in (16).

Let us now consider an inhomogeneous system (13a); explicitly

$$\begin{aligned} ax'_1 + bx'_2 &= f \\ cx'_1 + dx'_2 &= g \end{aligned} \qquad (17)$$

(this mimics higher order problems in a perturbation calculation). It has solutions only if the two equations are proportional, that is to say when

$$ag - cf = 0. \qquad (18)$$

The homogeneous adjoint problem (14) reads here

$$\begin{aligned} \bar{a}\tilde{x}_1 + \bar{c}\tilde{x}_2 &= 0, \\ \bar{b}\tilde{x}_1 + \bar{d}\tilde{x}_2 &= 0. \end{aligned}$$

A2. Differential Calculus

Its kernel is spanned by $\tilde{X} = (-\bar{c}, \bar{a})$, so that the compatibility condition (18) actually reads

$$\overline{(-\bar{c})}f + \overline{(\bar{a})}g = \langle \tilde{X}|F\rangle = 0$$

according to the definition of the scalar product (11). When this condition is fulfilled, the solution exists but is not unique and we can add a supplementary condition to remove the indeterminacy, e.g., X' orthogonal to the kernel of M,

$$\langle V|V'\rangle = 0, \qquad (19)$$

which has the advantage of presenting the solution of the inhomogeneous problem as a correction to the solution of the lowest order problem in a perturbation expansion (see Chapter 8).

2.2. Adjointness for Boundary Value Problems

Up until now we have considered finite-dimensional vector spaces. We now turn to vector spaces whose elements are functions (more generally finite sets of functions) defined on an interval, i.e., components indexed by a continuous variable, $t \in [a, b]$, instead of integers belonging to a finite set, $i \in \{1, 2, \ldots d\}$. For simplicity, we consider the case of a single function. The natural extension of the sum appearing in the definition of the scalar product is an integral, so that we define

$$\langle Y \mid X \rangle = \int_a^b \bar{y}(t) x(t)\, dt.$$

We are interested in the case where functions are solutions of a linear differential equation of order n, $LX = F$, with a differential operator L written as

$$LX = \sum_{m=0}^{n} a_m(t) x^{(m)}, \qquad (20)$$

where $x^{(m)}$ denotes the mth derivative of x with respect to t $\left(x^{(m)} \equiv d^m x/dt^m,\ x^{(1)} = x' = dx/dt,\ \text{and}\ x^{(0)} \equiv x\right)$. Note that coefficients a_m are supposed to be variable. To obtain a well-posed problem, we must add n boundary conditions of the general form $U_j X = u_j$ ($j = 1, \ldots, n$) with

$$U_j X = \sum_{m=0}^{n-1} \alpha_{jm} x^{(m)}(a) + \beta_{jm} x^{(m)}(b). \tag{21}$$

In many practical cases, boundary conditions apply separately at the two ends of the interval, i.e., $\beta_{jm} = 0$ when $\alpha_{jm} \neq 0$, and reciprocally. The case of homogeneous boundary conditions corresponds to $u_j = 0$ for all j.

Considering the definition of the adjoint operator in (12), we see that we must determine $L^\dagger$ such that:

$$\int_a^b \bar{y}(t) L^\dagger x(t)\, dt = \int_a^b \overline{\bar{x} L\, y(t)}\, dt = \int_a^b x \overline{L\, y(t)}\, dt$$

In order to remove $\bar{y}$ from the action of the differential operator on the r.h.s., we have to perform a series of integration by parts

$$\int_a^b x \bar{a}_m \bar{y}^{(m)}\, dt = \int_a^b \left(\left(x \bar{a}_m \bar{y}^{(m-1)} \right)' - \bar{y}^{(m-1)} \left(\bar{a}_m x \right)' \right) dt$$

$$= \left[x \bar{a}_m \bar{y}^{(m-1)} \right]_a^b - \int_a^b \bar{y}^{(m-1)} (\bar{a}_m x)'\, dt.$$

This procedure progressively lowers the order of derivatives involving $\bar{y}$ down to zero. The formal expression of $L^\dagger$ is then easily obtained:

$$L^\dagger X = \sum_{m=0}^n (-1)^m (a_m x_m)^{(m)}.$$

However, integrations by parts lead to the formation of a complicated *boundary form* involving derivatives of x, $\bar{a}_m$, and $\bar{y}$ (remember that a_m may depend on the independent variable t).

Inserting each boundary condition (21) of the direct problem into this boundary form and forcing it to vanish yields the boundary condition on the adjoint functions. As an example, consider the now familiar linearized Swift–Hohenberg model (here with x as the

A2. Differential Calculus

function and t as the independent variable to ensure the continuity of notations)

$$LX = r\,x - \left(\frac{d^2}{dt^2}+1\right)^2 x = (r-1)x - 2x'' - x^{(4)} \qquad (22)$$

with boundary conditions

$$x(a) = x'(a) = 0 = x(b) = x'(b)\,. \qquad (23)$$

Since L has constant coefficients and contains only even derivatives, $L^\dagger$ is identical to L, i.e., L is formally self-adjoint. The boundary form reads

$$[\bar{x},y] = -\left[\bar{x}'''y\right]_a^b + \left[\bar{x}''y'\right]_a^b - \left[\bar{x}'y''\right]_a^b$$
$$+ \left[\bar{x}y'''\right]_a^b - 2\left[\bar{x}'y\right]_a^b + 2\left[\bar{x}y'\right]_a^b.$$

Taking into account boundary conditions (23), we obtain

$$[\bar{x},y] = -\left[\bar{x}'''y\right]_a^b + \left[\bar{x}''y'\right]_a^b,$$

and since there are no boundary conditions on x''' and x'' we see that the boundary form vanishes for all x when

$$y(a) = y'(a) = 0 = y(b) = y'(b)\,.$$

Here, the adjoint functions fulfill the same set of boundary conditions as the original functions, but this is not always the case. From an algebraic point of view, the boundary value problem is self-adjoint if the differential operator is formally self-adjoint, $L^\dagger \equiv L$, and the boundary conditions on the functions, direct and adjoint, are identical.

Linear problems with variable coefficients frequently occur in the study of secondary instabilities. Consider, for example, a steady nonlinear solution to

$$\partial_t w = r\,w - (\partial_{x^2}+1)^2 w - w^3$$

(x is restored as the independent variable without risk of confusion).
Let us call this solution $w_0(x)$; then the resulting stability problem
reads

$$s\,\delta w = (r - 3w_0^2)\,\delta w - \left(\frac{d^2}{dx^2} + 1\right)^2 \delta w\,,$$

which is easily checked to be self-adjoint since the derivative does not act on variable coefficients. Now, replacing w^3 by $w\,\partial_x w$ we obtain

$$s\,\delta w = (r - w_0')\,\delta w - \left(\frac{d^2}{dx^2} + 1\right)^2 \delta w - w_0\,\delta w'\,,$$

which is no longer formally self-adjoint since $-w_0\,\delta w'$ gives $+(w_0\,\delta v)' = +w_0\,\delta v' + w_0'\,\delta v$, so that the adjoint reads

$$s\,\delta w = r\,\delta w - \left(\frac{d^2}{dx^2} + 1\right)^2 \delta w + w_0 \delta w'\,.$$

In concrete cases, complications frequently arise from the fact that physical states require more than a single function for their characterization (e.g., velocity components and temperature fluctuations). Examples considered here give the spirit of calculations to be performed.

3. Differential Equations with Delay

The multiplication of degrees of freedom studied in Chapters 8 to 10 was associated with the weakening of confinement effects in physical space. Here we are concerned with another possibility related to memory effects. Examples can be found in physics, electronics, biology, and ecology, etc. Finite-dimensional dynamical systems studied up until now were all of the form (1) with X, the argument of F, evaluated at time t. Introducing delayed interactions, we now assume that the evolution velocity dX/dt is a function of X evaluated at some earlier time $t - t^*$:

$$\frac{dX}{dt} = F\big(X(t - t^*)\big)\,. \qquad (24)$$

A2. Differential Calculus

The nature of the degrees of freedom involved in this new system is best understood from a comparison of the simplest finite difference approximations to the initial value problems (1) and (24):

$$X(t_{k+1}) = X(t_k) + \delta t\, F(X(t_k)), \qquad (25a)$$
$$X(t_{k+1}) = X(t_k) + \delta t\, F(X(t_{k-n})), \qquad (25b)$$

where $t_k = t_0 + k\,\delta t$ and $t^* = n\,\delta t$. Clearly, iteration (25a) starts with a single initial condition $X(t_0) = X_0$, while iteration (25b) requires a series of $n+1$ values $X(t_0), X(t-1),\ldots, X(t_{-n})$, since $X(t_0)$ and $X(t_{-n})$ serve for the calculation of $X(t_1)$, $X(t_1)$ and $X(t_{-n+1})$ for X_2, etc.

The number of degrees of freedom being the number of values required to specify a trajectory of the system, we can say that the differential system with delay is infinite-dimensional since n tends to infinity at given t^* when the limit $\delta t \to 0$ is taken; i.e., the initial data is a function defined on the interval $[-t^*, 0]$.

In fact, the effective number of degrees of freedom is much smaller. The reduction results from the temporal coherence introduced by the differential operator itself in much the same way as for partial differential equations. For simplicity, we discuss this property in a simple case where (24) can be written in the form

$$\frac{dX}{dt} + X(t) = G(X(t - t^*)). \qquad (26)$$

The l.h.s. accounts for a linear relaxation of X towards the equilibrium value $X = 0$ with a relaxation rate normalized to unity for convenience. $G(X)$ is a forcing term, delayed and essentially nonlinear with a typical humped shape sketched in Fig. 2.

We consider the case where X is a single real variable and first look for the linear stability of steady-state solutions, i.e., fixed points of $X_f = G(X_f)$. Inserting $X = X_f + \delta X$ into (26), and linearizing the expansion in powers of δX, we obtain

$$\frac{d\,\delta X}{dt} + \delta X = g\,\delta X(t - t^*) \qquad (27)$$

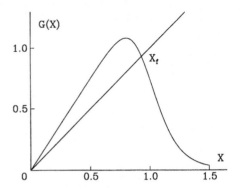

Fig. 2. Profile of the forcing term G for the Mackey–Glass model of blood pathology: $G(X) = aX/(1 + X^c)$, here with $a = 1.5$ and $c = 10$.

with $g = dG/dX$ at X_f ($g < 1$ in the situation of Fig. 2). The solution can be sought in the form

$$\delta X(t) = \delta X(0) \exp(st)$$

with s a solution of

$$s + 1 = g \exp(-st^*). \tag{28}$$

In spite of the fact that we have a single variable, the system is infinite-dimensional so that s must be taken as complex. Therefore, with $s = s' + is''$, Equation (28) reads

$$s' + 1 = g \exp(-s't^*) \cos(s''t^*), \tag{29a}$$
$$s'' = -g \exp(-s't^*) \sin(s''t^*). \tag{29b}$$

We are primarily interested in the existence and the number of neutral or unstable modes when t^* increases. It is first easily seen that there cannot be stationary unstable modes with $s'' = 0$ since, for $g < 1$, Equation (29a) has only negative roots. More generally, for complex roots Equation (29a) implies $|s' + 1| < |g| \exp(-s't^*)$, so that there cannot be roots with $s' > 0$ if $|g| < 1$. By contrast, if $g < -1$ unstable oscillatory modes can exist. The bifurcation points are given by $s' = 0$, i.e., $1 = g \cos(s''t^*)$ and $s'' = -g \sin(s''t^*)$ whose solutions read:

$$s''t^* = \pm \cos^{-1}(1/g) + 2k\pi \quad \text{with } s'' = \mp g \sin\left(\cos^{-1}(1/g)\right).$$

A2. Differential Calculus

At given g, the condition $t^* > 0$ leads to retaining a single series of solutions

$$t_k^* = \frac{\cos^{-1}(1/g) + 2k\pi}{|g|\sin(\cos^{-1}(1/g))}. \tag{30}$$

Each of these neutral modes is the starting point of an unstable branch. The growth rate of unstable modes is most easily obtained by varying g while keeping $t^* = t_k^*$ constant and large. Indeed, for t^* large, solutions to (28) corresponding to unstable modes should be such that $s't^*$ remains roughly constant to avoid a "killing" of the r.h.s. of (29a) by the exponential term, since we would then have $s' \sim -1$, which would contradict the assumption $s' > 0$. Up to corrections of order $1/t^*$, the problem for $g' = g + \delta g$ and $s' \neq 0$ is then the same as that for g except that g is replaced by $(g + \delta g)\exp(-s't^*)$. The persistence of a solution $(g+\delta, s' > 0, s'', t_k^*)$ corresponding to a given neutral mode $(g, s' = 0, s'', t_k^*)$ directly leads to the condition $g = (g + \delta g)\exp(-s't^*)$, which gives $s't^* \simeq \log(1 + \delta g/g) \simeq \delta g/g$.

From (30), we see that, at given g, the number of unstable modes grows linearly with the delay t^*. This is comparable to the linear growth with ℓ of the number of unstable modes for a one-dimensional partial differential equation defined on an interval of length ℓ. However, the behavior beyond threshold is drastically different since in the PDE case the growth rate is essentially independent of ℓ while here it decreases as $1/t^*$. This is due to the fact that the dynamics described by (24) puts a severe constraint on the solution in forcing a relation between values of X at distant times, which hinders the divergence.

This increase of the number of unstable modes with the delay shows us that the coupling between $X(t)$ and $X(t - t^*)$ cannot be as rigid as what one could have expected by neglecting dX/dt when compared with X in Equation (26), which would have led to the determination of $X(t)$ simply from the knowledge of $G(X(t - t^*))$. When $g < -1$, such a naive "adiabatic elimination" is therefore not legitimate and velocity fluctuations are amplified by the nonlinear coupling.

Another point of view on the same property is obtained by considering the integral representation of the solution

$$X(t) = X(0)\exp(-t) + \int_0^t dt' \exp(t'-t)\, G(X(t'-t^*)).$$

For t large enough, the first term can be dropped and we can replace the integral by an estimate that takes advantage of the short range of the integral kernel since it corresponds to an exponential damping at a rate $\tau = 1$. X at time t is therefore the result of an averaging of $G(X)$ over a time interval of order τ at time $t - t^*$. This fact is at the origin of the loss of temporal coherence and one can imagine that the delay t^* is split into sub-intervals of duration $\sim \tau$. The *effective* number of degrees of freedom then roughly corresponds to the number of "sub-harmonic vibration modes" of a string of length t^* with periods down to $\tau = 1$. At given g the bifurcation to time dependence is then expected for $t^* \sim 1$ and chaos for $t^* \sim 2\text{--}3$, which is indeed the case. Nonlinear effects should be properly taken into account but the orders of magnitude remain.

The bifurcation diagram, the Lyapunov exponents, and the entropy can be determined experimentally on explicit models. In the large t^* limit, a linear growth of the number of positive Lyapunov exponents is observed. These exponents scale as $1/t^*$, and the associated tangent solutions are seen to oscillate with an average number of periods on an interval of length t^* of the order of the index of the corresponding Lyapunov exponent. All these features can be heuristically related to the properties of the linearly unstable modes examined above. For example, the argument giving the growth rate is straightforwardly adapted to the case of the Lyapunov exponents since they derive from a linear stability analysis of trajectories more general than stationary states (fixed point), whereas the tangent dynamics remains governed by an equation of the form (27), except that g is now a complicated function of time (it depends on the solution that is followed). An equation of the form (28) should hold *on average*, from which the $1/t^*$ dependence would follow. Accordingly, the entropy is observed to remain roughly constant with t^* since it is the sum of $\sim t^*$ positive exponents all scaling as $1/t^*$, which is

A2. Differential Calculus

again in strong contrast with the partial differential case for which the entropy is generically an extensive quantity growing with the size of the system.

4. Bibliographical Notes

For initial value problems, a good reference is the book by Hirsch and Smale BN 5 [2] which spends several chapters to discuss linear algebra. A study of algebraic properties of boundary value problems can be found in Chapters 7 and 11 of:

[1] E.A. Coddington and N. Levinson, *Theory of ordinary differential equations* (McGraw-Hill, 1955).

For recent studies of the growth of chaos in differential equations with delays, consult, for example:

[2] J.D. Farmer, "Chaotic attractors of an infinite-dimensional dynamical system," Physica **4D**, 366 (1982).

[3] K. Ikeda and K. Matsumoto, "Study of a high-dimensional chaotic attractor," J. Stat. Phys. **44**, 955 (1986).

[4] B. Dorizzi, B. Grammaticos, M. Le Berre, Y. Pomeau, E. Ressayre, and A. Tallet, "Statistics and dimensions of chaos in differential delay systems," Phys. Rev. A **35**, 328 (1987).

Appendix 3

Numerical Simulations

1. Introduction

In mathematics and theoretical physics, numerical methods are mostly devoted to searching for numbers, e.g., roots of equations, eigenvalues, values of a function, etc. Here we shall be concerned with a different viewpoint, the search for specific approximate solutions to an evolution problem defined by a governing equation and some initial conditions (when the solution cannot be obtained in closed form by analytic means).

Present capabilities of modern supercomputers allow the *numerical simulation* of physical models based on primitive equations in nearly realistic situations. This has led to the emergence of *computational physics* as a field in itself. However, our main interest for simulations will lie rather in the flexibility of models and the complete control on initial conditions, thought to give useful hints on the nature of nonlinear processes and the approximations to be developed for their theoretical understanding.

In the spirit of the rest of this book, we shall not discuss in detail sophisticated numerical methods for large-scale simulations, but rather present some elements of standard simulation schemes for evolution problems in a sufficiently self-contained manner. We shall limit our ambition to situations that remain well within the range of numerical capabilities of personal computers (speed, accuracy, memory). Numerical problems to be considered can often be solved using routines either already present in computer libraries or at least described in introductory courses or reference books. Rather than developing such a conventional numerical-analysis type of approach, we shall stress the *physical meaning* of the proposed algorithms. Moreover, we shall present only simplified versions customized to specific cases of interest, so that their translation into short BASIC programs will be straightforward.

Here we will be mostly interested in deterministic initial value problems for differential equations, ordinary or partial. Methods to be developed can include weak extrinsic stochasticity, e.g., to test the sensitivity to random noise, but we will not discuss statistical (*Monte Carlo*) methods used for problems described in terms of fully stochastic models.

All numerical methods lead to the replacement of the actual solution of the primitive problem, say $F(\partial_t, \partial_x, v) = 0$, by a set of numbers with a finite (and usually quite small) number of digits representing the approximate solution $\underline{v}$ conveniently understood as the exact solution (up to round-off errors) of an approximate problem generally derived by truncation from the Taylor expansion of the original problem. In the following, we will drop the distinction between $\underline{v}$ and v and neglect round-off errors since they are usually much smaller than the truncation errors (at least when *double precision*, i.e., much more than six-digit accuracy, is used).

In general, the time dependence is evaluated at a sequence of regularly spaced times $t \to t_n = n \, \Delta t$; corresponding methods will be examined in Section 2. Space dependence, if any, is most naively described in terms of values taken at the nodes of a discrete lattice, i.e., $x \to x_j = j \, \Delta x$. Replacing partial derivatives by their *finite difference* approximations leads to the integration schemes examined in Section 3.1. Another description can be obtained in terms of components on some functional basis; corresponding methods, called *spectral* will be briefly presented more specifically in the case of Fourier modes (Section 3.2).

A3. Numerical Simulations

2. Finite Differences for Time Stepping

Let us consider the following first order initial value problem:

$$\frac{dv}{dt} \equiv v' = f(v) \quad \text{with} \quad v(t_0) = v_0. \tag{1}$$

It is written here as an ordinary differential equation, with primes denoting time derivatives, but the case of partial differential equations is identical as far as time evolution is concerned.

At some generic time t_n, expanding the solution in Taylor series we obtain

$$v_{n+1} = v_n + \Delta t\, v'_n + \frac{\Delta t^2}{2} v''_n + \frac{\Delta t^3}{6} v'''_n + \ldots, \tag{2}$$

where $v'_n = f(v_n)$ and $v''_n, v'''_n, \ldots$ can be evaluated from (1). Truncating this expression beyond the first-order term, we get the classical point-slope formula (Euler, see Fig. 1):

$$v_{n+1} = v_n + \Delta t\, f(v_n). \tag{3}$$

This integration scheme is called *first order* since it is exact up to corrections of order Δt^2 and *explicit* since one can go forward in time by simply knowing the value of v (and $f(v)$) at past times.

A second obvious first-order scheme is obtained from the evaluation of v_n in terms of v_{n+1}:

$$v_n = v_{n+1} - \Delta t\, v'_{n+1} + \frac{\Delta t^2}{2} v''_{n+1} - \frac{\Delta t^3}{6} v'''_{n+1} + \ldots, \tag{4}$$

which leads to formula

$$v_{n+1} = v_n + \Delta t\, f(v_{n+1}),$$

again first-order accurate. This scheme is called *implicit* since v_{n+1} is given by an implicit equation

$$v_{n+1} - \Delta t\, f(v_{n+1}) = v_n. \tag{5}$$

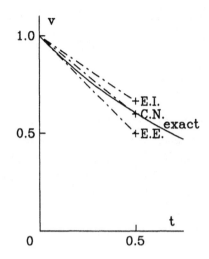

Fig. 1. Solution of $v' = -v$ at $t = \Delta t = 0.5$ with $v(0) = 1$: exact result compared with estimates from explicit Euler (E.E.), implicit Euler (I.E.), and Crank–Nicolson (C.N.) formulas, (3), (5), and (6), respectively.

Subtracting side-to-side Equation (4) from (2), noticing that $v''_{n+1} = v''_n + \Delta t\, v'''_n + \ldots$, we see that the term $\mathcal{O}(\Delta t^2)$ disappears, so that

$$v_{n+1} = v_n + \frac{\Delta t}{2}\bigl(f(v_n) + f(v_{n+1})\bigr). \tag{6}$$

This formula, which is nothing else than the *trapezoidal rule*, is called the *Crank–Nicolson* scheme (see Fig. 1). It is *second-order* accurate and *implicit*.

Accuracy considerations are best understood from the comparisons of error terms on a simple example. Taking $f(v) = -v$ with $v_0 = 1$, we have $v = \exp(-t)$. After a single time step, we have

Exact result: $\quad \exp(-\Delta t) = 1 - \Delta t + \frac{1}{2}\Delta t^2 - \frac{1}{6}\Delta t^3 + \ldots$

explicit Euler (3): $\quad (1 - \Delta t)$

implicit Euler (5): $\quad \frac{1}{1+\Delta t} = 1 - \Delta t + \Delta t^2 - \Delta t^3 + \ldots$

Crank-Nicolson (6): $\quad \frac{1 - \Delta t/2}{1 + \Delta t/2} = 1 - \Delta t + \frac{1}{2}\Delta t^2 - \frac{1}{4}\Delta t^3 + \ldots,$

which clearly shows the improvement brought by the Crank–Nicolson formula.

Up until now we have derived *single-step* formulas involving only

A3. Numerical Simulations

the knowledge of the solution at a single time. A simple way to improve formula (3) is by inserting

$$v_n'' = \frac{1}{\Delta t}(v_n' - v_{n-1}') + \mathcal{O}(\Delta t)$$

into (2), which yields the second-order *Adams–Bashford* formula

$$v_{n+1} = v_n + \frac{\Delta t}{2}(3f(v_n) - f(v_{n-1})), \qquad (7)$$

accurate up to corrections of order Δt^3. This second-order explicit formula is one of the simplest possible examples of a *multi-step* scheme, i.e., with v_{n+1} determined from the knowledge of v at several previous times, here $n-1$ and n.

An alternative is obtained by writing (4) at times $(n-1, n)$ instead of $(n, n+1)$, which gives

$$v_{n-1} = v_n - \Delta t\, v_n' + \frac{\Delta t^2}{2} v_n'' - \frac{\Delta t^3}{6} v_n''' + \ldots,$$

so that, upon subtraction from (2) we get the second-order accurate formula

$$v_{n+1} = v_{n-1} + 2\,\Delta t\, f(v_n), \qquad (8)$$

called the *leap-frog* scheme.

In addition to accuracy requirements, the choice of an integration formula is subjected to stability considerations. Indeed, the algorithm must not amplify round-off errors but instead *dissipate* fluctuations so that the solution followed is physically relevant and not parasitic. The problem is important since using for example two-step formulas for a first-order differential equation leads to two-dimensional *iterations* that depend on two initial conditions instead of one for the continuous time problem. Clearly, the added mode must be damped for the simulation to make sense. In this respect, let us compare the behavior of the Adams–Bashford and leap-frog schemes, again using $v' = -v$.

Iteration (7) reads

$$v_{n+1} = v_n + \frac{\Delta t}{2}(3v_n - v_{n-1})$$

or, preferably

$$u_{n+1} = v_n,$$
$$v_{n+1} = (\Delta t/2)u_n + (1 - 3\Delta t/2)v_n,$$

from which, by standard linear stability analysis, we derive the characteristic equation

$$s^2 - s\left(1 - \frac{3\Delta t}{2}\right) - \frac{\Delta t}{2} = 0.$$

At lowest order in Δt, the two roots are easily found to be $s_{(+)} = 1 - \Delta t$ and $s_{(-)} = -\Delta t/2$. The first one obviously corresponds to the physical solution and the second one to the numerical mode. The latter will be damped if $|s_{(-)}| < 1$, which is achieved provided Δt is small enough (here < 2); the Adams–Bashford scheme is said to be *conditionally stable*.

Iteration (8) reads

$$v_{n+1} = v_{n-1} - 2\Delta t\, v_n,$$

or, in terms of a two-dimensional iteration

$$u_{n+1} = v_n,$$
$$v_{n+1} = u_n - 2\Delta t\, v_n,$$

which yields

$$s^2 + 2\Delta t\, s - 1 = 0,$$

so that $s_{(+)} = 1 - \Delta t$ as before, but now $s_{(-)} = -1 - \Delta t$. Since $|s_{(-)}| > 1$, the numerical mode grows without bound; the leap-frog scheme is *unconditionally unstable*. Notice that since $s_{(-)}$ is negative, iterates corresponding to this numerical mode alternate from

A3. Numerical Simulations

positive to negative values. This sub-harmonic instability is due to the fact that the scheme leaves even and odd time steps uncoupled. Such an alternating behavior is often the signature of a *numerical instability*. Using the leap-frog scheme therefore implies some sort of control of the error growth. The numerical mode can be artificially damped by averaging the solution over two successive iterates every p iterations, but p must be kept small enough so that the amplitude of the numerical mode remains negligible with respect to truncation errors.

A high-order accuracy can be obtained from the use of more sophisticated multi-step formulas. Their general drawback comes from the need of generating a *numerical* initial condition involving several times from a single *physical* initial condition. The same problem arises when the time step has to be changed. To get out of these problems, one must use high order explicit formulas of the Runge–Kutta type.

The idea is to use the point slope formula but with a better estimate of the slope. Second-order accuracy has already been obtained by taking the mean of the slopes at t_n and t_{n+1} but this results in an implicit scheme (Crank–Nicolson). To avoid difficulties linked to this feature, one can first guess an approximate value of v_{n+1} from the explicit first order formula

$$\hat{v}_{n+1} = v_n + \Delta t\, f(v_n), \tag{9a}$$

calculate the slope at this point $f(\tilde{v}_{n+1})$, and take the average with $f(v_n)$ as a final estimate. This gives

$$v_{n+1} = v_n + \frac{\Delta t}{2}\big(f(v_n) + f(\tilde{v}_{n+1})\big). \tag{9b}$$

The correction is sufficient to achieve second-order accuracy while leaving an explicit scheme (Fig. 2a).

Another way to obtain roughly the same accuracy is to introduce an intermediate time $t_{n+1/2}$, to determine $\tilde{v}_{n+1/2}$ from the first-order formula

$$\tilde{v}_{n+1/2} = v_n + \frac{\Delta t}{2} f(v_n), \tag{10a}$$

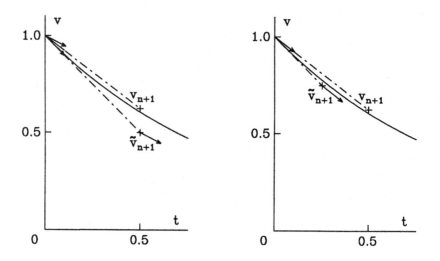

Fig. 2. Two different ways to ensure second-order accuracy using formulas of the Runge–Kutta type: a) from Equations (9); b) from Equations (10).

and to use the slope calculated at this point to extrapolate the solution at time t_{n+1} (Fig. 2b):

$$v_{n+1} = v_n + \Delta t\, f(\tilde{v}_{n+1/2}). \tag{10b}$$

Higher order formulas combine these two basic procedures to get a better evaluation of the slope. The popular fourth-order Runge–Kutta formula

$$v_{n+1} = v_n + \frac{\Delta t}{6}\left(f\left(\tilde{v}^{(1)}\right) + 2f\left(\tilde{v}^{(2)}\right) + 2f\left(\tilde{v}^{(3)}\right) + f\left(\tilde{v}^{(4)}\right)\right)$$

involves four intermediate steps given by

$$\tilde{v}^{(1)} = v_n,$$
$$\tilde{v}^{(2)} = v_n + \frac{\Delta t}{2} f\left(\tilde{v}^{(1)}\right),$$
$$\tilde{v}^{(3)} = v_n + \frac{\Delta t}{2} f\left(\tilde{v}^{(2)}\right),$$
$$\tilde{v}^{(4)} = v_n + \Delta t\, f\left(\tilde{v}^{(3)}\right).$$

Of course, this requires four successive evaluations of f but the scheme remains explicit; it starts with a single initial condition and the time step can be changed easily. The main formula can also be used in the non-autonomous case, just by keeping track of the correct time at the intermediate steps, i.e., $\tilde{y}^{(2,3)} = y_n + (\Delta t/2) f\bigl(t_{n+1/2}, y^{(1,2)}\bigr)$ and $\tilde{y}^{(4)} = y_n + \Delta t\, f\bigl(t_{n+1}, y^{(3)}\bigr)$.

Predictor-corrector methods are multi-step schemes often preferred to Runge–Kutta methods because they require a smaller number of evaluations of the function f to achieve the same accuracy, but their description goes beyond our purpose.

Explicit methods can be extended in a straightforward way to simulate differential systems coupling several variables (usually in a nonlinear way, e.g., the Lorenz model, Chapter 5). In the next section, we consider the case of partial differential equations which yields a different class of systems where the coupling between variables emerges from the action of spatial derivatives and takes a specific form justifying the recourse to implicit schemes.

3. Treatment of Partial Differential Equations

To be specific, we shall explain the methods taking the *diffusion equation* in one space dimension

$$\partial_t v = \partial_{x^2} v \qquad (11)$$

as a working example. Boundary conditions are a part of the physical problem considered. Usually, either the value of v (Dirichlet condition) or $\partial_x v$ (Neumann condition) have to be specified at the two ends of some interval of length ℓ, but periodic boundary conditions can also be imposed. Spectral methods to be presented later can be understood as a practical implementation of the reduction procedure to finite-dimensional dynamical systems. As such, they may appear less intuitive than finite difference methods which we consider first.

3.1. Finite Difference Methods

The numerical integration of initial and boundary value problems expressed in terms of partial differential equations involves two basic steps:

(1) the derivation of a consistent approximation of differential operators, and
(2) the stability analysis of the spatio-temporal scheme.

A third step would be the study of convergence of the solution of the fully discretized problem towards that of the continuous problem (it is ensured by theorems for linear problems such as (11)).

In one dimension, the solution is searched only at nodes of a periodic lattice $\{x_j\}$ with given spacing $\Delta x = \ell/(N-1)$, where ℓ is the length of the domain and N is the number of nodes, endpoints included. We shall denote the numerical solution as $v_{n,j}$ with subscript n and j for time and space, respectively but we shall drop the time subscript when it is not necessary.

3.1.1. Space Discretization and Consistency

Finite difference approximations of partial derivatives are easily obtained in the same way as time derivatives. For example, combining the two off-centered expressions of the first-order derivatives,

$$(\partial_x v)_j = \frac{1}{\Delta x}(v_{j+1} - v_j) \quad \text{and} \quad (\partial_x v)_j = \frac{1}{\Delta x}(v_j - v_{j-1}), \quad (12)$$

accurate at order Δx, one obtains the centered expression

$$(\partial_x v)_j = \frac{1}{2\Delta x}(v_{j+1} - v_{j-1}) \quad (13)$$

accurate at order Δx^2.

In the same way, the expression of the second-order derivative at order Δx^2 reads

$$(\partial_{x^2} v)_j = \frac{1}{\Delta x^2}(v_{j+1} - 2v_j + v_{j-1}), \quad (14)$$

A3. Numerical Simulations

while a better accuracy requires more points, e.g., at order Δx^4 five successive points:

$$(\partial_{x^2} v)_j = \frac{1}{12\,\Delta x^2}\left(-v_{j+2} + 16v_{j+1} - 30v_j + 16v_{j-1} - v_{j-2}\right). \quad (15)$$

Finally, the expression of the fourth-order derivative at order Δx^2 reads

$$(\partial_{x^4} v)_j = \frac{1}{\Delta x^4}\left(v_{j+2} - 4v_{j+1} + 6v_j - 4v_{j-1} + v_{j-2}\right). \quad (16)$$

In fact, the representation of partial differential operators in terms of finite differences is rather poor since the truncation error decreases algebraically as some power of Δx only (spectral methods are much better in this respect, see Section 3.2 below). This is a supplementary reason to take care of the *consistency* of the approximation. The order of consistency is the value of the exponent that controls the decrease of the error terms when Δx is lowered. For example, considering

$$\partial_t v = \varepsilon v - (\partial_{x^2} + 1)^2 v - v\partial_x v, \quad (17)$$

and using (14) for ∂_{x^4}, (15) for ∂_{x^2} and (12) for ∂_x yields a scheme consistent at order Δx only, even though a very accurate expression is used for ∂_{x^2}.

3.1.2. Boundary Conditions

Consistency considerations also affect the treatment of boundary conditions especially when they involve derivatives (Neumann condition). Since it is generally easy to achieve consistency at order Δx^2 in the bulk (e.g., using (13) rather than (12) for ∂_x), it would be a pity to spoil the quality of an approximation by a poor account of boundary conditions (though their effect might not propagate into the bulk but remain confined in a thin "numerical" boundary layer).

Assuming that condition $\partial_x v = 0$ is imposed at one end of the interval, using (12) yields an expression accurate at order Δx, which implies $v_j = v_{j+1}$. This can be realized in two ways, either with $j = 0$ at the boundary and $j = 1$ the first interior point, or else with $j = -1$ the first (fictitious) exterior point and $j + 1 = 0$ at the boundary. To preserve second-order consistency, we must take the off-centered formula

$$(\partial_x v)_j = \frac{1}{2\Delta x}(3v_j - 4v_{j+1} + v_{j+2})$$

accurate at order Δx^2 with $j = 0$ at the boundary, and the two first interior points $j + 1 = 1$ and $j + 2 = 2$. The boundary condition then reads $3v_0 - 4v_1 + v_2 = 0$. Another possibility consists of adding a fictitious exterior point $j = -1$ and an associated variable v_{-1}, which allows us to use formula (13) and simply force $v_{-1} = v_1$.

3.1.3. Time Discretization and Stability

Let us now turn to the evolution problem. For specificity, we consider an approximation of the diffusion equation (11) consistent at order Δx^2 and discretized in time according to one of the recipes of Section 2 (subscript n for time is restored). The most important question now relates to the stability of the scheme, that is to say whether numerical fluctuations are damped or not. In this respect, let us consider the two simplest schemes of order Δt derived from (3) and (5). Choosing the explicit Euler scheme, we have to evaluate $\partial_{x^2} v$ at t_n, which gives

$$v_{n+1,j} = v_{n,j} + \frac{\Delta t}{\Delta x^2}(v_{n,j+1} - 2v_{n,j} + v_{n,j-1}). \tag{18}$$

For the implicit Euler scheme, $\partial_{x^2} v$ is evaluated at t_{n+1}, so that we obtain

$$-\frac{\Delta t}{\Delta x^2} v_{n+1,j+1} + \left(1 + \frac{2\Delta t}{\Delta x^2}\right) v_{n+1,j} - \frac{\Delta t}{\Delta x^2} v_{n+1,j-1} = v_{n,i}. \tag{19}$$

The explicit method is straightforward and requires apparently less work than the implicit method since the latter involves the inversion

A3. Numerical Simulations

of a tridiagonal linear system. However, the stability study shows that more work *per* step is rewarding on the whole.

Stability analysis for finite difference schemes goes parallel to the study of physical instabilities. Forgetting boundary conditions, we analyze the evolution of a disturbance $\delta v_{n,j}$ to a given numerical solution $v_{j,n}$ by means of a discrete Fourier transform:

$$\delta v_{n,j} = \delta \tilde{v}_{n,k} \exp(ikj).$$

Inserting this expression into the discrete evolution Equations (18) and (19), we obtain

Explicit: $\quad \delta \tilde{v}_{n+1,k} = \left(1 - \dfrac{2\Delta t}{\Delta x^2}(1 - \cos(k))\right) \delta \tilde{v}_{n,k},$

Implicit: $\quad \left(1 + \dfrac{2\Delta t}{\Delta x^2}(1 - \cos(k))\right) \delta \tilde{v}_{n+1,k} = \delta \tilde{v}_{n,k},$

both of the general form $\delta \tilde{v}_{n+1,k} = \xi(k)\, \delta \tilde{v}_{n,k}$ and, as we already know, the perturbation grows when $|\xi(k)| > 1$ and decays when $|\xi(k)| < 1$.

For the explicit scheme, we get

$$1 - \frac{4\Delta t}{\Delta x^2} \leq \xi_{(e)} = 1 - \frac{2\Delta t}{\Delta x^2}(1 - \cos(k)) \leq 1.$$

The upper bound, reached for $k = 0$, is irrelevant. The lower bound is reached for $k = \pi$, so that the stability condition $\xi_{(e)}(\pi) > -1$ yields

$$-1 < 1 - \frac{4\Delta t}{\Delta x^2} \quad \Rightarrow \quad \frac{4\Delta t}{\Delta x^2} \leq 2 \quad \Rightarrow \quad \Delta t \leq \frac{\Delta x^2}{2}.$$

The explicit scheme is *conditionally stable*.

By contrast, the growth rate of perturbations in the implicit scheme

$$\xi_{(i)} = \left(1 + \frac{2\Delta t}{\Delta x^2}(1 - \cos(k))\right)^{-1}$$

is easily seen to be positive and smaller than 1 in all circumstances since $(1 - \cos(k)) \geq 0$. This scheme is therefore *unconditionally stable* and Δt can be chosen independently of Δx by consideration of temporal accuracy only.

Note that for the explicit scheme the most unstable perturbations are those with $k = \pi$. Going back to physical space we have $\delta v_{n,j} = \delta \tilde{v}_n (-1)^j$, which shows that the behavior of odd-numbered nodes is opposite to that of even-numbered nodes, which is again typical of numerical instabilities. At this stage, the only way to cure a numerical instability is to reduce the time step until the stability requirement is fulfilled or to turn to an implicit scheme. Higher order time stepping can be studied along parallel lines.

3.1.4. Efficient Solution of Implicit Schemes

The better stability properties of implicit schemes derive from the fact that they perform an averaging over the system. In the present case, the amount of supplementary work required is rewarding since the linear system to be solved at each time step is tridiagonal and direct elimination is made efficient by *L-U* decomposition.

The system to be solved can be written as $MV = F$, and the matrix M as the product LU of two triangular matrices L and U, respectively lower and upper triangular. M being tridiagonal ($M_{ij} \neq 0$ for $j = i$ and $j = i \pm 1$), L and U have a band structure with only two nonvanishing diagonals ($L_{ij} \neq 0$ for $j = i$ and $j = i - 1$; $U_{ij} \neq 0$ for $j = i$ and $j = i + 1$ with $U_{jj} = 1$). The problem which now reads

$$LUV = L(UV) = LW = F$$

is therefore replaced by two matrix equations

$$LW = F \quad \text{and} \quad UV = W$$

and, owing to the structure of matrices L and U, the solution is obtained from two simple first order recurrences, a *forward recurrence* to get W in terms of F:

$$L_{11}W_1 = F_1,$$
$$L_{i\,i-1}W_{i-1} + L_{ii}W_i = F_i, \qquad i = 1,\ldots,N,$$

A3. Numerical Simulations

and a *backward recurrence* for V in terms of W:
$$V_N = W_N,$$
$$V_i + U_{i\,i+1}V_{i+1} = W_i, \qquad i = N-1,\ldots,1,$$
where the required coefficients L_{ij} amd U_{ij} are determined once and for all by two easily derived set of recurrences. Everything can be easily implemented, even on small computers.

L-U decomposition is efficiently used also when fourth-order derivatives are involved, as for models of Chapter 8. The matrix M then becomes pentadiagonal and is further decomposed into two tridiagonal matrices L and U and the solution obtained by second-order forward and backward recurrences.

For two-dimensional problems, implicit schemes yield linear systems with *sparse* matrices having tridiagonal or pentadiagonal block structures. The solution can be obtained by the generalization of the previous algorithms to block matrices, i.e., by replacing all coefficients by square matrices and all operations such as products or inverses by the corresponding matrix operations. Though the method is very efficient, direct elimination by the block *L-U* algorithm requires a lot of memory for the storage of elementary matrices L_{ij} and U_{ij} involved in the recurrences, which forbids its use on small computers.

Since implicit schemes should be preferred to explicit ones for stability reasons, the solution of large linear systems then resorts to iterative methods that require practically no more memory than for the storage of the solution itself. The corresponding algorithms, e.g., over-relaxation, steepest descent, or conjugate gradients, are easy to implement, but their description goes beyond our purpose.

3.1.5. Treatment of Nonlinear Terms

Up until now we have considered linear problems only. In most cases of interest, nonlinearities are present so that we can write formally $\partial_t v = Lv + N(v)$. The linear term, now denoted as L, is usually the most dangerous since it contains the highest degree partial derivatives, which are most sensitive to numerical instabilities in explicit schemes. By contrast, the nonlinear terms are often more gentle. To keep second-order accuracy in time while avoiding the fully implicit treatment of the nonlinear term, one can develop a quasi-linearized Crank–Nicolson scheme by inserting $v_{n+1} = v_n + (v_{n+1} - v_n)$ into the scheme and keeping the first term of the expansion of the nonlinear term in powers of $v_{n+1} - v_n$ supposed to be small enough:

$$\bigl(N(v)\bigr)_{n+1} \approx \bigl(N(v)\bigr)_n + Q_n\bigl(v_{n+1} - v_n\bigr),$$

where Q_n denotes the resulting coefficient involving v evaluated at time t_n. If the nonlinearities are not dangerous from the numerical point of view, another possibility can be found by treating separately linear terms with a Crank–Nicolson scheme and the nonlinear term with an Adams–Bashford scheme:

$$v_{n+1} = v_n + \frac{\Delta t}{2}\Bigl[\bigl((Lv)_{n+1} + (Lv)_n\bigr) + \bigl(3(N(v))_n - (N(v))_{n-1}\bigr)\Bigr].$$

3.2. Spectral Methods

Spectral methods rest on a conversion of the partial differential equation into an infinite-dimensional differential system for a set of amplitudes corresponding to coordinates on some functional basis. We consider here only the case of periodic boundary conditions which can be treated by Fourier decomposition: $v(x,t) = \sum_k \hat{v}(k,t) \exp(ikx)$. Imposing $v(x + \ell, t) = v(x, t)$ implies $k = k_m = 2\pi m/\ell$ with m an integer ($\hat{v}(k,t) = \hat{v}_m(t)$, $-\infty < m < \infty$). In addition, for $v \in \mathbb{R}$, we have $\hat{v}(-k) = \overline{\hat{v}(k)}$.

For the diffusion equation (11), the solution becomes trivial:

$$\frac{d\hat{v}_m}{dt} = -k_m^2 \hat{v}_m \quad \Rightarrow \quad \hat{v}_m = \hat{v}_m(0) \exp\bigl(-k_m^2 t\bigr).$$

The approximation comes in when the series is truncated above some maximum value $N_{\max}$ so that $\sum_{-\infty}^{+\infty}$ is replaced by $\sum_{-N_{\max}}^{+N_{\max}}$. As a result, the spectrum v_m becomes periodic with period $2N_{\max}$ and the dynamics of structures with wavelengths shorter than $\lambda_{\min} =$

A3. Numerical Simulations

$2\pi/k_{\max} = l/N_{\max}$ is not resolved, which can be seen as the result of a sampling of the solution on a regularly spaced grid with spacing $\Delta x = l/2N_{\max}$.

The method can be said *of infinite order* since the error term decreases exponentially with $N_{\max}^2$ (as $\exp(-k_{\max}^2 t)$ at given t), i.e., more rapidly than any power of Δx.

Notice that here the spatial differential operator is diagonal in Fourier space. This may not be the case for different boundary conditions and different basis functions (e.g., the popular Chebyshev polynomials) but, usually, an exact expression of the differential operator can be found in spectral space.

The main difficulty then appears with nonlinearities since products in physical space become convolution products in spectral space, e.g.,

$$\widehat{v^2}_m = \sum_{m'} \hat{v}_{m-m'} \hat{v}'_m,$$

which comes from the Fourier analysis of

$$v^2 = \left(\sum_{m'=-N_{\max}}^{N_{\max}} \hat{v}_{m'} \exp(ik_{m'}x) \right) \left(\sum_{m''=-N_{\max}}^{N_{\max}} \hat{v}_{m''} \exp(ik_{m''}x) \right)$$

$$= \sum_{m',m''} \hat{v}_{m'} \hat{v}_{m''} \exp(i(k_{m'} + k_{m''})x).$$

The sum $k_{m'} + k_{m''}$ contains wavevectors corresponding to structures with $|k| > k_{\max}$, which are not resolved by the simulation. This *aliasing phenomenon* will be negligible if the amplitude of the corresponding modes is small enough not to perturb the dynamics of the modes kept in the truncation.

Even if the aliasing problem is correctly treated, the evaluation of convolution products is numerically expensive. However, since usually nonlinear terms are more easily computed in physical space and differential terms in spectral space, the idea is then to adapt the type of computation to the nature of the terms by passing from spectral to physical space and reciprocally. Such changes of representation involve $2N_{\max} \times 2N_{\max}$ matrices, so that the method is efficient only if matrix products can be evaluated by fast algorithms. This is precisely the case for Fourier transforms, especially when $2N_{\max}$ is a power of 2. The *fast Fourier transform* (FFT) takes advantage of relations between trigonometric lines evaluated at regularly spaced points and organizes the flow of data in order to decrease the number of operations by a factor of order $\log(N_{\max})/N_{\max}$. Considering model (17), noticing that $v\,\partial_x v = \frac{1}{2}\partial_x v^2$, we would obtain the general scheme:

$$\tilde{v}_n \begin{cases} \to \widehat{Lv_n} = \bigl(\varepsilon - (k_n^2 - 1)^2\bigr)\widehat{v}_n \\ \to v_j \to v^2(x_j) \to \left(\widehat{v^2}\right)_n \to \left(\widehat{v\partial_x v}\right)_n = \dfrac{ik_n}{2}\left(\widehat{v^2}\right)_n \end{cases} \to \tilde{v}_{n+1},$$

where the last step involves one of the temporal schemes studied in Section 2.

All the methods presented in this appendix require a rather light investment and should be used without hesitation to get a concrete feeling of how complicated nonlinear dynamics can become.

4. Bibliographical Notes

An introduction to numerics can be found in:

[1] A. Ralston and P. Rabinowitz, *A first course in numerical analysis*, 2nd edition (McGraw-Hill, 1978).

but people interested in concrete implementation of algorithms should consult:

[2] W. H. Press, B. P. Flannery, S. A. Teukolsky, and W. T. Vetterling, *Numerical recipes, the art of scientific computing* (Cambridge University Press, 1986)

which completes the description of methods by program listings; the source codes of the library in FORTRAN and Pascal are available on diskettes for personal computers.

A brief summary of useful formulas can be found in chapter 25, "Numerical interpolation, differentiation, and integration" of

[3] M. Abramowitz and I. A. Stegun, eds., *Handbook of Mathematical functions* (National Bureau of Standards, 1972).

A3. Numerical Simulations

The integration of differential equations is introduced in [1] above, and further in:

[4] B. Thomas, "The Runge–Kutta methods," Byte, April 1986.

For accurate simulations, predictor-corrector methods may be desirable; they are analyzed in detail by:

[5] L. Lapidus and J. H. Seinfeld, *Numerical solution of ordinary differential equations* (Academic Press, 1971).

One can begin the study of partial differential equations with:

[6] F. S. Acton, *Numerical methods that work* (Harper and Row, 1970).

where a pragmatic viewpoint is taken, especially for the treatment of boundary conditions.

Several interesting schemes and their stability properties are further presented in:

[7] R. D. Richtmyer, and K. W. Morton, *Difference methods for initial value problems* (Interscience, 1967).

For a summary of iterative algorithms adapted to the solution of sparse linear systems as they appear, e.g., in two-dimensional problems, see:

[8] J. H. Wilkinson, "Solution of linear algebraic equations and matrix problems by direct methods," in *Digital computer user's handbook*, M.K. Klerer, G.A. Korn, eds. (McGraw-Hill, 1967), Chapter 2.2.

A nice introduction to fast Fourier transforms with a short program example is presented in:

[9] R. J. Higgins, "Fast Fourier transform: an introduction with some mini computer experiments," American Journal of Physics 44, 766 (1976).

which also gives an interesting discussion of pitfalls in spectral analysis of interest to beginners. More advanced material can be found in:

[10] D. Gottlieb, S.A. Orszag: *Numerical analysis of spectral methods: theory and applications* (SIAM Publications, 1977).

[11] C. Canuto, M. Y. Hussaini, A. Quarteroni, and T. A. Zang, *Spectral Methods in fluid dynamics* (Springer-Verlag, 1988).

Subject Index

A

Absolute instability, 23, 405–408, 417
Absorbing state, 399–401
Adiabatic elimination of slaved modes, 10, 20, 41, 64, 137, 140–144, 188
Adjoint operator, 39, 446
  in boundary value problems, 447–450
  in perturbation expansions, *see* Fredholm alternative
Advection, 2, 387, 403, 405
Amplitude, 6, 10, 18–19, 34, 140, *see also* Envelope
Arnold map, 232–235
Arnold tongue, 230
Aspect ratio, 9, 38, 286, 287
Asymptotic behavior (permanent regime), 28
  wandering point, nonwandering set, 52, 175
Attractor, 48–49
  absorbing zone, attracting part, 45–47, 48
Autonomous system, 26–27, 28, 254

B

Balance equations, 419
Basic state, 2, 5, 28
Basin of attraction, 48

Bénard–Marangoni convection, 86–89
  competition with Rayleigh–Bénard convection, 90–93
Bendixon criterion, 176
Benjamin–Feir instability, 349, 390
Bifurcation, 3, 7, 49–50
  in one dimension, 157–165
  in two dimensions, 177–181
Bimodal instability, 119
Boundary conditions, 26
  on director field (nematics), 70, 75, 434–435
  numerical treatment, 467–468
  on roll pattern, 9, 133, 301, 363, 383
  on temperature field, 99–102
  on velocity field, 102–103
Boundary layer
  envelope, 350, 351–352, 362
  thermal, 8, 119, 120–121
Boundary value problem, 26, 35, 138, 444–450
Boussinesq approximation, 95, 96
Boussinesq equations, 96–99
Box counting, 261
Busse balloon, 114, 121, 124
Busse oscillations, 123, 134

C

Cantor set, 16, 266–267
Cascade of bifurcations, 3, 12, 125–126, 202

Catastrophes, 166, 182
Cellular automaton, 400, 416
Cellular instability, 37, 286
Center (elliptic point), 170, 172–174
Center manifold, 11, 44, 144–145, see also Adiabatic elimination
Central mode, 11, 41
Centrifugal instabilities, 80–82
Chaotic regime, 3, 203
Chaotic transient, 223, 240, 401
Closed flow, 22
Codimension, 69, 151
Coherence length, 20, 109, 286, 364, 393
  longitudinal/transverse, 290
  variation, 20, 287
Compatibility condition, 59, see also Fredholm alternative
Confined system, 9, 134
Confinement effects, 4, 8, 9, 286
Conservative system, 167–168
Constitutive equations, 426–428
Continuity equation, 61, 385, 386, 420–421
Continuous systems, 26
Continuous-time systems, 26
Control parameter, 6, 25
Convection, see Bénard–Marangoni; Rayleigh–Bénard; Fluid Mixture; Nematic liquid crystal
Convective instability, 23, 405–408
Core (of defects), 356, 364, 366
  from experiments, 278–279
Coupled map lattices, 400
Crises, 17, 217, 239–241, 243, 393
Critical exponents, 162–163, 401
Critical phenomena, 401
Critical point
  of map, 205
  of vector field, 42
Critical wavelength, wavevector, 6
Cross-roll instability, 115, 330–333
Curry–Yorke map, 236

D

Defects, 9, 134, 299, 352–359, see also Textures
Degenerate modes, 139, 146, 443
Degree of freedom (generalized coordinate), 10, 25
Determinism, 27–28
  evolution operator, 28, 34, 439–440
  existence and uniqueness, 28
Deterministic chaos, 16
Developed turbulence, 23, 404, 408, 417
Devil staircase, 234–235
Diadic map, 15, 250, 257–258, 260, 266
Differential equation with delays, 450–455
Diffusion, 56, 421
  general equation, 56, 465
  of modulations, 20, see also Envelope formalism
Diffusivity, 56
Dimension, 267–275
  correlation, 272
  embedding, 276–277
  fractal, 17, 268, 271, 272, 277
  Hausdorff, 269
  information, 271, 272
  Lyapunov (or Kaplan–Yorke), 273
  pointwise, 270
  Renyi (generalized), 270–273
  topological, 267, 277
Director field
  for nematics, 70, 431
  for textures, 364–365, 377

Subject Index

Discrete system, 25
Discrete-time system, 12, 26
Dislocation, 9, 132, 356–359, 364–366, 388–389
  motion, 133, 358, 359, 382
Dispersion relation, *see* Growth rate
Dissipative dynamical system, 4, 42, 168, 207
Dissipative standard map, 236
Dissipative structure, 3, 5
Divergence of trajectories, *see* Strange attractors; Lyapunov exponents
Drift flow, 122, 383–389
  and frustration, 389
  and phase dynamics 383–385
Dynamical system, 42
  discrete *vs.* continuous time, 12, 25–26

E

Eckhaus instability, 116, 345–346, 374
Eigenvalue, 441–443
  discrete *vs.* quasi-degenerate, 38, 140, 288
Elimination of nonresonant terms (normal forms), 147–150
Energy method, 30
Ensemble average, 247, 260
Entropy in information theory, 257–259
  K-entropy (Kolmogorov–Sinai), 259, 454
  and Lyapunov exponents, 266–267
  topological, 257–258
Entropy production, 29, 428
Envelope, 19, 289, 318–319
Envelope formalism
  stationary rolls, 19–20, 319–326
  waves, 334–336
Equilibrium point, 42
Equilibrium state, global/local, 1–2
Ergodic theory, 247
Extended system, 9, 20, 134–135

F

Fick law (molecular diffusion), 66, 430
Fixed point, 7, 11, 42, 153, 175–176
First Principle (energy continuity), 423–425
First return map, 13, 174, 184, 190
Floquet stability theory, Floquet exponents, 189–190
Fluid interfaces, 430
  boundary condition, 87, 91–92, 431
Fluid mixture, 429
  convection in binary mixtures, 64–70
Fluid particle, 419–420
Focus (spiral point), 170
Forced system, periodically, 26
  example, EHD in nematics, 78–80
Fourier modes, 35
Fractal structure, 14, 15, 207, *see also* Cantor set
Fredholm alternative, 446
  in perturbation expansions, 304, 306, 325, 372
Frustration, 302, 383, 389

G

Galerkin method, 109–112
Galilean invariance, 122–123, 384

Ginzburg–Landau equation, 337, 390
Gradient flow system, 166, 167
Grain and grain boundary, 9, 113, 298–299, 352–356, 362–363, 382
Growth rate
  of modes, 7, 34, 36, 286, 287, 290, 292, 320, 334–335
  of phase-space elements, 252–253, 273–274

H

Heat equation, 58, 72, 96
Hénon map, 14, 206–207
Homoclinic/heteroclinic orbit, 45, 172, 246
Homogeneous (or uniform) instability, 37, 390
Hopf bifurcation
  of fixed point, 179–181, 451–453
  of limit cycle, 194–197
    subcritical, and type-II Intermittency, 226
Howard oscillations, 120
Hyperbolic set, 49–50, 176
Hysteresis, 160

I

Imperfect bifurcation, 162–164
Initial conditions, 26
  controlled, e.g., thermal printing, 113
  uncontrolled, see Textures
Initial value problem, 26, 439–440, 443
  and stability analysis, 33
Instability of trajectories, 13–14, 203
Intermissions (laminar), 220, 395

Intermittency, see Spatio-temporal intermittency; Temporal intermittency
Intermittency factor, 395, see also Turbulent fraction
Invariant measure, 260–265
  for type-III intermittency, 263–265
Inverse (subcritical) bifurcation, 157–158, 180, 195
Inverse cascade, 215–217

K

Kinetic energy, 30
  generalized, 30, see also Lyapunov function; Energy method
Kolmogorov cascade (K41 spectrum), 23, 408–410
Kuramoto–Sivashinsky equation, 255–256, 274, 390–391, 394, 395

L

Laminar regime, 3
Landau–Hopf scenario, 12, 202
Large-dissipation limit, 187, 207, 236
Lebesgue measure, 261
Limit cycle, 11, 51, 174, 175–176, 203
Limit set, 48–49, 175
  exceptional, 48
Linear algebra, 440–443, 444–447
  eigenmodes, 39, 442
Linear differential systems, 440–441, 443–444
  in two dimensions, 169
Linear response, 1, 162, 427
Locking, 177, 228–235
Logistic map, 13, 208, 250

Subject Index

Lorenz map, 186–187
Lorenz model, 11–12, 185–186, 242–243
Lyapunov exponents, 16, 256
  computation from experiments, 279–281
  d-dimensional maps, 250–254
  delayed equations, 454
  differential systems, 254–256
  largest, 251–252, 280
  one-dimensional iterations, 16, 248–250
  relation with K-entropy, 266
Lyapunov function, 29, 30

M

Macroscopic system, 1–2
  continuous description, 419–431
Macro/meso/microscopic scale, 2
Marangoni effect, 86–87, see Bénard–Marangoni convection
Marangoni number, 89
Marginal mode, 11, 34, 143
Marginal stability curve, 36, 286
  from dimensional arguments, 60–61
  no-slip convection, 107, 112
  stress-free convection, 105
Matching (asymptotic), 349, 353–356
Material point, 419–420
Method of time delays, 275–276
Modulation of periodic pattern, 288, 318
  longitudinal/transverse, 290, 345–346
Mœbius band (and sub-harmonic bifurcation), 199, 200
Multiple-scale analysis, 319–321, 326–327, 335

N

Natural measures, 266
"Nature" of turbulence, 13, 52, 202–203, 411
Navier–Stokes equations, 96, 428
Nematic liquid crystals, 70, 431–438
  electrohydrodynamic (EHD) convection, 74–80
  thermal convection, 70–74
Neutral mode, 11
  and translational invariance, 191, 254, 300–301, 372
Newell criterion, 348
Newell–Whitehead–Segel equation, 328, 341
Newton's law for a continuous medium, 421
Node, 169
  improper, 172, 174
Nonlinear pendulum 172–174
Normal (supercritical) bifurcation, 157–158, 180, 195
Normal form
  for bifurcation of limit cycle, 192–194, 197–199
  saddle-node (or tangent), 198
  sub-harmonic, 198
  of linear operator (Jordan), 170, 442–443
  of nonlinear dynamical equations, 144–147
  and symmetries, 150
Normal modes, 33–38
Normal mode decomposition, 39–40, 140
No-slip boundary condition, 102
Number of degrees of freedom
  and aspect ratio, 38, 130, 287–288
  in delayed equations, 451, 453

in developed turbulence, 411
Numerical simulation
  of space dependence
    finite differences, 466–472
    spectral, 473–474
  of time evolution 459–465
    Adams–Bashford, 461–462, 472
    Crank–Nicolson, 460, 472
    Euler, 460
    first/second order, 459–461
    implicit/explicit, 459, 468–670
    leap-frog, 461–462
    Runge–Kutta, 463–465
    single/multi-step, 460–461, 465

O

Onset time, 295–296
Open flow, 22, 403
Order parameter, 6
Ordinary differential equations (ODE), 26, see also Numerical simulation
Oscillatory instability, 35, 390–392
  in binary-mixture convection, 67–69
Out-of-equilibrium state 2, 423–428

P

Parameter space, 6, 25
Partial differential equations (PDE), 26, see also Numerical simulation
Pattern, 9, 297
  hexagonal, 114, 297, 311–312
  roll, 6, 114, 297, 311, 333
  square, 114, 297, 311, 333
Pattern formation
  example, 293–302
  modeling, 291–293
Pattern selection, 359–366
  linear, 294–295
  nonlinear
    and frustration, 302, 383, 389
    long-term, 299–302, 360–366
    short-term, 297, 298, 330–334
Peixoto theorem, 176
Percolation, 398–400
  directed, 398, 402
Period doubling, see Sub-harmonic bifurcation
Periodic regime, 11, 51, 174, 177, 202
Permanent regime, 28
Perturbation expansion
  for convection, 313–318
  nonlinear, 39–40, 302–304
  regular vs. singular, 154–155
  uniform vs. modulated solutions, 325
Phase portrait, 42, 44, 50, 53, 333
Phase space, 10, 25
Phase transitions, 401
  Landau theory and bifurcations, 162
Phase variable (patterns)
  diffusion, 21, 300, 342, 370, 375
  diffusivity, 21, 373
  dynamics (generalized), 376–379, 387
  formalism, 370–374
  instabilities, 21, 374–376
  turbulence, 21, 391
Phase-winding solutions, 341–343, 347
Poincaré–Bendixon theorem, 174
Poincaré map, 12–13, 174, 184, 190
Poincaré section technique, 12, 174, 183–187
Potential, see Gradient flow system

Subject Index

for Newell–Whitehead equation, 360
for one-dimensional system, 153
for Swift–Hohenberg model, 293, 379
Prandtl number, 62, 63, 68, 113, 383–385
Pressure, 97, 424, 429
 elimination, 97
Primary instability, 3, 7

Q

Qualitative dynamics, 42–52
Quasi-periodic regime, 176–177, 196, 202, 228
Quasi-periodic scenario, 202
 example from convection, 127–129

R

Random forcing (noise), 27, 29
Random walk, 15,16
Rayleigh–Bénard convection
 linear analysis
  approximate, no-slip, 109–113
  exact, no-slip, 106–108
  exact, stress-free, 104–105
  simplified, 59–62
 mechanism, 5–6, 56–59
 nonlinear analysis
  no-slip, 316–318
  stress-free, 313–316
Rayleigh number, 62, 67, 113
Reduction of ODEs to iterations, 12, 13, see also Poincaré map; Poincaré section
 stroboscopic analysis, 12
 time-one map, 276

Reduction of PDEs to ODEs, 40–41, 140, see also Galerkin method
Regular point, 42, 175
Relaxational behavior, 11, 153, 387–388
Relaxation time, 3, 20, 109, 112, 288
 slowing-down, 20, 158
Renormalization of sub-harmonic cascade, 211–215
Resonance (frequency), 188, 192
 and locking, 230–231
 strong resonance, 191, 193, 197
 weak resonance, 193
Resonance (wavevector), 139, 304, 308–311, 322–324, 329
Reynolds number, 404, 408
Ruelle–Takens–Newhouse theorems, 202
Ruelle–Takens scenario, 13, 202, 237–238

S

Saddle–node bifurcation, 160, 178–179, 198, 216
 and type-I intermittency, 220
Saddle point, 43, 167, 170, 175
Scalar product, 39–40, 444–445
Scenario, 4, 9, 12, 13 202
Schmidt number, 68
Schwartzian derivative, 222
Second Principle (entropy source), 29, 425–426
Secondary instability, 3, 7
Secular term, 172, 318, 444
Sensitivity to initial conditions, 4, 203
Separatrix, 172

Singular point, 42
Skewed varicose instability, 123
Source/sink, 43–44
Spatial chaos *see* Textures
Spatio-temporal chaos, 18–22, 127, 383, 403
 example from convection, 131–134
Spatio-temporal intermittency, 22, 395–398
 modeling, 398–401
Specific secondary instabilities (convection), 118–124
Stability, 3, 28–33
 asymptotic, 30
 conditional, 31, 32, 159
 global, 29–33
 linear (local), 31, 33
 monotonic, 31
 numerical, 461–463, 468–470
 structural, 49–50, 174, 176
 uniform, 29, 31
Stable manifold, 44
Stable mode, 34, 143
Stationary instability, 35, 286, 390
 in binary-mixture, 67, 69
 in plain convection, 60
 in Taylor–Couette flow, 85
Stokes law (diffusion of vorticity), 427
 viscosity, kinematic viscosity, 56, 427, 428
Strange attractor, 4, 13, 203
Stress-free boundary condition, 103
Stress tensor, 422
Subcritical (inverse) bifurcation, 157–158, 180, 195, 303
Sub-harmonic bifurcation, 198–201, 216
 subcritical, and type-III intermittency, 222

Sub-harmonic cascade, 13–14, 207–215, 217
 example from convection, 130–131
 modeling, 204–207
Supercritical (normal) bifurcation, 157–158, 180, 195, 303
Superstable orbit, 208, 209, 211, 212
Swift–Hohenberg model, 292, 333, 448–450
 Gertsberg–Sivashinsky variant, 330, 333
 model "b", 138, 320, 394, 450
 modified, 293, 305, 333
 with vorticity, 386–389
Symmetries and envelope equations
 rotational, 326–328
 translational (for waves), 336
Symmetries and phase equations, 374, 378–379

T

Tangent space, tangent operator, 42, 143, 280
 stable/central/unstable space, 42, 44, 143–144
Taylor–Couette instability
 Rayleigh mechanism, 82–83
 simplified model, 83–86
 turbulence, 393–394
Temporal chaos, 4, 9, 10–17, 127
Temporal intermittency,
 modeling, 218–219
 type-I, 219–222
  example from convection, 129
 type-II, 226–228
 type-III, 222–226
  invariant measure, 263–265

Subject Index

Tent map, 187, 239, 250
Tertiary instability, 3, 8
Textures, 9, 291, 298, 301, 363–366, 370
Thermal boundary condition, 99–102
Thermal diffusion (Fourier law), 56, 427
Thermodynamic branch, 2, 5, 28
Thermodynamic potential, 1, 29, 423
Threshold, 6, 36, 62, *see also* Marginal stability
Time average, 247, 260
Topological turbulence, 391
Torus, 12, 176, 203, 267
  breakdown of, 235–237, 238, 246
Trajectory (orbit), 27
Transcritical (two-sided) bifurcation, 160–162, 165, 303
Transient regime, 28
Transition to turbulence in convection, 8–9, 125–126
Turning point, 155, 159–160
Turbulent
  burst, 220, 395
  fraction, 22, 396, 398, 401
  transient, 388, 398

U

Unfolding
  and codimension 151
  linear, 151–153
  nonlinear, 163–165
Uniform (or homogeneous) instability, 37, 390

Universality, 401–402
Universal secondary instabilities, 115–118, 344–349
Unpredictability, 15–16, 203
Unstable manifold, 44
Unstable mode, 34, 143

V

Vague attractor, 49, 154
van der Pol oscillator, 51, 148
Variational problem, *see* Potential; Relaxational behavior
Vector field, 27, 42
Vorticity, 97
  vertical, 98, 112–113, 122, 385

W

Wave, 286, 390
  left/right, standing/traveling, 335–337
  modulated, 334–335
  stability of, 346–349
Wavelength selection
  in convection, 126
  criteria, 350, 364, 379–383
Weak turbulence, 4, 126
Winding (rotation) number, 232
Windows of periodicity, 216–217

Z

Zigzag instability, 117, 122, 346, 374

Perspectives in Physics

Huzihiro Araki, A. Libchaber, and Giorgio Parisi, editors

Frances Bauer, Octavio Betancourt, Paul Garabedian, Masahiro Wakatani, *The Beta Equilibrium, Stability, and Transport Codes: Applications to the Design of Stellarators*
Pierre Pelcé, editor, *Dynamics of Curved Fronts*
Petre Dita and Vladimir Georgescu, editors, *Conformal Invariance and String Theory*
Paul Manneville, *Dissipative Structures and Weak Turbulence*

CPSIA information can be obtained at www.ICGtesting.com
Printed in the USA
BVOW06*0020050216

435543BV00004B/25/P